BUSINESS/SCIENCE/TECHNOLOGY DIVISION
CHICAGO PUBLIC LIBRARY
400 SOUTH STATE STREET
CHICAGO, IL 60605

R 14

CHICAGO PUBLIC LIBRARY
HAROLD WASHINGTON LIBRARY CENTER

R002.877844

NO SPECIAL LOAN SZ-BST 7-17-04

REFERENCE USE ONLY

R541.224
R912h
cop.1

FORM 125 M

APPLIED SCIENCE &
TECHNOLOGY DEPARTMENT

The Chicago Public Library

Received   May 8, 1972

THE HISTORY OF
VALENCY

TO SHIRLEY

THE HISTORY OF VALENCY

C. A. RUSSELL

HUMANITIES PRESS

1971

First published in 1971
by Leicester University Press

Distributed in North America
by Humanities Press Inc., New York

Copyright © Leicester University Press 1971

All rights reserved
No part of this publication may be reproduced,
stored in a retrieval system, or transmitted
in any form or by any means, electronic,
mechanical, photocopying, recording or otherwise,
without the prior permission of
Leicester University Press

Set in Monotype Times
Printed in Great Britain
by Alden & Mowbray Ltd at the Alden Press, Oxford
ISBN 0 391 00033 0

CONTENTS

			page	
	List of Plates			ix
	Preface			xi

PART ONE: The Origins of Valency

I	The Historical Background to the Theory of Valency		3
	1	The English Background	4
	2	The German Background	12
	3	The French Background	16
II	From Radicals to Valency		21
	1	The Rise of the Radical Theory	22
	2	Kolbe's Revival of the Radical Theory	27
	3	Frankland's Work	34
III	From Types to Valency		44
	1	The Rise of the Theory of Types	44
	2	The Concept of Polyatomic Radicals	49
	3	Recognition of the Tetravalent Carbon Atom	61
IV	Nomenclature		81
	1	Names used in Group Classification	81
	2	Origins of the Word "Valency"	83
	3	The Word "Bond"	89
V	Valency Notation		92
	1	Superscript Dashes	92
	2	Parentheses	94
	3	Brackets	95
	4	Touching or Intersecting Circles	96
	5	Linear Representation of Bonds	100
VI	The Question of Priority		108
	1	The Claims of Kekulé	110
	2	Contested Matters of Priority	113
	3	The Influence exerted by the Founders of Valency Theory	125

CONTENTS

PART TWO: Early Applications of Valency

- VII Valency and the Classification of the Elements — 137
  1. Classifications based upon Valencies — 138
  2. Periodic Variations in Valency — 139
- VIII Valency and the Theory of Structure — 142
  1. Factors opposing the Structure Theory — 142
  2. The Founding of the Theory of Structure — 146
  3. The Equivalence of Valencies — 152
  4. The Concept of Tautomerism — 156
- IX Valency and Stereochemistry — 159
  1. Recognition and Application of directed Valencies — 159
  2. Opposition to directed Valencies — 165

PART THREE: Variations in Valency

- X The Problem of Variable Valency — 171
  1. Theories involving fixed Valencies only — 176
  2. Molecular Compound Theory — 189
  3. The Admission of Variable Valency — 197
  4. The Conception of Auxiliary Valency — 201
- XI Valency in Unsaturated Compounds — 224
  1. "Gap-Theory" — 225
  2. Multiple Bond Theory — 231
  3. Reduced Valency Theory — 235
  4. Strain Theory — 237
  5. Partial Valency Theory — 240
- XII Valency in Aromatic Compounds — 242
  1. The Origin of the Hexagon Formula — 242
  2. The Diagonal Bond Concept — 247
  3. The Centric Formulae — 252
  4. The Oscillation Theory — 254
  5. Thiele's Theory — 256

CONTENTS

PART FOUR: Valency and Electricity

XIII The Renaissance of Electrochemistry 261
    1 The Electrochemical Theory of Berzelius 261
    2 The Decline of Dualism 262
    3 New Links between Chemistry and Electricity 265

XIV The Recognition of Electrovalency 270
    1 Developments from the Periodic Law 270
    2 The Impact of Atomic Spectra 274

XV Extensions of the Electronic Theory to Non-electrolytes 276
    1 Attempts to apply Dualistic Ideas to Non-ionic Compounds 276
    2 Introduction of the Concept of Electron-sharing 281
    3 The Recognition of Bond Polarization 287

XVI The Coming of Wave Mechanics 292
    1 The New Quantum Theory 292
    2 Structural Uncertainty in Organic Chemistry 296
    3 Orbitals and Bonds 302

XVII Some Conclusions 313
    1 The Internal Structure of the History of Valency 313
    2 The External Relations of the Theory of Valency 319

Periodical Publications Consulted 325
Notes and References 327
Indexes 365
    Index of Secondary Sources 365
    Index of Names 366
    Subject Index 370

PLATES

The Founders of Valency   *facing page* 160
1. Edward Frankland
2. August Kekulé

Some British Contributors   161
3. Archibald Scott Couper
4. Alexander Crum Brown
5. Alexander William Williamson
6. William Odling

Early Molecular Models   192
7. Hofmann's croquet-ball models
8. Dewar's benzene models
9. Kekulé's benzene model
10. van't Hoff's tetrahedral models

The Birth of Organo-metallic Chemistry   193
11. Putney House, *c.* 1810, later the College of Civil Engineering
12. Organo-metallic compounds prepared by Frankland

PREFACE

Of all the concepts used in modern chemistry, atomism apart, that of valency is one of the most fundamental. It is therefore curious that few efforts until now have been made to chronicle its development as a whole, although much excellent work has recently been done to illumine selected aspects or periods of growth. So far as I know, the subject has eluded attempts at historical analysis in the light of both developments within chemistry itself and also of the more profound changes, intellectual and social, that have characterized the last 150 years. The present work arose from a realization of this gap in our understanding and is based upon a London University Ph.D. thesis having the same title (1962).

Since undocumented accounts are useless to a historian, full references to primary sources are given. It is hoped, however, that they will be sufficiently unobtrusive for the ordinary chemist to find something of interest in the text. Partly with my fellow-chemists in mind I have included also a large number of references to secondary sources that are mostly fairly accessible and have given English renderings of foreign source material. I must accept responsibility for all such translations (chiefly from French and German).

Permission from the following to reproduce illustrations is gratefully acknowledged:

> The Chemical Society Library for the following from their collection of portraits: Plates 1, 4, 5, and 6.
> Institut für Organische Chemie, Technische Hochschule, Darmstadt for Plate 9.
> Rijksmuseum voor de Geschiedenis der Natuurwetenschappen, Leiden for Plate 10.
> The Manchester University Museum for Plate 12.
> The figure on p. 272 is reproduced from *Valence and the structure of Atoms and Molecules* by G. N. Lewis, copyright 1923 Reinhold Publishing Corporation, with permission from Litton Educational Publishing, Inc.

It will be obvious that I am heavily indebted to many of my colleagues in the history of science. I hope that they will regard their appearance in numerous footnotes as suitable acknowledgment not only of their scholarship but also of the friendly encouragement and help so often given. Here I must specially mention my gratitude to the members of the Department of History and Philosophy of Science at University College, London, and particularly my former supervisor, the late Professor Douglas McKie. I am also glad to record my thanks to Dr W. H. Brock, whose friendly interest has expressed itself in many practical ways during the production of the book, though in no way can he be held responsible for the opinions expressed. To Leicester University Press I am grateful for much patient co-operation, particularly to Mrs Susan Marshall who in seeing the book through the press has valiantly coped with the intricacies of a difficult text.

Most of this work has been carried out while a member of the chemistry staff of the Harris College, Preston. I am particularly grateful for the encouragement (and tolerance!) of my chemical colleagues and for the extensive assistance given by past and present members of the Library staff. Gratitude is also due to the Librarians of the Open University, Patent Office, British Museum, Chemical Society, Royal Society and Royal Institution Libraries in London; to those of the Universities of London, Manchester and Lancaster; to the reference librarians at the Harris Library, Preston and the Central Public Library, Lancaster; and especially to the staff of the physical sciences library at University College, London.

Finally I thank my long-suffering family who, at different times and in different ways, have borne with splendid fortitude the consequences of my preoccupation with this book.

C. A. RUSSELL

Open University, 1971

PART ONE
The Origins of Valency

CHAPTER I

The Historical Background to the Theory of Valency

THE INTERNATIONAL CHARACTER of science is guaranteed by its antecedents. No scientific theory today is built upon a substratum of facts collected in one country alone, or of concepts developed exclusively by one national group. Except where political considerations or commercial interests have regrettably led to the "classification" of knowledge, science has provided one of the best modern examples of international co-operation and interdependence. For this reason it is no longer meaningful to speak of French, or Russian or Indian characteristics in science, though obviously rates of progress in different branches of knowledge will vary enormously between nations. Even the barrier of language has been largely surmounted, science itself providing a partial *lingua franca*.

Chemistry has been particularly fortunate in these respects, although in the more distant past it shared in the segmentation of its sister-sciences. Thus the late eighteenth century saw the great argument over the phlogiston theory which "like the controversy between Newton and Descartes... produced something like a national division".[1] In the following century the most prolific and influential writer that chemistry had yet produced had cause bitterly to complain of the difficulties of getting his native Swedish translated and understood by those in other countries.[2]

Later still, German insularity played strangely upon the feelings of the Russian pioneer of structural chemistry, A. M. Butlerov:

One feature of these German congresses is particularly striking to us foreigners, so queer that I cannot pass over it in silence; it is an aspiration to assert their own nationality at every opportunity.... And there can be no doubt that this hypertrophy of the national feeling does plenty of harm to the Germans: it keeps them from duly recognizing every alien nationality.[3]

No doubt Butlerov was over-sensitive on this point, disappointed at a lack of recognition. Yet even if his analysis of the situation was deficient, its existence was undeniable.

While regretting all kinds of scientific nationalism – even in a minor key like these instances – it is unwise to ignore it as a fact. This is particularly so since valency emerged out of a conflict of ideas (radicals *v.* types) that in its early stages smacked strongly of national traits. As the following account will show, it is too facile to identify the warring factors with German and French schools respectively, but one cannot pass over the early developments in Germany and France which were, at first, quite independent.

Together with England, these two countries played the dominant role in the pre-history of valency. They were also the leaders in the European chemical industry of the nineteenth century. Their fortunes in this field also varied, and by the 1870s France and Britain had yielded the lead to Germany.[4] Political and social factors are obvious enough here, but they must also be taken into account in the more abstract realm of ideas, such as the prehistory of valency. Indeed, the further this subject is investigated the more clearly emerges the importance of the individual man of science, seen against the back-cloth of contemporary events. To take one illustration: in 1856 Kekulé left England for Germany, and was followed nine years later by Hofmann. These events contributed directly to the great expansion of the German chemical industry, and it is certain that their influence on the development of a theory of valency was at least comparable. A wholly English concept, or one that at least came to birth in England, would have had a very different pattern of growth from that which history records. Science is a human activity, and broad generalizations on national movements and attitudes cannot obscure the significance of the individual or of his experiences. If this is true of Kekulé and Hofmann, it is more completely true of J. J. Berzelius who, neither English, French nor German, bestrode like a colossus the whole field of European chemistry for thirty critical years. Regarded by many as the greatest chemist of his century, he wielded an influence on the course of events that may still be felt, and which in his own time did much to create the intellectual climate in which the theory of valency was born.

1. THE ENGLISH BACKGROUND

The year 1807 is an important one in the history of chemical ideas

THE ENGLISH BACKGROUND

and discoveries, particularly in England. The country was again at war with France, engaged in a struggle for existence that even Nelson's victory off Cape Trafalgar in 1805 had not terminated. Yet the Napoleonic war had as yet made little impact on the ordinary lives of the islanders, and science itself seems to have been affected even less. In London, Rumford's eight-year-old Royal Institution was gaining rapidly in social prestige, and its scientific lectures were growing in popularity with the élite to such an extent that their original educative purpose was in danger of being forgotten: "Philanthropy was out. Fashion for science was in."[5] Since the Royal Institution saw some highly important developments in chemical science, it will be convenient to look at it more closely.

Seven years earlier the science of electrochemistry had been born with the discoveries of voltaic electricity[6] and of electrolysis.[7] Something of the potentialities of the new science had been seen by the young Cornishman, Humphry Davy, who had been appointed to the Royal Institution in 1801, and who became its first Professor of Chemistry. Within two days of his arrival in London from Bristol he had commenced experiments on galvanism,[8] and within a few weeks his lectures had given electrochemistry the distinction of being the first new science to be directly and systematically reported to the public at large.[9]

Davy's lectures were thus no dreary rehearsal of long-established facts or theories. Even his textbook, years later, remains a kind of chemical autobiography.[10] Such an authoritative and personal presentation, in combination with Davy's natural eloquence and charm, assured the success of his lectures. Yet this rebounded upon Davy in a way that did disservice to his science, for the demands of a society that so lionized him became increasingly hard to reconcile with those from the laboratory bench. This fact, together with the requirement by his Managers that he pursued a study of the "useful" arts of tanning and of agriculture, diverted him for about four years from experimental electrochemistry. This situation lasted until the autumn of 1806, when less than five weeks of intense work on electrolytic transference equipped him with the basis of his first Bakerian Lecture to the Royal Society.[11] This astonishing discourse had shattered a whole array of misconceptions on the nature of electrolysis of aqueous solutions, had shown it to be a simple decomposition of water into hydrogen and oxygen alone, and (amongst much else) had proposed a view of chemical affinity in terms of electrical attractions. Now, in 1807, this lecture was published in the *Philosophical Transactions*. Its effect was immediate, and "it won him the kind of praise from a learned audience which his lecture of 1802 had won from a popular one".[12] Even Napoleon, though at war with England, paid his homage to Davy; the Institut de France awarded him its prize of 3,000 francs for the most important electrical research of the year.

It is tempting to ascribe to Davy's intuition a prophetic quality that it did not possess. His identification of chemical and electrical effects was not, and could not have been, the germ of our electronic theories of valency; one need reflect for a moment only on the anti-atomic philosophy of Davy to see that. In so far as Davy did have an effect on the rise of a theory of valency, that effect was obstructive rather than otherwise, at least in England. Yet this is not to deny the great value of his electrochemical theory. The autumn of 1807 was occupied with experiments conceived under its stimulus,[13] culminating in the triumphant isolation of potassium and sodium.[14] The use of electrolysis as a preparative and analytical tool began, for all practical purposes, in Davy's laboratory in the autumn of 1807.

One evening at about this time Davy was dining at the Royal Society Club at the *Crown and Anchor* in the Strand. With him at the meal and during a protracted conversation afterwards were Thomas Thomson, a Scottish chemist and author, and William Wollaston, then Secretary of the Royal Society. Both of these gentlemen had felt the impact of the mounting body of facts that we might now call stoichiometric regularities. Recent discoveries were pointing towards the Law of Multiple Proportions, and beyond it to an atomic interpretation. The Cumberland Quaker, John Dalton, had not yet committed to print his immortal generalization, but Thomson had met him three years earlier (1804), and now in 1807 had issued a third edition of his popular *System of Chemistry*[15] in which Dalton's atomic theory made its first public appearance. Indeed, the accumulation of empirical facts was now so compelling that Thomson believed that if Dalton had not produced an atomic theory others, particularly Wollaston, would have done so.[16] Thomson and Wollaston were thus to become ardent Daltonians, but it was otherwise with Davy. His ultimate units were the particles of eighteenth-century physics rather than the atoms of the new century's chemistry; of Boscovich rather than Dalton, perhaps. Yet his Notebooks reveal that his well-known reluctance to be committed to a *detailed* theoretical scheme was never more apparent than here.[17]

Thus on this occasion Davy was confronted with two redoubtable opponents, and for an hour and a half their conversation was of nothing other than the atomic theory. Yet Davy was unconvinced; "more prejudiced against it than ever", Thomson asserted.[18] Certainly Thomson himself, with John Davy later,[19] believed that Humphry Davy was to change his mind. But his Notebooks belie this,[20] and so do his letters.[21] He accepted Dalton's empirical generalization in the Law of Multiple Proportions, but refused to be committed by an atomic interpretation. Quite possibly this incident in 1807 made a greater impression on Thomson than it did on Davy, but the anecdote makes one thing very clear: Davy's opposition flew straight in the face of the very facts that had convinced his contemporaries the other way. Whatever may have been the reasons for his attitude (and one cannot exclude a reluctance to appreciate ideas produced by a contemporary rival[22]) it was certainly not a quick reaction based on superficial knowledge. Hence it was all the more firmly held, and therefore all the more obstructive. To understand the slow development of ideas on valency one must sense the feeling that was to become increasingly strong in nineteenth-century England that Dalton's atomic theory was more ingenious than correct. There were, of course, many factors contributing to this situation, but one may well reflect upon the opinion that it was Davy who gave to the whole century a cue "to a stream of disbelief in atoms".[23]

But Davy was more than an "influence". For all his individual genius and eccentricities, he remained typical of much of his age. In appreciating Davy's underlying attitude to atoms, therefore, we may see more clearly why the theory of valency was still half a century away.

Davy was above all a practical chemist. He had a deep-seated aversion to all theoretical schemes that seemed to him not rendered inevitable by empirical facts. There were personal reasons for his excessive caution in these matters; he had been severely treated in his youth for some rather wild speculations that he had unwisely made public, and thereafter that mistake was not to be repeated.

But it was more than a question of public relations. His private papers testify that he genuinely did abhor anything that might savour of a premature commitment to a theory. Atomism was just one example. It seems that, like many of his contemporaries, he was strongly moved by a constructive belief in progress, and particularly economic progress based upon science. Davy never ceased to *apply* his knowledge, whether to agriculture, tanning, medicine, the arts, the miners' safety lamp or even the electrochemical protection from corrosion of the ships of the Royal Navy. In this he was both a child of his age, and also a pioneer. Sir Llewelyn Woodward has described Davy's 1802 Discourse at the Royal Institution[24] as "one of the earliest and most remarkable examples of the constructive belief in progress".[25] Thus he became one of the first exponents of an attitude adopted by many English scientists in the nineteenth century. The rapid mechanization of industry and the coming of the railways illustrated the practical possibilities of progress in other fields. Had Davy lived another twenty years one may be sure he would have approved.

English chemistry was thus to be marked by a hale and hearty empiricism for half a century. No doubt the absence of involved theoretical schemes contributed to what recent writers have called "one of the most striking facts about British science during most of the nineteenth century",[26] its amateur standing. Enlightened laymen were often attracted to chemistry, especially, by its obvious practical utility. Undeterred by the complexities of the atomic weight controversy, they saw the science as a means to an end, and were often moved less by philosophy than by philanthropy. The short-term effects were good for those sections of society that could benefit from them, but chemistry inevitably suffered. When Liebig visited England in 1837, his assessment was not flattering: "England is not the land of science; there is only a widely dispersed amateurishness (weitgetriebener Dilettantismus)."[27] The justice of this observation is only too apparent.

The controversy over atomic weights was not, of course, a source of trouble only in Britain. The problem of measurement perplexed chemists of all nationalities. No certainty could be attached to figures suggested as atomic weights since for a long time no method was agreed for determining the relation between them and equivalents. Hence Davy and others after him acknowledged the great value of the latter and kept their own counsel on the former. Today we recognize that for any element the relation holds,

$$\text{atomic weight} = \text{equivalent} \times \text{valency}.$$

Only the equivalent could be measured in the early part of the century, and until some *independent* means of finding the atomic weight could be devised the valency of an element was not only inaccessible, it was almost without meaning. Not until the 1860s did the utility of Avogadro's hypothesis[28] dawn upon men generally, and not until then did a theory of valency make appreciable headway. Even then, however, the leaders of English chemistry were slow to take advantage of the new insights. In March 1869 a lecture was given at the Royal Institution by Crum Brown, a Scotsman who played a prominent part in the history of valency. Entitled "Chemical Constitution and its relation to Physical and Physiological Properties",[29] it used the new notation to represent molecular structure and was a model of clarity and conciseness. Yet *The Engineer's* review dismissed it as "of an abstruse character".[30] More astonishing still, perhaps, is the report of two years earlier that English chemists generally, including Edward Frankland who has been widely considered as having discovered valency in 1852, had suspended belief in atoms.[31] In these circumstances it is hardly surprising that we must look elsewhere than to the English chemical leaders for the chief advocates of the new concept.

Meanwhile we return to Davy in 1807, and to his contretemps at the *Crown and Anchor*. In one respect the trio meeting that night was a perfect cross-section of eminent British chemists: in the nature of their chemical education. Wollaston had received a medical training at Cambridge and had entered chemistry *via* medicine. Thomson had also studied medicine, but in Scotland where university teaching of chemistry was far more advanced than in the English universities. At Edinburgh he had profited greatly by the lectures of Joseph Black. These were the two main courses open to one who wished to study chemistry: a medical training in England or a spell at a Scottish university. Humphry Davy had neither of these, and was without any formal chemical education. In this respect he also illustrated a trend: Dalton and Faraday were two others whose genius compensated for their deficiencies in formal scientific education, and there were very many lesser men who shared the same misfortune without the same compensation. And the misfortune was not only theirs personally. The lack of facilities for a systematic scientific education in England goes far to account for the sterility of much chemical thinking far into the century, and for the late appearance of any clear ideas on valency. It also explains why Frankland and other key figures in this story made such efforts to train abroad. But that was later; Davy would have been denied even that opportunity.

Gradually the need became felt in England. On one hand the Mechanics' Institutes sprang up to provide the rudiments of a scientific education for those debarred from university by their status in life, and often entirely without adequate schooling behind them. On the other hand Harcourt and others laboured with success to create the British Association for the Advancement of Science in order "to give a stronger impulse and more systematic direction to scientific inquiry", and to give the national lead that had not been provided by the Royal Society, which, "scarcely labours itself, and does not attempt to guide the labours of others".[32] The activities of the British Association were complemented by those of the (Royal) Society of Arts and other organizations.

Unfortunately for England the realization of its scientific plight did not come early enough, and the first serious attempts to establish means of systematic education lagged behind similar efforts on the Continent. Yet when the first institutions were established they soon produced work of much value to science, and most of them had a significant part to play in the development of the theory of valency.

University College, London, founded after strenuous opposition from the older universities, received its Charter in 1836. Not restricted to Anglicans, it also departed from tradition in giving much fuller instruction in the sciences. From this foundation sprang the present University of London. For ten years before the Charter was granted active teaching was carried on at Gower Street, rivalled by King's College in the Strand from 1829. At University College, Alexander Williamson was professor from 1849 and an important figure in the early history of valency.

Two other institutions in London figure in the early development of this concept. In 1845, A. W. Hofmann became first professor of chemistry at the new Royal College of Chemistry with laboratories in Oxford Street. His achievement here "which had never been done before in Britain was to found a *school of research*".[33] At last England was offering a training in chemical research methods. Meanwhile, in the year 1851 the Royal School of Mines opened in Jermyn Street, and Lyon Playfair became its first professor of chemistry. It was a descendent of a museum of geology that had first appeared ten years before but was now transferred to the Office of Woods, Forests and Works. On Playfair's resignation in 1853, Hofmann assumed responsibility for both chemistry departments, and retained this until he returned to Germany in 1865. These two institutions owed much to the farsightedness of Prince Albert; it has been suggested that "one of the main claims to fame of the Prince Consort (and he has many) is that he saw clearly the link between scientific education and the well-being of industry and the nation".[34] Tragically, the impetus he gave in this respect was not maintained after his early death in 1861. As a result, the great surge of activity in the middle of the century, holding such splendid promise of atoning for the inertia of earlier decades, gradually spent itself, and there passed with it Britain's scientific leadership for many a long year. Yet it was in those momentous middle years, the 1850s, that the concept of valency was born. It is surely not coincidental that the fever of intellectual activity that preceded and accompanied its birth took place above all in London.[35]

From this digression well into the Victorian era we return once more, finally, to Humphry Davy. It would be ungrateful not to notice that in two ways, at least, he gave positive assistance to the chemical thinking of his successors. The first of these consisted of a clearer conception of a chemical element. Again the year 1807 is a critical one. On 19 October he and his assistant, Edmund Davy, became the first to set their eyes upon an alkali metal – potassium.[36] Electrolysis of fused potash gave the world a new metal and a triumphant vindication of Davy's own electrochemical theory.[37] A few days later he isolated sodium. The results were reported to the Royal Society in the following month.[38] Having decomposed the "fixed alkalis" and removed them from the list of elements, he later was able to demonstrate that chlorine *was* elementary, and that oxygen is *not* the essential principle of acidity since it is absent in hydrochloric acid.[39] Thus a good deal of dead wood was cleared away as two major errors about the elements were corrected.[40]

Davy's electrochemical theory continued to influence the course of chemical history long after Davy himself had left electrochemistry. It was, of course, the inspiration of Michael Faraday, Davy's protégé. More significant even than this was its stimulus on the great Swedish chemist, J. J. Berzelius, who used his extension of it as the cornerstone of a vast edifice of codified knowledge, destined to influence every chemist in Europe for thirty years and more. The electrochemical theory of Berzelius[41] gained advantage over its English predecessor by its whole-hearted commitment to the atomic theory. It was probably early in 1808 that Berzelius heard about Davy's achievements and views,[42] and also made his first acquaintance with Dalton's theory.[43] His fusion of these two lines of thought constituted the basis of his electrochemical theory. It is therefore not inappropriate to point out his indebtedness to the chemists of England.

2. THE GERMAN BACKGROUND

By 1807 the German states had suffered some of the consequences of hostility to France, although worse was to come before 1815 brought the downfall of Napoleon. From then on Germany was able to profit from its share of the thirty-three years' peace enjoyed by the Great Powers in Europe. Perhaps the rule of the Austrian Metternich proved to be "barren and sterile and unimaginative",[44] but at least the absence of major conflicts permitted science to get on with its task relatively undisturbed. More important, it made possible an influx of foreigners anxious to gain experience of German chemistry and eager to take home with them the benefits that they had gained.

It may be truly said that synthetic organic chemistry was born in Germany in this period. The reasons for this development are not very clear, and are probably complex. The account that is offered here may give a partial explanation of this important fact in the history of science.

The characteristic contribution of the German chemists in the previous century had been the theory of phlogiston – a thing the nineteenth century was anxious to forget.[45] From about 1800 there seems to have been a kind of theoretical vacuum in German chemistry; people seemed to be waiting for something to happen, and nothing to compare with the solid achievements of the English chemists was recorded. Yet during these years of relative inactivity the German romantic movement was taking shape in the form of a *Naturphilosophie*, to the growth of which Kant, Goethe and others had directly contributed.[46] This view of the world stressed its underlying oneness, and laid upon its adherents the necessity for looking for connections between natural phenomena that had hitherto been regarded quite separately. Poetry, philosophy and even music[47] are demonstrably in debt to the *Naturphilosophen* at this time, and natural science was not exempt. Oersted's work on electricity and magnetism[48] and Seebeck's on electricity and heat[49] owed their impetus largely to this influence. In chemistry it has been suggested[50] that Davy's work can be profitably reinterpreted in the light of his friendship with Coleridge, who had spent the winter of 1798–9 with Wordsworth in Germany and had returned full of enthusiasm for the German philosophy. Davy corrected the proofs of the second edition of *Lyrical Ballads*[51] in 1800, and thereafter reflected certain of the traits of the *Naturphilosophie* as mediated to him by Coleridge, and in particular the Idealism of Schelling.[52] It is certainly true that "when Davy is most emotionally involved and least scientific, when he is ready to talk of unity and affinity, that he comes nearest in sympathy to Coleridge and to a side of Wordsworth's poetry",[53] but his science and poetry were not totally detached, and his chemistry bears unmistakable marks of this German influence.[54]

If, therefore, the influence of *Naturphilosophie* may be detected as far away as England it is unlikely that chemistry in Germany was exempt from it. There are at least three focal points of German chemistry at this time that appear to be so coincident with those of this system of thought that one must at least admit that a connection is probable.

Kant had expressed the view that chemistry was a branch of applied mathematics, and this seems to be reflected in the researches performed by J. B. Richter who had studied under him at Konigsberg.[55] Richter, who died in 1807, had become preoccupied with the measurement of combining quantities, and is justly regarded as the founder of stoichiometry. In 1792 he had arrived at a generalized statement of the law of reciprocal proportions, from which it is but a step to a chemical atomic theory. Richter did not take that step, however, and when Dalton came to his own conclusions it was in complete ignorance of Richter's earlier work. Even in Germany this was very little known. However, his fellow-countryman, E. G. Fischer, summarized Richter's views in a German translation of Berthollet's *Recherches sur les Lois de l'Affinité*,[56] and Berthollet himself referred to them in the following year in his *Statique Chimique*.[57] In this way they first gained a wider recognition.

Yet Richter had not written entirely in vain. For some reason a copy of his *Anfangsgrunde der Stoichyometrie*[58] was to hand in 1807 when Berzelius was preparing his first textbook of general chemistry.[59] Acknowledging that it was "not generally read" he said that this encounter led him, ultimately, to the atomic theory itself.[60] Berzelius was to become the great opponent of *Naturphilosophie*, yet it seems that even he was unconsciously in debt to the system that he was to attack so vigorously.

Kant's mathematization of nature was perhaps not altogether typical of other philosophers of this school. Yet all shared to some extent in a second characteristic that can best be described as a romantic vitalism. The unity between man and nature[61] called for a new investigation into the organic world, a world which, for all the unity in nature, might be governed by totally different laws from its inorganic counterpart. Life (whatever that was) and living things became invested with a new fascination. It does not seem too far-fetched to see a common feature in Wordsworth's eulogy of the daffodils (or Davy's praise of Cornwall, for that matter) and the first determined attack on the nature of organic compounds in Germany. Nor is this to decry the analytical achievements elsewhere, the catholicity of Berzelius's chemical researches, and the manipulative advances of Chevreul and Gay-Lussac in France. And it must not be overlooked that a complex theoretical framework was unnecessary in the early stages of synthetic organic chemistry.

The prime mover of this development was Justus Liebig (1803–73). Dissatisfied with his chemical training in Germany, he spent two years in Paris, which was then (1822–4) a leading centre of organic research. Yet he had met the *Naturphilosophie*, for he afterwards complained that it had wasted his time.[62] Back in his own country as Professor of Chemistry at Giessen, Liebig spent twenty-eight years as Director of the most famous chemical laboratory in Europe at the time. Its research output alone would have won it world acclaim, but it had also the distinction of being one of the first teaching laboratories in Germany. In the peaceful years before

1848 it attracted students from many countries, and was a major factor in the subsequent "internationalism" of chemistry. Amongst the men who received training at Giessen were Hofmann and Kekulé (Germany), Gerhardt and Wurtz (France) and Frankland, Playfair, Stenhouse and Williamson (Great Britain). Every one of these chemists figures in the history of valency. When Liebig's additional influence through his *Annalen der Chemie*[63] is recalled, it becomes apparent that his rôle in shaping the whole development of chemical ideas (valency included) was comparable in importance to that of Berzelius.

Yet Liebig was not the only leader in the new organic chemistry. His friend Wöhler and a number of others were adding to the fame of Germany as a place for organic research. There seems to have been a parallel development here. Side by side with the new synthetic organic chemistry went the novel emphasis on practical chemical instruction in general. If the one owed something to the spirit of *Naturphilosophie* it is hard to accept the conclusion that the other was quite independent. We may perhaps infer that the new progress made in hacking a path through the "primeval forest"[64] of organic chemistry led to the realization that so complex a task demanded a new degree of practical training. Such a conclusion is hard to justify from documentary evidence, and obviously other factors were involved; but it seems permissible to hold it as a tentative explanation for this twofold achievement of nineteenth-century Germany. It is certainly true today that most of the specialized techniques of organic chemistry are hardly acquired without considerable instruction; one may cite the difficulty encountered by generations of students in the cases of recrystallisation and fractionation, as important now as in the 1830s. The one great chemist in the romantic tradition who made almost no contribution to organic chemistry was also the one who received no practical training: Humphry Davy. The conclusion that the new organic chemistry was both a partner and an offspring of practical teaching methods seems irresistible. Other countries than Germany lagged far behind in both respects. England, for instance, began to make good the deficiencies only in 1845. As late as 1866 an English reporter could write enviously of the new chemical laboratories being erected in Bonn and Berlin, and lament the lack of English counterparts.[65]

One further connection between the *Naturphilosophen* and German chemistry remains to be suggested. During the first half of the century Berzelius was setting about his prodigious task of rationalizing the whole of chemistry by his electrochemical scheme. Even he increased his debt to Germany, for the translation into German of many of his writings was performed by Wöhler, once a student with him, and this proved an indispensable means of broadcasting Berzelius's ideas. Although the electrochemical theory owed much to Davy, an important part of it had been a vague component of the ill-defined mass of chemical belief circulating in the Germany of the early nineteenth century. Berzelius's scheme was fundamentally a dualism, based upon positive and negative electricity, and as such it found an echo in the metaphysical dualism of some of the *Naturphilosophen*. Very early on, von Humboldt[66] and Ritter[67] had discussed a possible connection between electrical and chemical phenomena, and the latter had used the category of polarity in his view of matter. It has been said that Ritter "stands especially near the romantic nature philosophers with his dynamic conception of matter in which the category of polarity plays a decisive role".[68]

An English author, nearly contemporary with Berzelius, wrote of this dualism: "He considered that in all chemical combinations the elements may be considered as electro-positive and electro-negative, and made this opposition the basis of his chemical doctrines; in which he was followed by a large body of the chemists of Germany."[69] Thus the home of *Naturphilosophie* became the cradle of chemical dualism.[70]

3. THE FRENCH BACKGROUND

The claim of England and Germany to an important place in the prehistory of valency is as assured as it is in the later growth of the subject – probably more so in the case of England. Yet for chemical progress in general the laurels for the previous century had gone to neither of these countries, but to France. Here Lavoisier and his fellow-countrymen had overthrown the myth of phlogiston. That other myth voiced by Wurtz ("Chemistry is a French science; it was constituted by Lavoisier of immortal memory")[71] may not be dismissed just as a patriotic extravagance, for it reflects how greatly subsequent generations of Frenchmen recognized Lavoisier's genius and were inspired by his memory. It is perhaps for this reason more than any other that French chemistry entered the nineteenth century relatively free from the worst features of its German counterpart, especially its more speculative aspects.[72] With the passing of the first two or three decades of that century, a school of chemistry appeared in Paris whose contribution to the forming of a theory of valency was as valuable as it was distinctive.

In the year 1807 Napoleon's power was at its maximum. Even five years earlier Englishmen taking advantage of the peace of Amiens to visit France had been impressed by its rapid recovery from the ravages of the Revolution. Napoleon had then been gratefully acclaimed as First Consul for life, but now his stature was far greater. Only the English command of the sea frustrated his desire for world dominion, and for a while the country that he had led basked in his reflected glory, and science enjoyed his patronage.

Napoleon had long realized, like the Revolutionary Council before him, that science was worth encouraging. Under C.-L. Berthollet the French industries had been thoroughly reorganized, and France was now almost self-supporting. Together with Laplace, Berthollet founded in 1807 the Société d'Arceuil as a means for exchanging scientific information, welcoming foreign workers to its fortnightly meetings.[73] Davy was one of these, and in the same year he was awarded the Institut de France prize.

Berthollet's usefulness to science extended beyond these services to his own country, however. He was to bring about the fruition of an idea that had antecedents traceable over a long period in eighteenth-century France. This was the study of "affinity", associated with the names of Geoffroy, Buffon, Macquer, Baumé, Guyton de Morveau, Fourcroy and many others. In 1798, Berthollet accompanied Napoleon on his expedition to Egypt and was surprised to notice large deposits of soda on the shores of the Egyptian salt lakes. He concluded that they arose from a reaction between salt and limestone. This was a reversal of the familiar laboratory process:

calcium chloride + soda = calcium carbonate + salt
(*i.e.*, $CaCl_2 + Na_2CO_3 = CaCO_3\downarrow + 2NaCl$)

and he attributed it to the abnormally high concentration of sodium chloride. From this beginning he developed the idea (nascent in earlier writers) that *mass* could be a determinative factor in all combinations and decompositions produced by "elective affinity".

In cases like this "an excess of quantity of the body whose affinity is the weaker compensates for the weakness of the affinity".[74] This realization of the importance of mass in chemical reactions was incompatible with any view of absolute "elective affinities" and led, in time, to the enunciation by Guldberg and Waage of the law of mass action.[75] Unfortunately, it also led Berthollet to a false conclusion on the composition of compounds. Since relative masses influenced the proportions in which substances reacted together, could not the products have an indefinite composition, with varying ratios of constituents? After all, solutions behaved in this way and our sharp distinction (if it is so sharp) between compounds and mixtures was far less clearly seen then. This incorrect extension to his own theory led Berthollet into much trouble, including the celebrated controversy with Proust, but the subject is hardly germane to the history of valency. What must be said is that the doctrine of continuous composition was scarcely reconcilable with an atomic interpretation of chemistry. This was especially obvious when the law of constant composition had been enunciated by Proust,[76] and Berthollet's ideas generally went into eclipse. Views that seemed to be at variance with Daltonian atomism found little favour amongst many English chemists who tended to pay Dalton lip-service, and on the Continent there was not much enthusiasm for them. But once again Berzelius was among the first to perceive the value of a new theory, one that replaced the affinity tables of his own fellow-countryman Bergman. Able to disentangle and reject the incorrect strands in Berthollet's reasoning, Berzelius used his concept of mass action in his own electrochemical theory, reconciling it with atomism and giving an atomic interpretation to affinity.[77]

It is in this respect that Berthollet played his chief part in the evolution of a theory of valency, as an inspirer of Berzelius. Indeed, posterity unkindly commemorates his name chiefly in connection with those inorganic compounds that are the despair of classical valency theory, the non-stoichiometric or "Berthollide" compounds. His great services to physical chemistry, of which he can claim to have been a founder, are not detailed here but are his most valuable legacy to us.

When Berthollet died in 1822 the war was over and an exhausted France was making a slow recovery from the consequences of protracted conflicts. Science itself shared something of the country's debility, and its earlier promise showed little hope of fulfilment. For a time, following the educational reforms of 1802, science "occupied a far more important place than in the schools of other countries, and the lycées and advanced colleges provided France with the young men who for a generation placed her ahead of the world in science and particularly in chemistry".[78] Yet the ambitious plans at the secondary school level for the training of scientists were not fully implemented, and with their failure went the inevitable decline of science itself.[79]

In contrast to Germany, France did not see any considerable alliance between speculative philosophy and natural science. The French chemists, many of whom were working under conditions of severe difficulty, continued to produce valuable experimental work without the stimulus of a *Naturphilosophie*, though also without the spectacular success of their German colleagues. It will be recalled that organic chemistry in the early years of the nineteenth century was a promising field of enquiry that did not require a deep theoretical insight. Under Chevreul, Dumas, Gay-Lussac and others much useful work was done here, and the first thirty years of the century were less barren than might have been expected.

It was from France that Berzelius received the first determined opposition to his dualistic theory. First Dumas, then in harsher vein Laurent and Gerhardt, led the revolt that at length replaced dualism by the Unitary Theory and the Theory of Types. At first the new views were a characteristically French product (and were often castigated by opponents as such); they did not enter Germany until 1855.[80] Since both dualism and its successors played such a fundamental part in the rise of valency they are treated at some length later.[81]

The insulation of French chemistry from German influence was a remarkable feature of the early nineteenth century. The recent animosity between the two nations was doubtless reinforced by the attachment to "freedom" in post-Revolutionary France which would have militated strongly against an acceptance of bondage to German idealism. Moreover, the intellectual climate of Paris had long been unfavourable to anything that might be construed as occultism in science. This aspect of French life became embodied, magnified and revitalized in the bizarre figure of Auguste Comte

(1798–1857). The founder of positivism, he became the leader of a small but influential group in Paris that sought to emancipate science from all traces of mysticism and superstition. Among those who came into the circle of his influence were Laurent, Gerhardt and Williamson, all of them distinguished chemists whose writings shared a sturdy agnosticism in the matter of chemical constitution.[82] Inhibited by their reluctance to proceed beyond the facts into speculation on the inner recesses of molecules, they were naturally denied more than a shadowy prescience of valency, though Williamson was less hampered in this respect than his French colleagues. Yet the Parisian chemists, through their Theory of Types and with their cherished independence of thought, kept alive the study of organic chemistry at just the time when the Radical Theory was beginning to crumble. As the older theoretical framework started to give way, the Type Theory provided the temporary scaffolding necessary to maintain the unity of the science until such time as more permanent measures were available. Thus both radicals and types may be seen as temporary expedients, ultimately to be replaced by the concept that sprang from them both: the classical theory of valency.

CHAPTER II

From Radicals to Valency

THE STRANGEST FEATURE of the development of the ideas leading to the formulation of the doctrine of valency is that two almost independent routes were followed and that both of these were *via* organic chemistry.

To the modern chemist nothing is more natural than to interpret in valency terms the stoichiometric laws as exemplified by simple inorganic compounds. Why then was it necessary to force a passage through what Wöhler called "the dark and impenetrable jungle" of organic chemistry? Fundamentally the reason would seem to be this: inorganic compounds on the whole failed to offer the incentive to an analytical view of their constitution by virtue of their very simplicity. It was believed by some, for example, that water consisted of two atoms of hydrogen and one of oxygen, and that was the beginning and end of it. What else remained to be discovered? Vivid mental pictures of Daltonian atoms often found no place in the thoughts of chemists whose literature was filled with suggestions that such atoms did not exist, or if they did were not important. Hence such questions as to how, or even where, the three atoms in water were connected had no need to be raised.

But in the field of organic chemistry matters were different. Similarities discovered to the inorganic branch led to that vast system of argument by analogy that constituted the Theory of Types. They pointed directly to the Theory of Radicals and cast doubts upon a "vital force". The compounds of carbon presented a challenge to the best minds of nineteenth-century chemistry, and once this was accepted there could be no respite until the problem was solved. Moreover, greater point was given to an enquiry into the nature of organic materials by the discovery of isomerism. After this it was never possible to avoid speculation on what was eventually to be called "structure" of carbon compounds. Clearly a molecular formula alone could not be a sufficient description of organic molecules if several of these could share the same formula.

So it came about that attempts were made to unravel the mysteries of organic chemistry, and, as is often the case in science, the results of relatively particular enquiries turned out to have a far more general significance than would have been expected when the investigations began.

It is probable that all scientific enquiries must be pursued within the framework of a theoretical system, however tentatively it may be held. Two such systems prevailed in the 25 years leading up to the emergence of a theory of valency. Always in opposition until then, they ultimately became merged in the theory they anticipated. The investigations that had as their mainspring the *Type Theory* will be considered in the next chapter. Meanwhile some attention will be given to the progress of the *Theory of Radicals* in so far as it led to the doctrine of valency.

1. THE RISE OF THE RADICAL THEORY

The idea of a radical seems to have originated in 1787 with Guyton de Morveau who spoke of it as "the simple substance [of an acid] which modifies the oxygen".[1] Lavoisier considered organic acids as oxides of radicals which contained at least the elements carbon and hydrogen.[2] This oxygen theory of acids received severe blows by the work of Davy,[3] Dulong[4] and Gay-Lussac. The last-named showed that prussic acid contained no oxygen, though he regarded it to some extent as a special case. His discovery of cyanogen and other derivatives of prussic acid led him to suppose that cyanogen was a compound radical, "a body which, though compound, acts the part of a simple substance in its combinations with hydrogen and metals".[5]

Berzelius took this idea of a radical, clarified it in terms of Dalton's atomic theory and made it an integral part of his electrochemical system.[6] This dualistic view of chemistry regarded all compounds, organic included, as being composed of two radicals, one electropositive, the other electronegative. The radicals themselves might also have a dualistic structure being composed, for example, of positive carbon and negative oxygen. It is difficult to exaggerate the influence which these views exerted over chemistry for twenty years and more; the resourcefulness with which Berzelius used them to unite widely different phenomena remained unabated until his death, even though his closing years were marred by a growing hostility to his views as evidence against them began to accumulate. The basis of the electrochemical system was the assumption that when elements or radicals combined together a neutralization of opposite electricities occurred; only that amount of "negative" component would be held in combination that saturated the electricity of the "positive" part. This idea of saturation has been described by Palmer as "the first germ" of a theory of valency,[7] and Sidgwick has stated that "the earliest valency theory was that of Berzelius".[8] But the conception of a degree of saturation with a constant value (or a small number of fixed values) was quite absent, as was its application to elementary atoms. It turned out to be far more important that Berzelius forced chemistry for a time into a dualistic mould, and this led to an analytical view of chemical compounds that was of great assistance in developing a theory of valency.

The relationships of alcohol, ether, muriatic ether (ethyl chloride) and other compounds caused much interest about this time, though a final settlement of even their molecular formulae was not possible owing to the neglect of Avogadro's hypothesis. Dumas and Boullay (1828) supposed them to be built upon the radical which in modern symbolism would be C_2H_4.[9] Berzelius termed this radical "etherin" and exchanged the formulae for alcohol and ether in 1832.[10] The analogy was thus with compounds of ammonia which can be written as $NH_3 + HCl$, and so on. By the next year, however, Berzelius had adopted a different view, and no longer used the analogy between etherin and ammonia, whose alkaline properties are absent in ethylene. Instead he regarded ether as an oxide corresponding to our $(C_2H_5)_2O$. This radical was also combined with acids in ethyl chloride, *etc.*, but was absent in alcohol whose simpler formula he acknowledged from vapour densities; this was regarded as an oxide of a radical that we should call C_2H_6.[11] This obscuring of the close relation between alcohol and ether caused Liebig the next year to regard *all* these compounds as being derived from a radical he named "ethyl" and wrote as C_4H_{10}.[12]

These differing views are summarized in Table I, where the symbols in italics are those reduced to the modern atomic weight system $C = 12$, $H = 1$ *etc.*

The change of view from etherin to ethyl was a definite step in the progress of radical theories, for whereas etherin could be identified

TABLE I

Author and date	Note Reference	Symbols Alcohol	Ether	Muriatic ether
Dumas & Boullay, 1828	9	$4C_2H_2 + 2H_2O$ $C_2H_4 + H_2O$	$4C_2H_2 + H_2O$ $2C_2H_4 + H_2O$	$2C_2H_2 + HCl$ $C_2H_4 + HCl$
Berzelius, 1832	10	$C^4H^8 + 2\dot{H}$* $C_2H_4 + 2H_2O$	$C^4H^8 + \dot{H}$ $C_2H_4 + H_2O$	$C^4H^8 + H^2Cl^2$ $C_2H_4 + 2HCl$
Berzelius, 1833	11	$CH^3 + O$ $C_2H_6 + O$	$C^2H^5 + O$ $(C_2H_5)_2 + O$	$C^2H^5 + Cl$ $C_2H_5 + Cl$
Liebig, 1834	12	$C^4H^{10}O.H^2O$ $C_4H_{10}O.H_2O$	$C^4H^{10}O$ $C_4H_{10}O$	$C^4H^{10}Cl^2$ $C_4H_{10}Cl_2$

* In Berzelian notation $\dot{H}$ stands for H^2O.
Italicized symbols are modern versions where $C = 12$, $H = 1$, $O = 16$, etc.

as the gas ethylene, ethyl was entirely hypothetical. Indeed it was a radical in a new sense and corresponded to the hypothetical ammonium. Kekulé attributed the success of the [ethyl] Radical Theory to the fact that in inorganic chemistry at that time it was becoming more usual to regard ammonium salts[13] as made from ammonium rather than ammonia:

> Since it was precisely at this time that that part of the ammonium salts analogous to potassium was called a composite metal, so the radical theory naturally carried the day over the aetherin theory in the same way as the ammonium theory displaced that of ammonia.[14]

Powerful support for the existence of radicals came from another quarter when in 1832 Liebig and Wöhler published their researches on the radical benzoyl.[15] This group of atoms, which they wrote as $C^{14}H^{10}O^2$, persisted through a variety of reactions and was present in several well-defined compounds (known to us as benzaldehyde, benzoic acid, benzoyl chloride, benzamide, benzoylformonitrile and others). These discoveries were hailed by Berzelius as a triumph for his viewpoint and were claimed as "the dawning of a new day in vegetable chemistry".[16]

This period marked the high-water mark for the "old radical theory", as it came to be called. The acknowledgment of methyl[17] and amyl[18] as well as ethyl, led to the recognition of a new series of compounds (the alcohols) and of radicals derived from them.

Particularly in Germany, the Radical Theory gained widespread approval, and its temporary ascendancy in France was marked by a paper from Liebig and Dumas, then in a kind of metastable union, in which radicals were upheld as "the true elements on which organic chemistry is founded". The only difference between the two main branches of the science is that "in mineral chemistry the radicals are simple; in organic chemistry the radicals are compound". The time was ripe for a clarification of the idea of radicals, and this was given by Liebig, who stated cyanogen to be a radical:

(1) Because it is the unchanging constituent of a series of compounds;
(2) because it is capable of replacement in these by simple substances; and,
(3) because, in those cases where it is combined with one element, this latter can be exchanged for its equivalent of another element.[19]

At least two of these conditions must be fulfilled for an atomic complex to be a radical.

Nevertheless, it was about this time that the simple dualism of the Radical Theory began to break down. The insistence by Berzelius that radicals do not contain oxygen was hardly compatible with his enthusiastic welcome of benzoyl, $C^{14}H^{10}O^2$, and shortly afterwards he renounced this in favour of the idea that benzoyl and benzoic acid were both oxides of the binary radical $C^{14}H^{10}$.[20] The facts of halogenation conflicted with the immutability of radicals and undermined the foundations of the electrochemical system. How these events affected the course of organic chemistry will be partly considered in the following chapter. Meanwhile we must follow the fortunes of the Radical Theory itself.

From 1837 to 1843 a series of papers appeared in Poggendorf's *Annalen*[21] in which Bunsen described his researches on certain organic compounds of arsenic, largely performed at Marburg. These centred round an evil-smelling liquid for which Berzelius proposed the name cacodyl and which we know now as tetramethyldiarsine. This is present in the product from the action of potassium acetate on arsenious oxide discovered by Cadet in 1760.[22] Bunsen's achievement was to show that this product, alkarsin, contained the oxide of the radical cacodyl which was capable of remaining unchanged through a long series of reactions. To clarify various points that will emerge in the following discussion, modern formulae are used to show the principal reactions studied by Bunsen:

Reaction scheme

$$CH_3COOK + As_2O_3$$

branches to:

(cacodyl): $(CH_3)_2As-As(CH_3)_2$

(cacodyl oxide): $(CH_3)_2As-O-As(CH_3)_2$ (formed via air from cacodyl)

From cacodyl with chlorine ⇌ (zinc) $(CH_3)_2As-Cl$

From cacodyl oxide with hydrochloric acid → $(CH_3)_2As-Cl$

From cacodyl oxide with moist air → $(CH_3)_2As(=O)(OH)$

$(CH_3)_2As-Cl$ with chlorine → $(CH_3)_2As^+(Cl)(Cl) \; Cl^-$

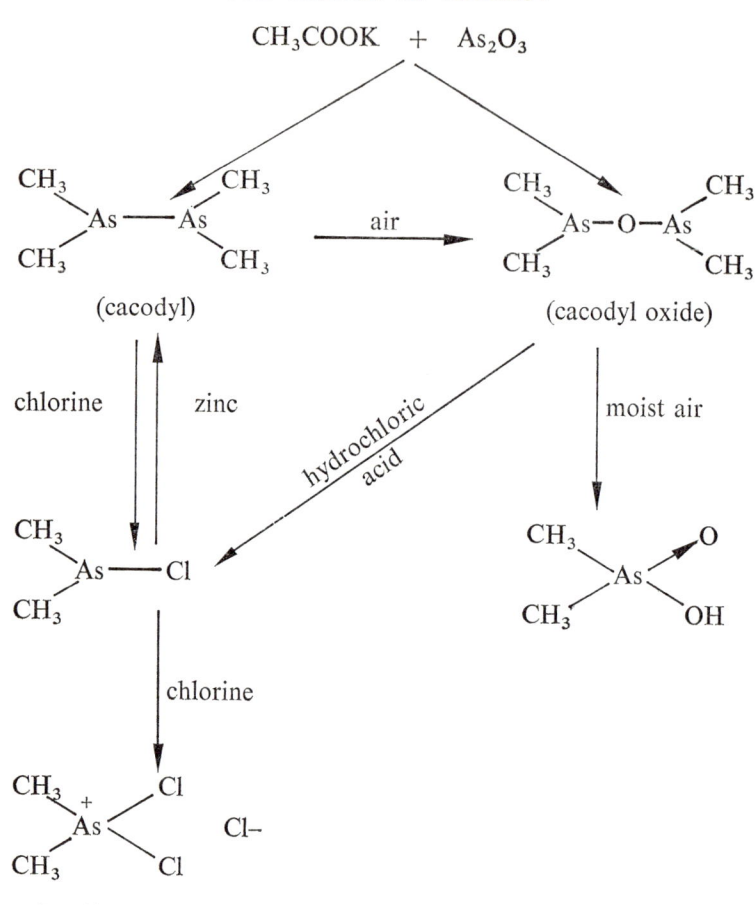

It will be noted that the empirical formula (C = 12, H = 1) for cacodyl is half the molecular formula as indicated by vapour densities. Hence until the application of these in the 1860s, most chemists wrote cacodyl as $As(CH_3)_2$, or in equivalent terms. It thus looked as though cacodyl was a genuine radical – the first to be isolated since cyanogen. Consequently the researches of Bunsen were hailed by the now hard-pressed Berzelius with the utmost joy,[23] and he went so far as to say:

> The research is a foundation stone of the theory of compound radicals of which kakodyl is the only one whose properties correspond in every particular with those of the simple radicals.[23d]

Certainly it enabled – or impelled – Kolbe and Frankland to re-animate the now declining theory of radicals. More than that, it was a discernible milestone on the journey to valency. As Roscoe writes in the Bunsen Memorial Lecture:

The cacodyl research claims our interest, not only because, as we have seen, it furnishes us with the first example of an isolable radical, but also because it assisted Frankland and Kekulé in more exactly illustrating the term "chemical valency".[24]

Admittedly this tribute is inaccurate in its omission of Gay-Lussac's cyanogen, but Bunsen's work came at the psychological moment and was worked out far more fully. Otherwise it is no over-statement, as will be seen. Frankland speaks of Bunsen's discovery as "one of the most important steps in the development of organic chemistry" and the behaviour of cacodyl as "a most remarkable confirmation of the theory of organic radicals, as propounded by Berzelius and Liebig".[25] It is interesting that Kekulé, not in the "radical" tradition, gives no mention to Bunsen in the historical section of his textbook.

2. KOLBE'S REVIVAL OF THE RADICAL THEORY

Hermann Kolbe (1818–84) was eminently fitted by temperament and background to be the protagonist of the New Radical Theory. Possessing literary and polemical skills of a high order, he was capable of the most ardent defence of a position which he considered right; it must be conceded that on occasion his writings were seasoned with a personal invective that now seems regrettable. His experimental and theoretical work were of a very high order, the latter being characterized, however, with a single-minded inability to see the significance of physical data for chemical constitution.

He was rooted in the tradition of German chemistry which, as Kekulé pointed out, tended to favour the concept of radicals.[26] In 1838 he became a pupil at Göttingen of Friedrich Wöhler who had studied under Berzelius and whose friendship with Liebig had already produced the benzoyl researches (see p. 24). This was followed by some years (1842–5) at Marburg, as assistant to Bunsen. At a similar post in London with Lyon Playfair he made the acquaintance of Edward Frankland; the two assistants then spent some months together in Marburg (1847) where they enjoyed the hospitality of Bunsen's laboratory at which the researches on cacodyl were in progress. After a short time in Brunswick editing the *Dictionary of Chemistry* begun by Liebig, Kolbe returned once more to Marburg in 1851, this time to succeed Bunsen who had been appointed to the chair at Heidelberg.

It is therefore not surprising that the doctrine of radicals should have made a deep impression on Kolbe by the late 1840s. As will be seen in the following chapter (p. 45), Berzelius eventually came to view acetic acid as a compound of "oxalic acid" with the copula C_2H_3:

$$C^2H^3 + C^2O^3 + HO$$

The term "copula" had been originally used by Gerhardt[27] to indicate the organic part of a molecule that might be joined to an inorganic part. Berzelius applied these ideas of "copulated" or "conjugated" compounds (*gepaarte Verbindungen*) to trichloracetic acid, and used all the force of his now declining authority to advocate them.

Amongst the earliest work by Kolbe was the successful synthesis of trichloracetic and trichlormethyl sulphonic acids, both of which could be reduced to the parent unchlorinated acids. It thus seemed reasonable to Kolbe to follow Berzelius and write them as

$$C_2Cl_3 + C_2O_3 + HO \quad \text{and} \quad C_2Cl_3 + S_2O_5 + HO$$

and their parent acids as

$$C_2H_3 + C_2O_3 + HO \quad \text{and} \quad C_2H_3 + S_2O_5 + HO$$

(In these formulae C, O and S have half their modern atomic weights.)

To Kolbe, therefore, Berzelius's dualistic formulae still had something to commend them. Impelled by a wider curiosity, he wondered if such views could be extended to the other fatty acids and was joined in this enquiry by Edward Frankland. Their work was started in London shortly before they left for Marburg in May 1847 and was reported in English and (at greater length) in German. The research was significant in several ways; it was Frankland's earliest publication, it used a Liebig condenser in the reflux position almost for the first time, and it marked a definite advance in the fortunes of the Radical Theory.[28]

This work centred round the nitriles of the fatty acids, and identified them with cyanides of alkyl ("alcohol") radicals. Fehling had successfully converted ammonium benzoate to benzonitrile[29] by a reaction closely analogous to the actions of heat on ammonium oxalate and formate. Now if benzoic acid were constituted in a similar way to acetic, benzonitrile must be the cyanide of the phenyl radical, and, moreover, benzonitrile would be analogous to "ethylic cyanide" [ethyl cyanide], so that the latter would presumably undergo hydrolysis in the same way. The test was therefore to attempt to hydrolyse ethyl cyanide with alkali. This was done, and the product on acidification identified as propionic acid.

In this way, propionic acid was identified as "ethyl oxalic acid", with the result that an economy of radicals was now possible in organic chemistry, a specific radical being no longer necessary for each acid; all could be regarded as copulated compounds of oxalic acid. Of this Frankland and Kolbe wrote (in almost identical terms in both papers):

It is probable that nature, in creating the innumerable and manifold products of the organic kingdom by a wonderful combination of those few elements which are at her disposal, likewise makes use of these extensive combining powers of the organic radicals as the simplest means of accomplishing her greatest works.[30]

Shortly after this Dumas was able to convert ammonium acetate to acetonitrile by the action of phosphorus pentoxide.[31]

Meanwhile, Kolbe was working by himself on a new approach to the matter. As an advocate of a theory whose roots were in the electrochemical system of Berzelius, it is not surprising that his fertile mind soon turned in the direction of electrolysis. As a result of what he accomplished, his name has become permanently linked with the electrolytic method of synthesis of hydrocarbons and other compounds. At this time, however, the researches were chiefly important for adding further weight to his view of the copulated nature of fatty acids.

On subjecting the potassium salt of "valerianic acid" to electrolysis,[32] he obtained at the positive pole:

(i) carbonic acid [CO_2];
(ii) an "ethereal oil", shown to be a mixture of:
   (a) "valyl", "C_8H_9" [C_8H_{18}, octane];
   (b) "a valerianic ether" formed from "valyl" + oxygen +

valerianic acid at the positive pole, written as C_8H_9O, $(C_8H_9)\,C_2O_3$ [$C_4H_9.COOC_4H_9$, butyl valerate];
(iii) acid;
(iv) an "odourous gas": carbonic acid; "valyl"; and an olefiant gas, either C_4H_4 or C_8H_8 [butene, C_4H_8].

Thus in isolating "valyl" he thought he had obtained the "radical" attached to the copulated oxalic acid. This was, of course, the dimer of this radical, but the error is less important than it may seem, for the discovery gave substance to the hitherto despised copulae and was the natural complement of the nitrile researches. Kolbe's own representation of what occurred is given thus:

$$\underbrace{HO.\,(C_8H_9)C_2O_3}_{\text{valerianic acid}} + O = C_8H_9 + 2CO_2 + HO$$

followed by two simultaneous but independent reactions:

$$C_8H_9 + O = 2(C_4H_4) + HO,$$
$$C_8H_9 + O + (C_8H_9)C_2O_3 = \underbrace{C_8H_9O,\,(C_8H_9)C_2O_3}_{\text{valerianate of oxide of valyl}}$$

[Present theory would indicate discharge of valerate ions at the anode with production of butyl radicals,

$$C_4H_9COO^- \rightarrow C_4H_9\cdot + CO_2 + e^-$$

followed by disproportionation of radicals as indicated by the overall reaction

$$2C_4H_9\cdot \rightarrow C_4H_8 + C_4H_{10}$$

and formation of butyl valerate from butyl radicals and (probably) the valerate anion.]

In a similar way, Kolbe obtained "methyl", C_2H_3 [actually ethane, C_2H_6], from potassium acetate.

What Kolbe had done by these researches was to give reality to the hypothetical copulae. If it is true, as Kekulé said, that "radicals had melted into copulae; from the radical theory had come a theory of copulae", this dissolution of the radicals had been only in respect of their unchangeability. By the time of Berzelius's death (1848), Kolbe had shown how these radicals were far more than figments of a great imagination, and indeed believed he had isolated them. To quote Kekulé again: "The influence of the copulated formulae maintained by Berzelius, and the union of these with the radical theory, produced the doctrine of paired radicals (Kolbe, 1848)."[33]

Thus when Kolbe settled down to writing his contributions to Liebig's *Dictionary of Chemistry* his own researches on nitriles and electrolyses of organic acids had firmly convinced him of the presence of radicals in organic compounds. Bunsen's work on cacodyl had already run its triumphal course, and Frankland, imbued with the same spirit, was about to begin his own search for radicals in a different field (see p. 35).

One of the most distinctive features attributed to copulae at this time was that their presence did not exert an appreciable influence on the rest of the molecule. Chlorination of the methyl in acetic acid produced a compound that was closely similar (see p. 45). Conjugation of various "alcohol radicals" [alkyl] to "oxalic acid" gave rise to acids of the same neutralizing power as oxalic, and altogether these copulae seemed to exert a very minor influence on the compound's behaviour. This fitted in very well with Kolbe's latest discoveries. But it also appeared to have a direct bearing on the cacodyl problem. Bunsen himself had expressed no opinion on the structure of this "radical", but he had noted that it formed an acid like arsenious acid. Cacodylic and arsenious acids were regarded by Kolbe as respectively

$$(C_2H_3)_2AsO_3 \text{ and } AsO_3$$

Hence the two methyl radicals did not exert much of an effect, and cacodyl could be written as $As(C_2H_3)_2$; *i.e.*, one atom of arsenic was conjugated with "two semi-molecules of methyl".[34] We now know that this argument is strewn with misconceptions, for not only must the cacodyl formula be doubled, but also the cacodylic acid is related to arsenic, rather than arsenious, acid or oxide; there is not much analogy between the two statements:

$$As(C_2H_3)_2 + 3[O] \longrightarrow (C_2H_3)_2 AsO_3$$

and

$$\begin{array}{c}CH_3\\ \\ CH_3\end{array}\!\!\!\!As\!\!-\!\!As\!\!\begin{array}{c}CH_3\\ \\ CH_3\end{array} + 3[O] + H_2O \longrightarrow 2\ \begin{array}{c}CH_3\\ \\ CH_3\end{array}\!\!\!\!As\!\!\begin{array}{c}O\\ \\ OH\end{array}$$

The fatty acids could also be viewed in this light. Up to 1848 Kolbe had regarded acetic acid as

$$HO.(C_2H_3).C_2O_3,$$

a view in full accord with his work on electrolysis, and hydrolysis of the nitriles. But what of its derivatives, acetyl chloride, acetamide, acetaldehyde and the like? By making a slight alteration in this formula he was able to explain the constitution and interconversion of these, and at the same time retain the advantages of a structure where methyl was clearly indicated. Moreover the series of acetyl compounds were brought into line with the cacodyl family. In short, he regarded acetic acid as an oxygen compound of the radical then known as acetyl ($C_2H_3C_2$; $C = 6$), first put forward by Liebig as C_4H_6,[35] but where "the last C_2 alone forms the connecting-link for the oxygen, the methyl being in some way only an appendix".[36] So the view was expressed, and emphasized in italics,

that in acetyl compounds there actually exists an acetyl radical, but that this cannot, as hitherto, be considered as an atomic complex composed of four equivalents of carbon and three equivalents of hydrogen, whose four carbon equivalents have equal functions, but it is much more likely to be composed of two equivalents of carbon, with methyl as the copula:

$$acetyl = (C_2H_3)C_2$$

in which C_2 serves exclusively as the point of attack of the binding power of oxygen, chlorine, etc.[36a]

This viewpoint would apply to a number of acetyl compounds as:

$HO.(C_2H_3)C_2O_3$ acetic acid
$HO.(C_2H_3)C_2O$ [acet-]aldehyde
$HO.(C_2H_3)C_2 \begin{cases} O_2 \\ NH_2 \end{cases}$ acetamide

Analogous constitutions would follow for other fatty acids as

$HO.HC_2O_3$ formic acid
$HO.(C_4H_5)C_2O_3$ propionic acid

and also the dithionic acids containing the conjugate sulphur radicals

$$(C_2H_3)S_2 \text{ and } (C_4H_5)S_2$$

Kolbe foresaw the existence of conjugated radicals with other elements as selenium, phosphorus and antimony, a prediction that Frankland and others were soon to bring to pass.

We can generalize all these suggestions in the form

$$R_nX$$

where R is an alcohol [alkyl] radical or hydrogen, n an integer, and X an atom of a non-metal (including the "double atom" C_2) to which everything else in the molecule is attached. These other addenda are all electronegative atoms or groups.

It is our contention that this idea of Kolbe's represents his final contribution to the birth of valency theory before the publication of Frankland's paper of 1852, and that this contribution was highly significant – more so than is often acknowledged. There are at least three seminal ideas of the greatest importance here, apart from the general recognition of the existence of radicals which was shared by others at the time.

In the first place there is the acknowledgment of a definite grouping of atoms and one that is accessible to experiment. This conflicted directly with the views of Gerhardt and his school (see p. 47) and came now at a time when "structural agnosticism" in chemistry was at its highest. Moreover it was the first theory for the structure of cacodyl, and to the adherents of dualism will have had a strong appeal. It seems, therefore, that the clarity and definiteness of these views must have heightened the barrier between the adherents of Radical and Type Theories.

Secondly, and of more importance, Kolbe's speculation on the structure of these molecules represents a new advance in an analytical view of radicals. For it was his constant aim to reduce organic structures to their lowest factors, and this included breakdown of complex radicals to simpler components. Thus, acetic acid having been "broken down" into methyl and oxalyl, it was further analysed into methyl, carbon and oxygen; acetyl was no longer the ultimate unit, and cacodyl, itself a radical, was expressed in terms of the simpler methyl. The logical consummation of this process was, of course, analysis into individual atoms, but this crowning achievement was not to be Kolbe's. However, his influence on Frankland could scarcely have been other than profound; and this refusal to accept a radical as the final point of analysis was to lead ultimately to the concept of radicals becoming a useful guide to structure but no longer a necessity.

Finally, in this work Kolbe focused attention for the first time on the individual atom in a molecular context. Here was an atom linked directly to an alkyl radical and capable of joining up with several atoms of oxygen or other electronegative element. This concentration on one atom "as the point of attack of the binding power" is getting very near to valency.

Nevertheless, Kolbe did not arrive at the concept of valency himself, though he eventually accepted Frankland's view of it (see p. 132). The greatest weakness of his arguments lay in his acceptance of a copula that had very little influence; in the case of cacodyl, as we have seen, this was founded on an entirely false analogy. His use of the double carbon atom and of copulated water were smaller points against him. Were it not for these disabilities it is probable that the Kekulé–Couper–Butlerov controversy on structure theory would have been rendered unnecessary by a general recognition that its founder was in fact Hermann Kolbe.

3. FRANKLAND'S WORK LEADING UP TO THE DOCTRINE OF SATURATION CAPACITY

Edward Frankland[37] was born in 1825 at Churchtown, near Garstang, Lancashire, spending most of his early life in that district and being apprenticed to a pharmacist in Lancaster. Apart from some informal tuition from medical friends and others, his years in Lancashire seemed scientifically barren. In 1845 he was able to leave the north for the London laboratory of Lyon Playfair, a chemist in the Government Department of Woods and Forests, and shortly became Playfair's assistant at the Putney College for Civil Engineers.

In London he made the acquaintance of a number of able and enthusiastic chemists at Playfair's laboratory and in the Chemical Society which he joined in 1847. Most momentous of his friendships was that with Kolbe, also an assistant to Playfair, who infected Frankland with his own zeal for discovering the constitution of organic compounds. Together they completed in London their work on the hydrolysis of ethyl cyanide,[38] Frankland's first published research. Kolbe was by now tired with the mineralogical analysis he was required to do, and for which he had a "supreme contempt".[39] He persuaded Frankland to accompany him for a short while to Bunsen's laboratory at Marburg. Here they continued their researches on nitriles,[40] but Frankland returned in the autumn of the same year (1847) to take up a teaching appointment at Queenwood College, Hampshire.

There can be no doubt that the three months spent in Bunsen's laboratory had a profound effect on Frankland's outlook. In particular he was impelled to seek for ways of isolating other radicals than cacodyl, and especially the alkyl radicals whose isolation Liebig had foretold years before.[41] He wrote that: "The isolation of the alcohol radicals was, at this time, the dream of many chemists, whilst others doubted or even denied their very existence. I was also smitten with the fever, and determined to try my hand at the solution of the problem."[42]

At the same time as their researches on nitriles at Marburg, Kolbe and Frankland turned their attention to the isolation of the radical ethyl. Their efforts were unsuccessful. In choosing their reagents, they were guided by the conditions Bunsen himself had pointed out for the separation of organic radicals by the action of metals or liquids: the temperature must be less than the boiling point of the liquid, and the inorganic compound formed must be soluble in the new radical.[43] They chose to act on potassium with ethyl cyanide, hoping the known affinity for cyanide possessed by the metal would enable them to release ethyl at a low enough temperature.[44]

They did not obtain ethyl. Instead they found a gas and a water-insoluble white solid. The gas, in its gravimetric composition, appeared to be "methyl", but on chlorination gave a gas corresponding to the expected chloride of ethyl but differing in physical properties. In other words it differed from the "methyl" that had been obtained previously by electrolysis of acetic acid. The situation appears to have been this: the "methyl" is of course our ethane $[C_2H_6]$, and the behaviour of the resultant chloro-compound must have been due to air being present. This Frankland admitted later,[45] but the confusion was made worse by his statement that

It is highly probable that the so-called methyl gas, generated by the decomposition of cyanide of ethyl by potassium is also the hydruret of ethyl, and therefore only isomeric [=polymeric] with the true radical methyl....[46]

In fact all so-called "methyl" samples were ethane, though this one must have been impure. It is certainly true that the ethyl cyanide was impure (as again Frankland admits[47]), although it had been dried with calcium chloride. Water or alcohol must have been present to account for the formation of ethane at all. The solid product, called by them kyanethine, was probably an aminopyrimidine, from trimerization of ethyl cyanide:

[Chemical structure: pyrimidine with CH₃, C₂H₅, H₂N, and C₂H₅ substituents]

This research, interesting though it was, did not succeed in its chosen object, the isolation of ethyl. Shortly afterwards, Frankland was recalled to England to take up his duties at Queenwood. He was soon to find that his teaching appointment left little time for research, but opportunities could be made, and on 28 July 1848 he performed an experiment that was to become a classic in the history of science.[48]

Undeterred by his previous failure, he decided to change his reagents. At first (10 April) he had tried to use potassamide and then potassium to remove oxygen from ether and to liberate ethyl. This was of course unsuccessful and he turned his attention to ethyl iodide. He claims he was the first to use this material in research, and that he was led to it by comparison with hydrogen iodide whose ease of decomposition was well known. Assuming the ethyl compound might be similar, he tried heating it with potassium, and analysed the gaseous products in a eudiometer. These were found to be ethylene and "ethylic hydride" only, in equal volumes (April and May).[49] Replacement of ethyl cyanide by the simpler iodide was thus not successful, but the possibility remained of exchanging the alkali metal with a less reactive one, and thus obtaining the ethyl radical undecomposed. In this way Frankland was led to heat together zinc and ethyl iodide in a sealed tube immersed in an oil-bath. The reaction was performed on 28 July following a trip to the Continent and some abortive attempts to prepare formyl from chloroform and potassium.

It was not possible to analyse the contents of the tubes at the time as Frankland's only eudiometer had exploded in May and he had no access to any other. Fortunately, however, Frankland's time at Queenwood was now nearly over, as he and Tyndall had been invited to join Bunsen at Marburg. They arrived there in October, Frankland still in possession of his sealed tube! Construction of a new eudiometer was speedily accomplished, but for various reasons the opening of the tube under water had to wait until the following

February (18th). An immediate and violent reaction occurred; a liquid, originally present in the tube, entirely volatilized, and produced a gas, about 40 times its own volume, inflammable and capable of absorption by alcohol. White crystals also present effervesced with the water to give zinc iodide and a gas similar to the first. The two gases were analysed eudiometrically, and each shown to be a mixture. The following were the results:[50]

TABLE II

Frankland's names and formulae	Modern names and formulae	% in first gas	% in second gas
elayl C_2H_2	ethylene C_2H_4	21.7	2.78
ethyl C_4H_5	butane C_4H_{10}	50.3	22.41
methyl C_2H_3	ethane C_2H_6	25.79	74.81
nitrogen —	nitrogen N_2	2.48	—

Frankland was of course using the atomic weight C = 6. Analysis of ethyl iodide on this basis gave ethyl not as C_2H_5 but as C_4H_5. Also butane will have the same *empirical* formula as ethyl, but owing to the disregard for Avogadro's hypothesis, a doubling of this to give the molecular formula was not then contemplated. Hence it was impossible for Frankland to observe any difference between an ethyl radical and its dimer, butane. Thus he believed he had isolated ethyl and so fulfilled the long-cherished ambition of a great many chemists of the day.

In the same paper[51] he gave the properties of ethyl and referred to the demonstration of its homogeneity by diffusion (an innovation in chemistry). He also described the production of "methyl" [ethane] from zinc and water and ethyl iodide, identical with that obtained from potassium and ethyl cyanide (p. 35). The action of zinc and ethyl iodide was tried in the presence of alcohol and of ether, and ethyl iodide alone was treated with several other metals and with arsenic.

Having completed his investigation on "ethyl" Frankland now turned to investigate "methyl", and on 12 July 1849 opened two tubes containing the reaction products of zinc and methyl iodide. The gas first liberated was identified as "pure light carburetted hydrogen" [methane], and this was accompanied by a poisonous, spontaneously inflammable gas formed here for the first time and shown by Frankland to be zinc methyl. This reacted with water as rapidly as potassium and evolved two more equivalents of methane. Analogously, crude zinc ethyl was made from ethyl iodide and zinc, and was shown to be closely similar. The formulae given by Frankland were respectively $(C_2H_3)Zn$ and $(C_4H_5)Zn$ [modern formulae $(CH_3)_2Zn$ and $(C_2H_5)_2Zn$]. These results were communicated to the Chemical Society in November of that year.[52]

These results were highly significant for chemistry. They gave the world the first organo-metallic compounds, and opened the way to the synthetic possibilities of Grignard's discoveries. The so-called radicals "methyl" and "ethyl", discovered in the home of cacodyl, provided added weight to the arguments in favour of the radical theory. When Frankland left the laboratory of Bunsen for a few months at that of Liebig, it was not surprising that he should extend his work to the preparation of amyl,[53] which, when successfully completed, confirmed yet further the view of Liebig (see p. 23) on the constitution of alcohol and ether.

Frankland's views on the constitution of his "radicals" gave rise to a lively controversy. Viewed in retrospect, the arguments seem rather sterile, and the neglect of Avogadro almost incomprehensible. Almost immediately Gerhardt and Laurent saw in Frankland's results a further illustration of the former's law of homology, and that the so-called radicals were probably homologues of marsh-gas, necessitating a doubling of Frankland's formulae. Frankland kept to his position, however, and in his paper on amyl compared his products with hydrogen, which is fairly unreactive in the free state, but active "*in statu nascenti*", giving rise to many compounds, such as "hydrurets" like "light carburetted hydrogen", C_2H_3+H, *etc.*[54] Hofmann attempted unsuccessfully to resolve the issue by experiment and took up the dubious implication that hydrogen was unreactive. At least it combined readily with chlorine, which could not be said of Frankland's radicals. But after a dispassionate survey of both sides of the question he came down, reluctantly it seems, "in favour of the lower formulae".[55] Frankland at once replied, dealt with Hofmann's objections to his contentions, and announced some eudiometric evidence of his own, some of which results are extremely hard to explain.[56] So at least he admitted years later, in saying that his results were then "nearly the only evidence in support of the non-identity of the two series of hydrocarbons", and suggesting a repetition of the experiments.[57] The crux of the problem – the issue of whether and when formulae should be doubled – was pointed out by Brodie in a paper that clearly set out the issues, and concluded on the side of Gerhardt.[58] But it was not until 1864 that Schorlemmer finally established the identity of Frankland's radicals with the paraffins (see p. 155).

Now this controversy as to the nature of Frankland's "radicals" raged strongly, but for only a few months. Its effect, however, was to reinforce that of Kolbe's paper of the same period,[59] to keep alive the concept of a radical, and to stimulate Frankland still further to experiment with organo-metallic compounds in the hope of making yet more "radicals". Indeed, the organo-metallic compounds themselves were regarded as radicals. As he wrote of one in 1849: "It is highly probable that this body which for the present I propose to call *Zincmethyl*, plays the part of a radical, combining directly with oxygen, chlorine, iodine, *etc*."[60] There was thus the double attraction of discovering more radicals of the cacodyl type, and using them to produce "alcohol radicals". In the same paper he predicted some twenty more compounds of the kind known today as metal alkyls.

Therefore when Frankland left the Continent for England in the winter of 1849/50 he came with every stimulus to pursue further his work on organo-metallic compounds. He was brought back to this country by a call to the Chair of Chemistry at the College of Civil Engineering, Putney. This was followed later the same year by the Chair at Owen's College, Manchester.

At Putney Frankland was able to bring about the exposure of various reaction mixtures to strong sunlight. On the roof of his laboratory, unscreened by neighbouring buildings, he made good use of the increase in daylight as the spring advanced in 1850 (see Plates 11 and 12). Various metals and various alkyl halides were reacted together, and the sun's rays were focussed upon them. Remarkable changes began to occur and a whole new tract of organic chemistry became gradually revealed. More important still, a consideration of some of the relationships disclosed set in motion a sequence of ideas that was to result directly in the theory of valency. In Frankland's own words:

I had not proceeded far in the investigation of these compounds at Putney before the facts brought to light began to impress upon me the existence of a fixity in the maximum combining value or capacity of saturation in the metallic elements which had not before been suspected.[61]

The paper in which these ideas were first announced was communicated to the Royal Society on 10 May 1852.[62] Its publication was, however, delayed by a year as it "was inadvertently laid aside in his private drawer by the Secretary (Professor Stokes)".[63] It was also published by the Chemical Society.[64]

Frankland describes first the preparation of "iodide of stanethylium", "C_4H_5SnI", from tin and ethyl iodide in strong sunlight. On successive treatment with alkali, dilute hydrochloric acid and zinc this gave rise to a yellow oil, "C_4H_5Sn", which he termed "stanethylium". It boiled, with decomposition, at 150°, and combined with air to form an oxide, and with hydrochloric, hydrobromic and hydriodic acids to form, respectively, chloride, bromide and iodide of stanethylium. The relationships that Frankland discovered may be summarized thus:

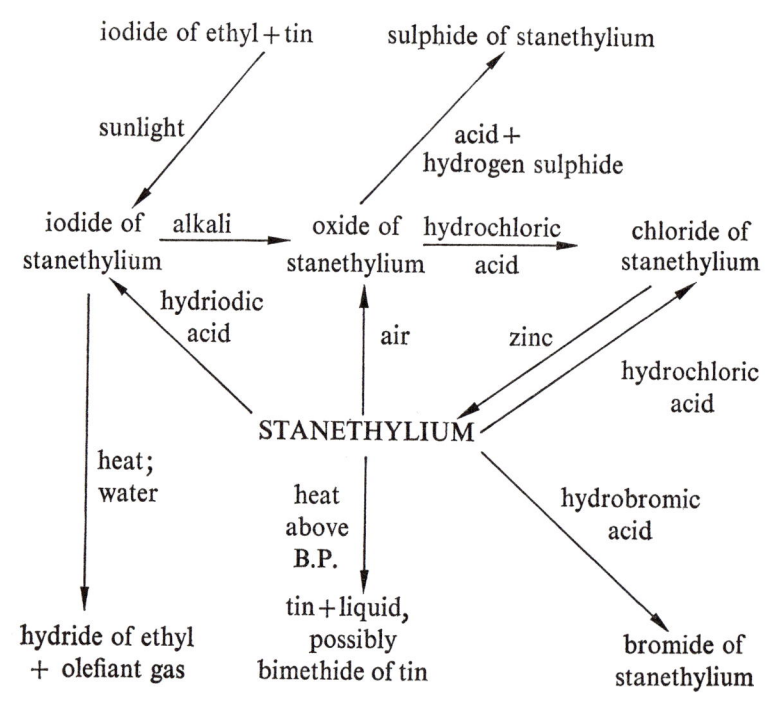

```
iodide of ethyl + tin           sulphide of stanethylium
            \                        ↗
    sunlight \                      / acid +
              ↘                    /  hydrogen sulphide
  iodide of   alkali   oxide of   hydrochloric    chloride of
  stanethylium ─────→ stanethylium ───────────→  stanethylium
      │  ↖ hydriodic       ↑           ↗
      │    acid           air   zinc ↗
      │                    │       ↗      hydrochloric
      │                    │     ↗            acid
      │              STANETHYLIUM
      │ heat;              │       ↘
      │ water             heat      ↘  hydrobromic
      │                   above       ↘    acid
      ↓                   B.P.          ↘
  hydride of ethyl         │              ↘
  + olefiant gas      tin + liquid,       bromide of
                      possibly            stanethylium
                      bimethide of tin
```

Here, then, was another substance like zinc ethyl capable of rapid combination with the non-metals; more important still,

> stanethylium perfectly resembles cacodyl in its reactions, combining directly with the electronegative elements and regenerating the compounds from which it has been derived.[65]

Frankland also reported the action of zinc and methyl iodide at 150°C to give a gas, perfectly resembling "the methyl procured by Kolbe for the electrolysis of acetic acid", some crystals, and a colourless liquid that was shown to be zinc methyl. Descriptions followed of zinc ethyl and amyl, and of methyl mercury iodide.

Having described his new products, Frankland then turned to theories upon them. The great reactivity of his organo-metallic compounds offered an obvious parallel to the behaviour of cacodyl, and this was duly noted. If Kolbe was right (see p. 31) in regarding cacodyl as "arsenic conjugated with two atoms of methyl", these must be deemed to have the "alcohol radicals" conjugated on to zinc, tin, *etc.* But conjugation, in Kolbe's sense, did not alter the "essential chemical character" of a substance. In that case, therefore, stanethylium should closely resemble tin, as indeed cacodyl should resemble arsenic.

> If therefore we assume the organo-metallic bodies above mentioned to be conjugated with various hydrocarbons, we might reasonably expect that the chemical relations of the metal to oxygen, chlorine, sulphur, *etc.* would remain unchanged.[66]

But in fact *this was not so*, even for cacodyl, and now the new organo-metallic compounds reiterated the difficulty. With the arsenic compounds one would expect two stages in oxidation:

cacodyl → protoxide of cacodyl → further product (X)
 and cacodylic acid

corresponding to

arsenic → protoxide of arsenic → arsenic acid
 and arsenious acid

In fact the second product (X) for cacodyl was not found, even when cacodylic acid was distilled with arsenic acid. Moreover, protoxide of cacodyl dissolves readily in hydrochloric acid, whereas Frankland believed that protoxide of arsenic [As_2O_3] had no well-defined basic character.

Stanethylium was as difficult. Like tin it should, if conjugated, give rise to a protoxide and peroxide, but Frankland was "quite unable to form any higher oxide than that described".[67] Only on removal of ethyl would another equivalent of oxygen go in. Stibethyl posed the same kind of problem.

So it was that the facts forced themselves upon Frankland with such directness that he was impelled to write:

When the formulae of inorganic chemical compounds are considered, even a superficial observer is struck with the general symmetry of their construction; the compounds of nitrogen, phosphorus, antimony and arsenic especially exhibit the tendency of the elements to form compounds containing 3 or 5 equivalents of other elements, and it is in these proportions that their affinities are best satisfied; thus in the ternal group we have NO_3, NH_3, NI_3, NS_3, PO_3, PH_3, PCl_3, SbO_3, SbH_3, $SbCl_3$, AsO_3, AsH_3, $AsCl_3$, *etc.*; and in the 5-atom group NO_5, NH_4O, NH_4I, PO_5, PH_4I, *etc.* Without offering any hypothesis regarding the cause of this symmetrical grouping of atoms, it is sufficiently evident, from the examples just given, that such a tendency or law prevails, and that, no matter what the character of the uniting atoms may be, the combining power of the attracting element, if I may be allowed the term, is always satisfied by the same number of these atoms.[68]

In the light of this Frankland considered that the radical and type theories could perhaps be reconciled. Instead of regarding the "conjugate organic radicals" as analogous to the metal, it would be better to deem them based on the metal's oxygen, sulphur or chlorine compounds as the true molecular types; he gave the following table:[69]

Inorganic types	*Organo-metallic derivatives*	
As $\begin{cases} S \\ S \end{cases}$	As $\begin{cases} C_2H_3 \\ C_2H_3 \end{cases}$	cacodyl
As $\begin{cases} O \\ O \\ O \end{cases}$	As $\begin{cases} C_2H_3 \\ C_2H_3 \\ O \end{cases}$	oxide of cacodyl
As $\begin{cases} O \\ O \\ O \\ O \\ O \end{cases}$	As $\begin{cases} C_2H_3 \\ C_2H_3 \\ O \\ O \\ O \end{cases}$	cacodylic acid

ZnO $\qquad$ Zn(C$_2$H$_3$)

$$\text{Zn}\begin{cases} \text{O} \\ \text{O}_x \end{cases} \qquad \text{Zn}\begin{cases} \text{C}_2\text{H}_3 \\ \text{O}_x \end{cases}$$

$$\text{Sb}\begin{cases} \text{O} \\ \text{O} \\ \text{O} \end{cases} \qquad \text{Sb}\begin{cases} \text{C}_4\text{H}_5 \\ \text{C}_4\text{H}_5 \\ \text{C}_4\text{H}_5 \end{cases}$$

$$\text{Sb}\begin{cases} \text{O} \\ \text{O} \\ \text{O} \\ \text{O} \\ \text{O} \end{cases} \qquad \text{Sb}\begin{cases} \text{C}_4\text{H}_5 \\ \text{C}_4\text{H}_5 \\ \text{C}_4\text{H}_5 \\ \text{O} \\ \text{O} \end{cases} \text{ and } \text{Sb}\begin{cases} \text{C}_4\text{H}_5 \\ \text{C}_4\text{H}_5 \\ \text{C}_4\text{H}_5 \\ \text{C}_4\text{H}_5 \\ \text{O} \end{cases}$$

binoxide of stibethine $\qquad$ oxide of stibethylium

SnO $\qquad$ Sn(C$_4$H$_5$)

$$\text{Sn}\begin{cases} \text{O} \\ \text{O} \end{cases} \qquad \text{Sn}\begin{cases} \text{C}_4\text{H}_5 \\ \text{O} \end{cases}$$

$$\text{Hg}\begin{cases} \text{I} \\ \text{I} \end{cases} \qquad \text{Hg}\begin{cases} \text{C}_2\text{H}_3 \\ \text{I} \end{cases}$$

With one slight and doubtful exception, all known cases are covered in this way. After one or two brief comments, and a suggestion for changes in nomenclature, he brings this classic paper to a close.

Thus the Theory of Radicals, winding its tortuous way through the dark forest of organic chemistry, with innumerable checks and many fundamental errors, emerged vindicated by the work of Frankland. But its triumph was short-lived. For not merely were Frankland's "radicals" soon to be shown to be not true radicals but dimers;[70] not only was the supposed similarity in reactivity between cacodyl and the zinc alkyls quite spurious, and their activity due to quite different causes; but in this actual culminating paper was the admission that the electrochemical (radical) viewpoint was only one of several, and the Type Theory was evoked as the climax of the argument. And from this reconciliation emerged one thing far greater than either: the notion that elements have a definite combining power.

CHAPTER III

From Types to Valency

1. THE RISE OF THE THEORY OF TYPES

THE ORIGINS of the theory of valency lie not only in the changing ideas of radicals. They may also be discerned in the doctrine of types, and we must now glance back at the developments taking place in this field of speculation.

To the twentieth century the Theory of Types has several curious features, not least of which seems to be the almost perverse refusal of its adherents to extrapolate just one stage further and arrive at once at Valency and Structure Theory. That they ultimately did so is not surprising; the astonishing thing is that it took so long. A chemist who held these views (as Kepler said of Copernicus) "failed to see how rich he was" and indeed without the final extension to rational formulae he was hardly in a better position than those adherents of the rival viewpoint, the Theory of Radicals. In the event both theories were to lead, almost independently, to valency, but the more spectacular contributions came from the "typists" whose views dominated chemistry in the middle of the last century.

The development of the Type Theory is well known, and for this reason only a brief summary is needed here.

Fundamentally the Type Theory was an attempt to bring some kind of order into the bewildering chaos of organic chemistry, especially in those cases where the Electrochemical (or Radical) Theory was unsatisfactory. The latter had been applied triumphantly by Berzelius, Liebig, Kolbe and others to the breakdown of molecules by electrolysis and to the ejection of groups of atoms from molecules by other atoms having the same electrochemical character, but to a greater degree. This included displacement of the "free radicals" themselves by the action of alkali metals on alkyl halides. But for a reaction

$$A + BC \longrightarrow AC + B$$

A and B had to be of similar electrochemical character. Consequently the theory received a severe blow when it was discovered that chlorine (electronegative) could replace hydrogen (positive).[1] In the light of dualism this was incomprehensible. The greatest difficulty arose with acetic acid, which Berzelius had written as an oxide of a hydrocarbon radical, *e.g.*, as

$$C^4H^6 + 3O + H^2O$$

(by a common error doubling the true molecular weight).[2] The chlorinated product he refused to write as

$$C^4Cl^6 + 3O + H^2O$$

but represented it instead as

$$H^2O + C^2O^3 + C^2Cl^6$$

so avoiding direct replacement of H by Cl.[3] These units were termed "copulae", the antecedents of radicals in the electrochemical sense.

Even at this time, however, formulae were regarded at least in some sense as expressing chemical behaviour, and the gross dissimilarity between these two was underlined by the discovery that in fact trichloracetic acid could be readily converted back to acetic acid by reduction, and that the two acids were closely related.[4] For this reason Berzelius resorted to

$$H^2O + C^2O^3 + C^2H^6$$

for acetic acid,[5] thereby admitting direct replacement of H by Cl in the copula.

Such a face-saving retreat gained little support from his contemporaries and won much bitter criticism,[6] although it is interesting that Japp, in his Kekulé Memorial Lecture,[7] defends it as "perfectly legitimate" since the copulae are not ionizable, and the electrochemical theory strictly applies only when ionization takes place, so that Berzelius is here merely withdrawing from a position of excessive zeal.

But the total effect of these discoveries was generally to discredit a dualistic, electrochemical analysis of molecules into copulae or radicals, and to focus attention instead on molecules as wholes or units. This tendency was emphasized by the incompetence of the Radical Theory – at least as Berzelius held it – to explain acids other than as oxides of radicals or (later) as hydrates of oxides. That these views were wrong was indicated by a growing body of evidence from Berthollet,[8] Davy,[9] Dulong,[10] Graham,[11] Dumas and Liebig,[12] and above all Gerhardt.[13] The last-named showed that acetic acid has the single formula $C_2H_4O_2$, and thus trichloracetic acid is $C_2HCl_3O_2$ which excludes altogether the possibility of water, H_2O, existing in it. Hence "there is no water in our acids and no oxide in our salts".

Into the theoretical vacuum created by the collapse of the electrochemical system came the new views of organic compounds known as the Unitary and Type Theories, at the centre of which was the chemical school at Paris.

J. B. A. Dumas (1800–84) arrived in Paris from Geneva in 1823. There he spent the rest of his life, establishing his own laboratory in 1832, and there he began his work on substitution reactions. In 1834[14] he published his laws of substitution and in 1839[15] his Theory of Types. The similarity in the properties of acetic acid and its chlorinated product led him to suppose that "in organic chemistry there are certain types which remain unchanged, even when their hydrogen is replaced by an equal volume of chlorine, bromine and iodine". By "types" he meant substances containing the same number of equivalents united in the same way and showing the same fundamental chemical properties, as acetic and trichloracetic acids, chloral and acetaldehyde, *etc.* To these "chemical" types he added "mechanical" types which were similar in formulae but not in chemical properties, an idea that he seems to have borrowed from Regnault.[16] The essence of Dumas' theory was the oneness of a chemical compound, and that its properties "are determined by the positions of the particles and far less by their chemical nature".

Dumas did not arrive at these conclusions at once. Following his papers of 1834, he did not extend his laws of substitution to the doctrine of types immediately. Instead, his assistant, Auguste Laurent, developed his ideas to such a degree that Dumas himself at first disowned them,[17] though later came to substantially the same opinion. Laurent's Nuclear Theory[18] regarded organic compounds as "fundamental nuclei", consisting of carbon and hydrogen, and "chemical nuclei" obtained by replacement in the former of hydrogen by other atoms or groups. The primary nucleus he compared to a prism whose edges were hydrogen and whose angles carbon. Removal of hydrogen would cause collapse (decomposition) unless it were replaced by another atom. Sceptical at first, Dumas was won over to these ideas by his own discovery of chloracetic acid, and his influence undoubtedly was greater than that of Laurent, whose theories were, however, adopted by Gmelin, and exerted a profound influence on Gerhardt. With Melsens' discovery[19] and Berzelius' reactions to it, the fortunes of the Radical Theory were at their nadir.

In 1850, Hofmann[20] and Williamson[21] were able to show respectively that amines can be regarded as substituted ammonia, and alcohols and ether are derivatives of water. Thus the "ammonia-" and "water-types" were established. But the chief architect of the victory of the type theories was Charles Gerhardt (1816–56). Having studied under Liebig he spent many years in Paris where he collaborated closely with Laurent. The systematization by Gerhardt with its interpretation and promulgation by his colleague was to have a profound influence on chemistry.

Gerhardt's theory of residues put forward in 1839[22] retained the unitary view of Dumas, but assumed that double decomposition reactions involve the rearrangement of "residues", or atomic complexes left over after reaction. In the nitration of benzene, for example, the hydrogen of the hydrocarbon combines with the oxygen of the nitric acid, and the two residues left ("phenyl" and "nitro") unite to give nitrobenzene. These residues differed from radicals in that they did not have an independent existence and were merely expressions of the mode of reaction. As a piece of information this is completely useless; as a linking of aspects of the Type and Radical Theories (substitution by radicals in types) it marked a definite advance.

Later, in 1853,[23] Gerhardt systematized existing speculation on types, and proposed four of these: ammonia, water, hydrochloric acid and hydrogen. Again substitution in the types by radicals was allowed, and these replacing radicals could be composed of simpler ones, *i.e.* they could be "conjugated". But in all this there was no hint that such radicals actually existed or that the internal structure of a molecule could ever be determined. Indeed the last possibility is explicitly rejected again and again in Gerhardt's writings. So persistent was this "structural agnosticism" that when Kolbe added a "carbonic acid type" in 1855,[24] although an opponent of Gerhardt's views generally, he was unable to admit any structural significance whatever for his own representations.

This aspect of the Type Theory was one of its most conspicuous (and damaging) features; chemical formulae merely indicated what reactions a compound might undergo, but said nothing of the ways in which its atoms were united. In a curious way it reflected a number of aspects of mid-nineteenth-century philosophy. It was, of course, essentially a French phenomenon; Dumas, Laurent, Gerhardt and others were all working in Paris where one of the most influential figures was the philosopher Auguste Comte (1798–1857). The founder of positivism, Comte taught that science progressed from a theological interpretation of nature to a metaphysical one and from there to a "positive" view where final causes are no longer sought and outward phenomena are all that matter. No philosophy could be more damaging to any search for explanations *behind* empirical facts, and it is surely more than a coincidence that its chief expositor was the friend and teacher of some of the leading chemists of Paris.[25]

Moreover, the use of analogy that is the basis of the Type Theory was common enough in nineteenth-century science. In fact it probably reached a maximum at that time, in biology, chemistry and geology. "It was not enough that compounds were; as with the flowers and fossils of the nineteenth century they had to be like."[26] Comparison of all organic compounds with four simple inorganic types was therefore quite in keeping with the general scientific outlook.

Yet again, the Type Theory was essentially a unitary hypothesis. A "Gestalt" approach to an entity as complicated as a molecule is not only psychologically understandable, we can now say that it is chemically defensible, for each molecule *is* a unit and is different from a "linear combination" of its atomic constituents. And although it was based essentially on a limited range of reactions (substitutions or double decompositions), the same criticism applies to the Radical Theory at the root of which are the facts of electrolysis.

Above all the Type Theory, like its electrochemical rival, represented a definite groping towards the doctrine of valency. Frankland has rightly attributed to Laurent and Dumas a glimpse of what he himself realized in 1852 (see p. 42), and suggested that it was on account of this that their Type Theories came into being.[27] Their influence would have been far greater if they had not insisted on the dependence of property on atomic position rather than nature. With this we may well agree.

2. THE CONCEPT OF POLYATOMIC RADICALS

As the Theory of Types was reaching its culmination in Paris with the work of Gerhardt, new developments were taking place in London. Conceived at the time merely as modifications to the Type Theory, they played a vital part in paving the way to valency. It will be convenient first to mention briefly the most important figures involved in these developments.

By the middle of the century the Chair of Chemistry at University College, London, was occupied by A. W. Williamson (1826–1904), who had been a pupil of Liebig and a friend of the philosopher Comte. Here, at University College, he produced some valuable results in the years 1850–60, but after this, till 1887, diverted nearly all his main energies to teaching, and thereby exerted a widespread influence for the ideas he held. Like many others in London at the time he accepted the principle of Types, but in a rather more liberal fashion than many and with more respect for radicals than would be expected from a thoroughgoing "Typist".

William Odling (1829–1921) had been a pupil of Gerhardt in Paris before he settled in London, first as Medical Officer of Health for Lambeth and then as lecturer in chemistry at St Bartholomew's Hospital in the year (1863) that Frankland exchanged this post for a Chair at the Royal Institution. An upholder of the Parisian Type Theory, he translated into English Laurent's *Chemical Method* in 1855 (correcting the atomic weights to Gerhardt's new values of $C = 12$, $O = 16$, *etc.*). In his books and papers he used typical formulae, but with the same degree of freedom that characterized Williamson, whose pupil and friend he was.

The chief participant in the developments that were to ensue in London was, however, neither of these. From December 1853[28] to the autumn of 1855 chemical circles in London were enlivened by the appearance of the youthful August Kekulé,[29] assistant to John Stenhouse at St Bartholomew's Hospital. Born in 1829 in Darmstadt, Kekulé abandoned an early inclination to study architecture and became a pupil of Liebig, at Giessen, following this with a year in Paris where he came under the spell of Gerhardt's teaching, and was even given the manuscripts of Gerhardt's famous *Traité* before it was published. This first-hand contact with the chief protagonists of the Type Theory influenced him for many years to come. His collaboration with Stenhouse in London (on Liebig's suggestion) was of very little value, and the work monotonous and dull. He was, however, able to direct his boundless energies into "unofficial" channels and conduct exploratory experiments that were to have great significance. Moreover, his acquaintance with Williamson broadened into friendship and respect, and the tedium of his work at the Hospital was frequently relieved by visits to University College. He spoke appreciatively of Williamson's clarity of formulae and thought, and comments: "This was excellent schooling for making the mind independent."[30] Odling, also, helped Kekulé in acquiring that open-minded attitude to types that characterized this group of chemists. The two men were of the same age and both published a major paper in the same year (1854).[31] When, in 1856, Kekulé had returned to Germany as Privatdocent at Heidelberg, he wrote to Williamson, speaking of his constant advocacy of Williamson's and Gerhardt's views in his lectures, and his dismay at the hold on inorganic chemists of "the swindle of electrochemical and dualistic addition".[32]

Mention must also be made of A. W. von Hofmann (1818–92), another former pupil of Liebig's, who had been Professor at the Royal College of Chemistry in London since 1845. Although his direct contribution to the theory of polyatomic radicals was slight, his experimental work on amines led to the recognition of the ammonia type, which with Williamson's water type, formed the basis of Gerhardt's theory.

Meanwhile one distinguished English chemist was absent from London for a period almost exactly co-extensive with that in which these ideas were circulating in the capital. From 1851 to 1857 Edward Frankland was occupying the first Chair of Chemistry at Owen's College, Manchester. Whether his absence had any effect upon the development of the new views will be discussed in a later chapter (p. 130), but he certainly played no direct part in the rise of the theory of polyatomic radicals.

These, then, were the chief instigators of the movement that was to culminate in the clear formulation of the theory of valency. Their contributions can now be considered.

In 1850, Williamson published his first paper on the ethers.[33] He attempted, in imitation of Hofmann's work, to replace hydrogen in alcohols by other radicals. But when he acted on ethyl iodide with potassium ethoxide he obtained ordinary ether instead. Realizing the significance of this he investigated matters further, and, as is well known, established that ether and alcohol contained respectively two and one ethyl radicals and gave to them their modern formulae. This had many implications for science, one of which was the formulation of the water type.

Echoing an earlier suggestion of Laurent,[34] Williamson related ether and alcohol to water, and asserted that for all inorganic and most organic compounds "one simple type will be found sufficient, ... water".[35] Thus he wrote:

$$\begin{array}{cccc} H & C_2H_5 & C_2H_5 & C_2H_3O \\ O & O & O & O \quad etc. \\ H & C_2H_5 & H & H \\ \text{water} & \text{ether} & \text{alcohol} & \text{acetic acid} \end{array}$$

Shortly afterwards Gerhardt was able to write that this ether theory was "now almost universally adopted".[36]

Nothing would be more incorrect than to infer from this method of writing formulae that Williamson was trying to suggest that the oxygen was *joining* two parts of the molecule. His types were still being used in Gerhardt's sense of "reaction-formulae". But the following year saw a publication marking a definite progression of thought, and the advent of the theory of polyatomic radicals (though not under that name).

In 1851 Williamson read a paper "On the Constitution of Salts",[37] the basic premiss of which was that all salts are based upon the water type. This marked a new departure, as did his view of hydrated oxides as belonging to the same type; potash was not $K_2O + H_2O$, but

$$\begin{matrix} H \\ K \end{matrix} O$$

More striking, however, was the following statement, made by a disciple of Comte and adherent of the Type school:

Formulae ... may be used as an actual image of what we rationally suppose to be the arrangement of constituent atoms in a compound, as an orrery is an image of what we conclude to be the arrangement of our planetary system.[38]

This is an extension, not an abandonment, of the Theory of Types.

Indeed of the latter he wrote "we owe to M. Dumas an idea which has already been the vehicle of many an important discovery in science, and which is undoubtedly destined to receive more general application". He was now giving it this "more general application" in the shape of greater physical realism. For a "Typist" this was a concession of the first magnitude.

In conformity with his suggestion that salts be referred to water, he regarded acids as derived from the same substance. Replacement of one hydrogen atom by a compound radical in one molecule of water gave a monobasic acid, *e.g.*,

$$\begin{array}{cc} C_2H_3O & NO_2 \\ O & O \\ H & H \end{array}$$

To account for dibasic acids he proposed a "multiple type", of two molecules of water in which two atoms of hydrogen were replaced by one diatomic radical, *e.g.*, CO, C_2O_2, SO_2, *etc.* The multiple types duly inaugurated, Williamson then elaborated the idea in the following passage, where he is speaking of the hydrolysis product of "cyanic ether" [ethyl cyanate] with potash:

One atom of carbonic oxide is here equivalent to two atoms of hydrogen, and by replacing them, holds together the two atoms of hydrate in which they are contained, thus necessarily forming a bibasic compound $\begin{array}{c}(CO)\\O_2,\\K_2\end{array}$ carbonate of potash.[39]

This formation of potassium carbonate he wrote as:

$$\begin{array}{cc} K_2 & K_2 \\ O_2 & O_2 \\ (H_2) & (CO) \end{array}$$

$$\begin{array}{ccc} C_2H_5 & = & C_2H_5 \\ CO & & (H_2) \\ N & & N \end{array}$$

Thus he illustrates his enlightened views of rational formulae by saying that the carbonic acid radical "holds together" the rest of the molecule. He had grasped the same idea that Frankland propounded the following year,[40] but in relation to polyatomic radicals rather than elementary atoms. Thus there is in this paper a recogni-

THE CONCEPT OF POLYATOMIC RADICALS 53

tion of one of the essential concepts of valency theory, namely the union of a definite number of radicals to one centre, a polyvalent group. These radicals were, like Gerhardt's "residues", entities of which we can have no direct experience; they must be distinguished from analogous radicals in the free state (*e.g.*, zinc is different in zinc sulphate and in the uncombined metal). This is a definite improvement on Frankland's early views, where no distinction of this kind was made.

This paper contains the first reference to the *necessity* of a dyad radical being present in a double type. The main ideas appear to have been "adapted by several eminent chemists in Great Britain and France",[41] though this does not necessarily include these premonitions of valency, for the most obvious thesis of the paper was the relegation of so many compounds to the water type.

However that may be, the concept of polyatomic radicals acting as a kind of "molecular cement" within the types remained at the front of Williamson's mind, and is further illustrated in two papers of 1854. In the first of these[42] he disproves an idea held by Gerhardt on the action of phosphorus pentachloride on sulphuric acid, to the effect that the dibasic acid is first dehydrated to the anhydrous acid, and *this* reacts with more pentachloride with replacement of 1 oxygen atom by 2 of chlorine. By obtaining as intermediate chlorsulphonic acid, Williamson shows the process to be a step-wise introduction of Cl in place of 1 O and 1 H:

$$
\begin{array}{ccc}
\begin{array}{c} H \\ O \\ SO_2 \\ O \\ H \end{array}
&
\rightarrow
\begin{array}{c} H \\ O \\ SO_2 \\ Cl \end{array}
&
\rightarrow
\begin{array}{c} Cl \\ SO_2 \\ Cl \end{array}
\end{array}
$$

The existence of this body, which we may call chlorhydrated sulphuric acid, furnishes the most direct evidence of the truth of the notion that the bibasic character of sulphuric acid is owing to the fact of one atom of its radical SO_2 replacing or (to use the customary expression) being equivalent to two atoms of hydrogen.[43]

In other words sulphuric acid is $SO_2(OH)_2$, and the (divalent) radical SO_2 unites two hydroxyl groups, thus giving two replaceable hydrogen atoms.

Four months later, with his student Kay, he was able to publish his findings on the action of sodium ethoxide on chloroform to produce orthoformic ester. The latter "may be equally well conceived to be a body in which the hydrogen of three atoms of alcohol is replaced by the tribasic radical of chloroform".[44] Thus chloroform could be regarded as the chloride of the (trivalent) radical CH.

Of these contributions, Kekulé's friend and colleague in Stenhouse's laboratory, Edward Divers, has written that Williamson "showed, for the first time, how chemical formulae might be so written as to represent chemical constitution" and again, "Williamson founded 'valency' as far as the combining power of a binary radical is concerned".[45] And from Baeyer comes the acknowledgment that "the fundamental idea of interpreting the types by the valence of the atoms was started by Williamson".[46]

It must be a matter of great regret that after this time Williamson's output of original research rapidly dwindled, largely owing to his heavy teaching commitments at University College. Nevertheless, he continued to exert an important influence for some thirty years, particularly by virtue of his post at University College and his Presidency of the London Chemical Society (*cf.* pp. 126, *etc.*).

Meanwhile, Williamson's ideas on polybasic radicals were being extended by William Odling. In his first paper, "On the Constitution of Acids and Salts" (1854),[47] he applied them to a large number of inorganic compounds. For example he showed that the phosphoric acids could be regarded as compounds of the triatomic radical PO. He indicated the number of hydrogen atoms to which one of the group or element concerned was equivalent by dashes. Thus he wrote:

Type	Compound		
$3H_2O$	PO''' / H_3	$3O'''$	phosphoric acid
$2H_2O$	PO''' / H	$2O''$	metaphosphoric acid
$5H_2O$	PO''' / PO''' / H_4	$5O$	pyrophosphoric acid
$3HCl$	$PO''' \; Cl_3$		phosphoryl chloride

Odling also allowed "mixed types", thiosulphuric acid being considered as a union of the types H_2O and H_2S:

Type	Thiosulphuric acid
	H
H_2S	S
	SO_2
H_2O	O
	H

Williamson's ideas were extended in four ways: by allowing a *mixed* type to be held together by a polyvalent radical, by the use of these notation marks, by the introduction of the idea of *replacement value* and by applying "polybasicity" to metals.

Nevertheless, Odling's position was in some ways a retreat from that of his friend. He reflected Gerhardt's influence clearly in his assertion that:

In the three best-known hydrocarbons ... the conception of self-existent constituent compound radicals is not only unnecessary but irrational. The particular groupings of atoms, which we denominate compound radicals, do not have an existence apart from the other constituents of the bodies into which they are said to enter.[48]

The same influence is discernible in his adherence to the Type Theory for many years to come (brackets included), and his reluctance to employ the new symbolism of chemical bonds:

Such a system, used to express with equal confidence the ascertained and the unascertained constitution of all bodies whatsoever, has I believe exerted and still continues to exert a most prejudicial influence on the study of Chemical Science, by making the fanciful sticking-together of variously prolonged discs of more importance than the investigation of phenomena.[49]

This is, of course, a protest against the abuse rather than the use of such a system, but it denotes a characteristically cautious attitude to rational formulae.

During this period, Kekulé was working in London, and by going beyond what was required of him by Stenhouse was able to carry out some research leading to the discovery of thiacetic acid. This he reported in 1854, in two slightly different versions, in English and German.[50]

He had been impressed with the similarity of action of phosphorus chlorides and sulphides on water, and had wondered if a type based on "sulphuretted hydrogen" existed, analogous to that of water. Accordingly he prepared phosphorus tri- and pentasulphides and acted with them on members of the water series. He wrote:[51]

> Experiment has proved that these combinations of sulphur and phosphorus act on the members of the series of water in the same manner (although less violently) as the corresponding compounds of chlorine and phosphorus; however, there is this difference that by using the chlorine compounds the product is resolved into *two* groups of atoms, while by using the sulphur compounds there is obtained only one group; a peculiarity, which, according to the bibasic nature of sulphur, must have been expected. By acting on these compounds of sulphur and phosphorus with water one atom of sulphuretted hydrogen is obtained; while the chlorides give two atoms of hydrochloric acid,
>
> $$6{H \atop H}\}O + P_2S_3 = 3{H \atop H}\}S + 2PO_3H_3$$
>
> $$6{H \atop H}\}O + 2PCl_3 = 6HCl + 2PO_3H_3$$

The corresponding passage in the German version reads thus in speaking of the reactions of alcohol:

> It is indeed seen that the decomposition is essentially the same; only when the chlorides of phosphorus are employed, the product breaks up into chloride of othyl and hydrochloric acid, *or* into two atoms of chloride of othyl, as the case may be; whereas, on employing the sulphur compounds of phosphorus, both groups remain united because the quantity of sulphur equivalent to two atoms of chlorine is indivisible. . . . It is not merely a difference in formulation but in actual fact that one atom of water contains two atoms of hydrogen and one atom of oxygen, and that the quantity of chlorine equivalent to one indivisible atom of oxygen is itself divisible by two, whereas sulphur, like oxygen, is dibasic so that one atom of sulphur is equivalent to two of chlorine.[52]

(This defence of Gerhardt's atomic weights is absent from the English version, which for some reason was less forthright in its maintenance of the Typical view as opposed to that of Kolbe.)

This probably represented Kekulé's private views on the matter before his experiments began; it is in fact largely an emphasis of what Williamson had shown in proving that oxygen can unite two dissimilar radicals. When he tried the action of the phosphorus sulphide on acetic acid, "acetic ether" and other members of the water type, these ideas were confirmed:

It will be seen that the action of tersulphide and pentasulphide of phosphorus above described produces sulphuretted organic compounds by substituting sulphur for oxygen. The compounds obtained in this way may also be formed by replacing one or two atoms of hydrogen in sulphuretted hydrogen (H_2S) or one or two atoms of metal in sulphide of potassium (K_2S), or in sulphide of hydrogen and potassium (KHS), by organic radicals.[53]

The general result of this publication was to impress contemporaries with the possibilities of further discoveries in related fields, as is indicated by the letters written by Gerhardt and Williamson in support of Kekulé's application for the Chair at Zurich later that year.[54] It was the first public acknowledgment from Kekulé of polybasic radicals, and it helped to consolidate the ideas put forward by Williamson. According to Japp, it showed "that the theory of the linking of atoms and groups by means of polyad radicals was already present to Kekulé's mind",[55] while more recently Hiebert has written: "His views in the 1854 paper were noticeably orientated in the direction of valency considerations, but they were certainly not explicit."[56] There seems no reason to quarrel with either of these assessments.

While the chemists in London were thus advocating the new ideas of polybasic radicals, their views were beginning to receive support elsewhere, particularly (as would be expected) in Paris. C. A. Wurtz in 1855 showed that glycerol was derived from the tribasic radical C_3H_5,[57] and in 1856 that ethylene and propylene glycols were respectively based on the dibasic radicals C_2H_4 and C_3H_6.[58] Generally chemists were "beginning to familiarize themselves with the idea of polyvalent compound radicals of different hydrogen-replacing power", though Gerhardt, who died in 1857, was never able to grasp this fully.[59]

Early in 1856 Kekulé left London for the post of Privatdocent at Heidelberg where for two years he was to teach organic chemistry under Bunsen who, however, diverted his own energies into other channels of chemistry. But the influence of his days in Paris and London was far from lost. Maturing in his mind were reminiscences of long conversations with Gerhardt and (still more, perhaps) with Williamson, while his famous "vision" on a London omnibus (see p. 70) was stored in his memory until the time was ripe for its consequences to be made public.

After some eighteen months at Heidelberg, Kekulé published a paper "On the So-called Copulated Compounds and the Theory of Polyatomic Radicals".[60] This was occasioned by the views of Limpricht and von Uslar[61] and of Mendius[62] on the nature of some sulphonic acids. Since there is an obvious parallel between sulphonation and nitration, they regarded these acids as having constitutions analogous to the nitro-compounds, *i.e.* as being substitution products of the parent substance.

The whole issue was bedevilled by the question as to whether sulphosalicylic acid was dibasic or tribasic (which of course depends entirely on how one defines basicity), and had ramifications that do not concern us here. But the controversy amounted to essentially this: Limpricht and his colleagues wrote sulphobenzoic acid as

$$C_{14}H_4(S_2O_4)O_2$$
$$O_4,$$
$$H_2$$

which with $C = 12$, $S = 32$, $0 = 16$ becomes

$$C_7H_4(SO_2)O$$
$$O_2$$
$$H_2$$

Kekulé, however, objected to the needless multiplication of large radicals like their $C_{14}H_4(S_2O_4)O_2$, merely to conserve a simple type. To bring it into line with Williamson's ideas on sulphuric acid, he advocated instead a formulation based on the mixed type $H_2/2H_2O$ where simpler radicals were united by the dibasic oxygen atom.

In fact, however, the paper made little of the attack on his opponents, who seemed to provide him with a platform for expressing his views on polyatomic radicals generally. As this paper is generally overshadowed by the more famous one of 1858, and seems of nearly comparable importance, a brief summary of its contents follows.

Kekulé begins by deprecating the use of bigger and more complex radicals, merely to "be able to refer to one simple type. One is then not far removed from the form of the radical theory that Berzelius represented in his last years, which knew nearly as many hypothetical radicals as there were compounds produced". He then expounds his theory under four main headings.

(a) Idea of Types

We are at once confronted with the assertion that "the number of atoms of an element or radical that combines with one atom (of an element or ... a radical) depends upon the basicity or relationship-size (*Verwandtschaftsgrosse*) of the component parts". Thus elements are of three kinds, with basicities of 1, 2 and 3, from which come the three chief types represented by HH, OH_2 and NH_3.

But, in addition to these, other types can arise by replacement of one atom by an equivalent of another element giving, for example, HCl, SH_2, etc. More important are the *multiple types* in which several molecules are linked by polyatomic radicals (*e.g.*, sulphuric acid, phosphoric acid, glycerol, *etc.*). Where two molecules of types are linked together and are different, *mixed types* arise, as H_2/HCl, *etc.*

The dehydration of dibasic acids is easily explained on this description as a kind of internal rearrangement. Thus:

$$\left. \begin{array}{l} H \\ SO_2'' \\ H \end{array} \right\} \begin{array}{l} O \\ \\ O \end{array} \quad \text{giving} \quad \begin{array}{l} SO_2'', O \\ H \\ H \end{array} \Big\} O$$

(b) Copulated Compounds

"The so-called copulated compounds are in no way constituted differently from other compounds; they can be referred in the same way to types in which hydrogen is replaced by a radical."[63]

This is his central thesis, and he proceeds to show by many examples how compounds previously regarded as copulated can be more easily understood in terms of multiple or mixed types. Thus the formula for sulphobenzoic acid he writes as

$$\left. \begin{array}{l} C_7H_4O \\ \\ SO_2'' \\ \\ H_2 \end{array} \right\} \begin{array}{l} O \\ \\ \\ O \end{array} , \text{ based on } \begin{array}{l} \left\{ \begin{array}{l} H \\ H \end{array} \right. \\ H \\ \\ \left\{ \begin{array}{l} H \\ H \end{array} \right. \\ \\ H \end{array} \begin{array}{l} \\ \\ O \\ \\ \\ O \end{array}$$

Admittedly this makes for complicated formulae that take up a good deal of space, but Kekulé deems this worth while since such representations "express the behaviour of bodies better than those in ordinary use".

(c) *Idea of Radicals*

"Radicals are nothing further than the remainder left unattacked after definite decompositions."

In the light of this definition Kekulé proceeds to give several illustrations. For example, sulphuric acid loses two hydrogen atoms in salt formation, so implying the radical SO_4; but the effect of phosphorus pentachloride is to replace additionally two oxygen atoms, leaving unscathed only the group SO_2. Similarly, sulphonation of benzene indicates the presence of a radical C_6H_5 in the resultant benzene sulphonic acid, while the action of phosphorus pentachloride on the latter suggests the group $C_6H_5SO_2$.

In other words Kekulé is applying a *reductio ad absurdum* to the views he challenges by showing how arbitrary and unnecessary is the multiplication of radicals, and how dependent are the formulae on the reactions chosen. The theory of polyatomic radicals obviated the necessity for inventing new radicals for each compound discovered.

(d) *Basicity of Radicals*

Finally, Kekulé shows how the overall basicity of a radical depends upon its constituents. For instance, loss of each hydrogen atom raises the basicity by one unit at a time, as C_2H_5, C_2H_4 and C_2H_3, which have atomicities of respectively 1, 2 and 3.

This kind of transition occurs when, *e.g.*, ethane is treated with chlorine. But increase in basicity of radicals often occurs in nitrations also. For benzene the two nitration products are clearly

$$C_6H_4(NO_2),H \qquad \text{and} \qquad C_6H_3(NO_2)_2,H$$

and so their reduction products can be assumed to be respectively

$$C_6H_4(NH_2),H \qquad \text{and} \qquad C_6H_3(NH_2)_2,H$$

But conventional type-theory regards both as belonging to the ammonia type, thereby completely obscuring their relationship:

$$N \begin{cases} C_6H_5 \\ H \\ H \end{cases} \quad \text{and} \quad N \begin{cases} C_6H_4NH_2 \\ H \\ H \end{cases}$$

Surely the assumption of a radical changing its basicity on nitration is preferable to such artificiality.

In this way Kekulé gave powerful advocacy to the conception of polyatomic radicals. His publication brought a prompt reply from Limpricht who dealt mainly with the issue as to whether sulphonation products were substitution products or not.[64] He admitted that his new radicals "are again composed of other radicals", and which one chose was "only a question of usefulness", the best being "that which is least easily misunderstood".[65] This paper in its turn prompted a further statement from Kekulé, but the emphasis in this now shifted from the relatively minor issues of sulphonic acids to far more profound matters. Therefore it will be considered in the next section.

Meanwhile, however, it is noteworthy that the paper of 1857 was a climax to the work and speculations of the London chemists. Indeed the debt to them is acknowledged; Kekulé's suggestions are "nothing more than a further explanation of early ideas which Williamson has communicated on occasion" and "which Odling first extended" and "whose suitability can most probably be questioned no longer".[66] Its fundamental achievement was that it emphasized the legitimacy of breaking down the complicated radicals involved in the formulation of compounds on the basis of simple types alone. If a large monatomic radical could be conceived to be made of other smaller radicals, some of them polyatomic, then the next logical step was to dissect the simple radicals themselves. And if this operation were carried to completion one would be left with the elements, naked and open to view for the first time in history. To the consummation of this analytical process we shall now turn.

3. RECOGNITION OF THE TETRAVALENT CARBON ATOM

It has already been noted (see p. 21) that the most fruitful ideas in the development of theories leading to valency were found in the organic branch of chemistry. Also, the whole effect of the doctrine of polyatomic radicals in the types was to focus attention on ever smaller units in those radicals. Consequently, it was inevitable that sooner or later the nature of the one kind of atom common to all organic compounds – carbon – should come under scrutiny. Indeed, without a clear understanding of this basic factor no real progress could be made.

So it happened in 1858 that the nettle was grasped, and one of the most basic of all chemical discoveries announced when Kekulé and Couper explained their views of the nature of the carbon atom. The two publications were almost simultaneous, and independent of each other. Their appearance within a month was partly coincidental, but these ideas were "in the air" at the time and had in one sense been anticipated by Frankland. It would perhaps be fair to say that they were the logical extension of Frankland's views, but now applied to the carbon atom. The mutual debts of these chemists in the matter of valency theory will be discussed in Chapter VI. Meanwhile let us consider Kekulé first.

(i) *The Contribution of Kekulé*

The name of Kekulé more than any other has been associated with this aspect of the development in theoretical chemistry, the recognition of tetravalent carbon. It was chiefly by a paper published in 1858 that his ideas became generally known, "Ueber die Constitution und die Metamorphosen der chemischen Verbindungen und über die chemische Natur des Kohlenstoffs".[67] The centenary of this publication was widely celebrated in 1958 (in the U.S.A., London, Ghent and elsewhere), and several symposia were held in honour of this part of Kekulé's work. Nevertheless it is possible to find distinct traces of the conception of tetravalent carbon in several earlier publications, the two chief examples being in the previous year and emanating from his Heidelberg laboratory.

First, then, came a paper "Ueber die Konstitution des Knallquecksilbers" ("On the constitution of mercury fulminate").[68] To a modern reader it looks distinctly unpromising; Kekulé has apparently done a *volte-face* and gone back to the old atomic weights that he used before his 1854 paper (*i.e.*, $C = 6$, $O = 8$); moreover, his conclusion that fulminic acid is what would be termed now nitroacetonitrile is quite wrong. It is needless to go into the reasons that led him to this latter mistake, and his retrograde step to the old atomic weight system is probably just a concession to current German usage. Hiebert thinks he may well have been the only chemist in Germany at that time holding the new views.[69] A concession of this kind would have enabled better comparison to be made with a paper on the same subject by his colleague Schischkoff,[70] and since he was not here primarily concerned with propagation of new fundamental views was probably worth making.

Having written fulminate of mercury as $C_2(NO_4)(C_2N)Hg_2$, he proceeded:

This formula shows at the first glance that mercuric fulminate exhibits in its composition the closest analogy with a large number of known compounds to which, for example, chloroform belongs:

C_2 H   Cl   Cl   Cl

We could regard it as nitrated chloroform in which the chlorine is replaced partly by cyanogen and partly by mercury.

The following compounds may be referred to this same type:

C_2	H	H	H	H	marsh gas
C_2	H	H	H	Cl	methyl chloride, *etc.*
C_2	H	Cl	Cl	Cl	chloroform, *etc.*
C_2	(NO_4)	Cl	Cl	Cl	chloropicrin
C_2	(NO_4)	(NO_4)	Cl	Cl	Marignac's oil
C_2	(NO_4)	Br	Br	Br	bromopicrin
C_2	H	H	H	(C_2N)	acetonitrile
C_2	Cl	Cl	Cl	(C_2N)	trichloracetonitrile
C_2	(NO_4)	Hg	Hg	(C_2N)	mercuric fulminate
C_2	(NO_4)	H	H	(C_2N)	hypothetical fulminic acid[71]

Thus to Gerhardt's four types (see p. 47) is now added a fifth, commonly called the "marsh-gas type". In view of the extension and modification that the Type Theory was undergoing at that time, Kekulé's suggestion is less remarkable than might appear. Multiplication of types was in the air, and in fact the marsh-gas type had already been suggested by Odling, though Kekulé probably had not seen a copy of this paper in Germany. It is unlikely that Gerhardt exerted much influence here, as his formula for marsh-gas, $C^4H^2H^6$, bore little resemblance to Kekulé's.

Kekulé goes on to say:

In assigning these compounds to the same type, I do not use the word in the sense of Gerhardt's Unitary Theory, but in that in which it was first employed by Dumas on the occasion of his fruitful investigation on the subject of types. I wish essentially to indicate the relations in which the said compounds stand to one another; that the one, under the influence of appropriate reagents, can be produced from, or transformed into, the other.[72]

In other words, these compounds were related together by mutual transformation, not by similarity of property. This was a break-away from the rather barren viewpoint of Gerhardt who, as we have seen (see p. 47), allowed his types merely the significance of underlining resemblances in reactions. It was in fact a reversion to an earlier idea that could be more enlightening when it came to considerations of composition.

Kekulé insisted on this point later when assailed by Kolbe on matters of priority (see p. 111). And it is evident that a "mechanical type" is envisaged from his inclusion of methyl chloride, which was obviously a member of the HCl type *in Gerhardt's sense*. Whether this meant that Kekulé wished to establish the marsh-gas type to go alongside Gerhardt's other four has been doubted by Meyer,[73] but the issue was not for long to remain in doubt.

Now the question that this paper raises here is the extent to which it contributed to the development of the theory of valency, and in particular to the doctrine of tetravalent carbon. One point that seems to have been generally overlooked is this shift of emphasis from "chemical" to "mechanical" types – from relationships of reaction to relationships of constitution, and indeed almost of structure. This was a healthy change from Gerhardt's structural agnosticism, and yet a further blow to that viewpoint from within the framework of the Type Theory itself.

But more particularly one may clearly see here the beginnings of a recognition of the tetravalency of carbon, just as the water type implied a divalent oxygen. Japp says that here we have "the earliest enunciation of the tetra-valency of carbon",[74] though he goes on to admit the incompleteness of the demonstration, since the class of compounds to which it could be applied is strictly limited to those formed from methane by replacement of hydrogen by monovalent radicals. Japp regards this as free from any suggestion of the law of mutual linking of carbon atoms, though Hiebert sees a glimmering of this idea in

$$C_2 \quad (NO_4) \quad H \quad H \quad (C_2N)$$

and other formulae with a substituent containing "C_2".[75] If Kekulé had regarded ethane as methane in which one hydrogen atom was replaced by a methyl radical, then his marsh-gas type would have led directly to self-linking carbon atoms. But although Frankland had already made this suggestion[76] there is no evidence that Kekulé had given it serious thought.

In the following year Kekulé published his second paper on fulminic acid.[77] He added three more members of the marsh-gas type (nitroform, trinitroacetonitrile and dibromonitroacetonitrile) and reiterated his views thus:

All these bodies of one series can be included in one mechanical type: all contain the same number of atoms, if one reckons the nitro-group and the cyanide as radicals analogous to the elements; at the same time they show great differences in the individual properties, occasioned by the differences in dynamic nature of the elements which have entered into them.[78]

Meanwhile, however, he had published his far more famous paper on the copulated compounds and theory of polyatomic radicals that we have already discussed (see p. 58). Apart from the main theme of this, there are two minor references in it that have bearing on the notion of tetravalent carbon. (It is also noteworthy in using the new atomic weights ($C = 12$, *etc.*).) The first reference is a footnote to his classification of elements into three groups according to their atomicity:

Carbon, as may easily be shown, and as I shall explain in detail later, is tetrabasic and tetratomic; *i.e.*, one atom of carbon $= C = 12$ is equivalent to four atoms of hydrogen.[79]

The second reference occurs in the main text of the paper where Kekulé speaks of the basicity of radicals (see p. 60), and observes:

The simplest hydrogen compound of the (tetratomic) carbon marsh-gas, CH_4, *e.g.*, behaves neither as a radical nor as the compound of a radical. The representation of marsh-gas as methyl hydride is only schematic; no reaction is known in which a methyl compound arises from marsh-gas.[80]

Here again is a reference to the "mechanical" nature of the type, but the parenthetical way in which the concept of tetratomic carbon is introduced is distinctly curious. All these references to the idea of a tetravalent carbon atom are preliminaries to the great paper of 1858 where Kekulé gave for the first time a systematic account of the doctrine.

In the controversy that had arisen between Limpricht and Kekulé over the sulphonic acids, the issue had resolved itself into whether it was right to regard them as substitution products. If they were so constituted, then (by Laurent's conception of substitution products at least) they must conserve the same type as that from which they were derived. Since this did not appear to be so, the argument became lively indeed. Kekulé's paper of 1857 brought a prompt rejoinder from Limpricht, and the paper now under discussion was Kekulé's reply to this. With the opening paragraphs we are not concerned as they dealt with the details of the case under consideration and were complicated by issues now long dead. But they did underline the complexity of the whole matter, and suggested the need for going back to first principles if that were possible. Accordingly, having acknowledged his debt for what follows to Williamson, Odling, Gerhardt and especially Wurtz,[81] Kekulé states:

I deem it necessary, and in the case of present chemical knowledge possible in many cases, for the explanation of the properties of chemical compounds, *to go back to the elements themselves* that compose the compounds.[82]

No more auspicious beginning could be devised for the introduction to the chemical world of a theory to encompass the whole of organic chemistry.

Four main topics are then discussed, as outlined below, the first three being in essence an introduction to the fourth. There is also a brief concluding section.

(a) *Chemical Metamorphosis: Combination and Decomposition*

Kekulé now strikes a direct blow at the Theory of Types as conceived by Gerhardt. The latter himself had stressed that it is based on double decompositions as the reaction type. But Kekulé shows that this is not sufficiently general, and that other kinds of reaction must be considered. He recognizes three main types of reaction:

Direct addition
   E.g., the reaction of ammonia and hydrochloric acid, of phosphorus trichloride and chlorine, *etc.*

Union of more molecules by attack of a polyatomic radical
   E.g., the action of water on anhydrides to give acids and imides to give amides. A case of the latter is that of succinimide:

$$\left.\begin{array}{c}C_4H_4O_2'' \\ H\end{array}\right\}N \qquad \left.\begin{array}{c}H \\ H\end{array}\right\}N$$

$$\left.\begin{array}{c}H \\ H\end{array}\right\}O \quad \longrightarrow \quad C_4H_4O_2'' \left.\begin{array}{c} \\ H\end{array}\right\}O \qquad \text{(succinamic acid)}$$

Double decomposition

Exchange of equivalent weights always occurs – a monatomic radical by a monatomic, a diatomic by a diatomic or by two monatomic, *etc*. A readjustment of affinities occurs, and in the simplest case can be represented generally thus:

a | b	a b	a b
a' | b'	a' b'	a' b'
(before)	(during)	(after)

(b) *Action of Sulphuric Acid on Organic Compounds*

Three kinds of reaction are considered, where the dibasic radical SO_2 has different roles. Representative reactions are:

$$\left.\begin{array}{c}C_6H_5 \\ H\end{array}\right\}O \quad \longrightarrow \quad \left.\begin{array}{c}C_6H_5 \\ SO_2 \\ H\end{array}\right\}\begin{array}{c}O \\ O\end{array}$$

$$\overline{SO_2,O}$$

$$\left.\begin{array}{c}SO_2,O \\ \hline H \\ C_2H_4 \\ H\end{array}\right\}O \quad \longrightarrow \quad \left.\begin{array}{c}H \\ SO_2 \\ C_2H_4 \\ H\end{array}\right\}\begin{array}{c}O \\ O\end{array}$$

(c) *Radicals, Types and Rational Formulae*

An empirical view of rational formulae is stated; they are "replacement formulae and ... cannot be otherwise". Therefore, choice of a rational formula is governed by considerations of utility. Thus of the following for benzene sulphonic acid, the first is "most rational".

68 THE ORIGINS OF VALENCY

$$\left.\begin{array}{c}C_6H_5\\SO_2''\\H\end{array}\right\}O \qquad \left.\begin{array}{c}C_6H_5SO_2''\\H\end{array}\right\}O \qquad C_6H_5SO_3,H$$

Having disposed of these preliminary points, Kekulé now turns to his fourth section: *The Constitution of Radicals: the Nature of Carbon.* His first step is to show that radicals may be considered as combinations of atoms in which "affinity units" (*Verwandschaftseinheiten*) have combined with each other. *E.g.*,

The radical of sulphuric acid SO_2 contains three atoms, each of which is diatomic, thus representing two affinity units. In joining together, one affinity unit of one atom combines with one of the other six affinity units, so four are used to hold the three atoms themselves together; two remain over, and the group appears to be diatomic; it unites, *e.g.*, with two atoms of a monatomic element:[83]

$$S''\left\{\begin{array}{c}O''\\O''\end{array}\right. \qquad S''\left\{\begin{array}{c}Cl\\O''\\O''\\Cl\end{array}\right. \quad \text{(sulphuryl chloride)}$$

It will, of course, be noted that this is completely wrong. It implies that the chlorine atoms are linked to oxygen and states that sulphur is divalent. The former error is a consequence of the latter, and this in turn arises from Kekulé's refusal to admit several different valency states for one element; and as sulphur is divalent in its simplest compounds, it was assumed that this is universally the case. Fortunately, this did not affect the main issue.

Kekulé now turned to carbon:

If only the simplest compounds of carbon are considered . . ., it is striking that the amount of carbon which the chemist has known as the least possible, the atom, always combines with four atoms of a monatomic, or two atoms of a diatomic element; that generally, the sum of the chemical units of the elements which are bound to one atom of carbon is equal to four. This leads to the view that carbon is tetratomic (tetrabasic).[84]

Various examples are now given of its various kinds of combination:

IV+4*I*: CH_4, CCl_4, CH_3Cl, $CHCl_3$
IV+(*II*+2*I*): $COCl_2$

IV+2II: CO_2, CS_2
IV+III+I: CNH

But what of substances containing more than one atom of carbon? In these cases,

... it must be assumed that at least part of the atoms are held just by the affinity of the carbon, and that the carbon atoms themselves are joined together, so that naturally a part of the affinity of one for the other will bind an equally great part of the affinity of the other.

The simplest, and therefore the most obvious, case of such linking together of two carbon atoms is this, that one affinity unit of one atom is bound to one of the other. Of the 2 ×4 affinity units of the 2 carbon atoms, two are thus used to hold both atoms together; there still remain six extra which can be found by the atoms of the other elements. In other words, one group of 2 atoms of carbon = C_2 will be hexatomic.[85]

After quoting a number of illustrations of this (ethyl chloride, acetonitrile, *etc.*), Kekulé extends his argument to a hydrocarbon with *n* carbon atoms united in this way, showing that the number of hydrogen atoms will be given by:

$$n(4-2)+2 = 2n+2$$

It is possible for polyvalent atoms to be joined to a carbon atom in such a way that only one of their affinities is united to the carbon atom; this may give rise to compounds as these:

$$\left.\begin{array}{l}C_2H_5\\ H\end{array}\right\}O \quad \left.\begin{array}{l}C_2H_5\\ H\\ H\end{array}\right\}N \quad \left.\begin{array}{l}C_2H_3O\\ C_2H_5\end{array}\right\}O \quad \left.\begin{array}{l}C_2H_5\\ C_2H_5\\ C_2H_5\end{array}\right\}N$$

Thus in some cases these polyatomic atoms are interposed in a chain of carbon atoms. If this is so, "the carbon group appears to be a radical, and it is said that the radical replaces one atom of hydrogen of the type".

Thus we have a reconciliation of Type and Radical Theories of the utmost clarity. Kekulé brings this section to a close by unifying yet more concepts: the existence of homologous series ("the carbon atoms ... are arranged together in the same way"); decompositions with and without fragmentation of the carbon skeleton; the existence of aromatic compounds with "a denser arrangement of carbon atoms" – all these are briefly but convincingly considered in the light of the new theory.

A final short section outlines "Principles of a Classification of Organic Compounds" based on the ideas he has enunciated, and brings the paper to a close.

To attempt to assess the importance of this paper before dealing with the developments that took place following its publication would be quite unwarranted. But one or two points are fairly clear merely from a perusal of its contents.

Kekulé's view is essentially a development from the Type Theory, and not an attempt to supplant it. The merit that Kekulé appears to be claiming for it is that it explains the types, even justifies them, but does not make them unnecessary. One might say that the language is the language of valency, but the thought-forms are those of types. This holds for the symbolism also. It is surely significant that in the first paper there is no attempt to picture a carbon chain in order to enunciate the concept of linking carbon atoms, or to represent a molecule with two carbon atoms joined together. Again, there seems to be a Gerhardtian reluctance to ascribe physical reality to the formulae even now. In fact Kekulé "thought his formulae were merely summaries of reactions; like Gerhardt's, only better".[86]

This much is clear. One other matter may be conveniently mentioned here – the origin of the theory enunciated in this paper.

It is true that F. Rochleder had implied self-linking carbon atoms in 1852,[87] but it is unlikely that Kekulé was influenced by this. According to his own account "my ideas on chemical valency and the manner of joining atoms... were already being formed during my stay in London".[88] Elaborating on this he told the famous story of his "vision", which will bear repetition here if only as a reminder of the diverse origins of different aspects of valency theory.

During my stay in London, I lived for a considerable time in Clapham Road near the Common. But I often spent my evenings with my friend Hugo Müller in Islington, at the other end of the great town. We spoke of many things, but chiefly of our beloved chemistry. One fine summer evening, I was returning by the last omnibus, "outside" as usual, through the deserted streets, at other times so full of life. I fell into a reverie. There before my eyes gambolled the atoms. I had often seen them moving before, each tiny being, but I had never succeeded in discerning the nature of their motion. This time I saw how frequently two smaller atoms united to form a pair; how a larger one embraced two smaller ones; how still larger ones kept hold of three or four of the smaller, whilst the whole kept whirling in a giddy dance. I saw how the larger ones formed a chain dragging the smaller ones after them. I saw what our past master Kopp, my honoured teacher and friend, has shown us so vividly in his "World of Molecules", but I saw it long before him. The cry of the conductor "Clapham Road" awoke me from my dream, but I spent a part of the night putting down on paper the sketches at least of these dream-forms. Thus began the structure theory.[89]

How far this was meant to be taken seriously is a matter for conjecture. But a symposium in one's honour is hardly a likely place for autobiographical inventiveness. Kekulé went on to say that at Heidelberg he wrote down these ideas and showed them to two of his closer friends; "both thoughtfully shook their heads"! As the two most closely agreeing with this description were Baeyer and Erlenmeyer it is tempting to suppose this reference is to them. However, deeming either the contents or the timing unsuitable, Kekulé withheld the paper for a year, until Limpricht's publication caused him to make his famous reply. This was a revised form of his original sketch, the more polemical parts having been considered, presumably by Liebig, unsuitable for publication.[90]

(ii) *The Contribution of Couper*

Archibald Scott Couper was born in 1831 at Kirkintilloch, Scotland, of a prosperous cotton-weaving family. From 1851-2 he read classics at Glasgow University, moving in 1852 to Edinburgh where he studied logics and metaphysics under Sir William Hamilton, a famous exponent of "commonsense" philosophy, and an admirer of Kant.

A frequent visitor to Germany, in 1855 he settled in Berlin and, for some unknown reason, began the study of chemistry. He made remarkable progress, and the following year was admitted to Wurtz's laboratory in Paris. Here the ideas of Gerhardt were freely circulating, and publications were freely flowing from the pen of Wurtz. The latter had just completed his preparation of ethylene glycol by hydrolysis of its diacetate obtained in turn from the iodide.[91] Couper's first task was to extend this to the benzene series, *via* brominated benzene. This would have been practically impossible, so it was as well that the operations were terminated by an explosion. In his report on this,[92] he used the system $C = 6$, $O = 8$.

His next research, initiated by a desire to throw light on the vexed question of its basicity, was on salicylic acid. He showed that with phosphorus pentachloride, a product was obtained that had the molecular formula $C_7H_4PO_3Cl_3$ (modern A.W.s), and which he termed "trichlorophosphate of salicyl". This on rapid distillation forms "monochlorophosphate of salicyl" ($C_7H_4PO_4Cl$), and both these materials with water give rise to the tribasic "phosphosalicylic acid" ($C_7H_7PO_6$).

These results he published in French and English in 1858.[93] The two papers differ somewhat. In the latter he changes from C = 6 to C = 12 (so this was almost certainly slightly later), though in both O = 8. But the most startling difference between the two versions is that the English one gives quasi-constitutional formulae according, he says, "to the rational theory which I seek to develop in another paper".[94] This other paper is the source of our knowledge on Couper's valency theory.[95] But before discussing this, it is worth observing that the work on salicylic acid has had a significant history.

Attempts were made to repeat Couper's work by Kekulé[96] and Kolbe,[97] but in both cases without success, and the "trichlorophosphate of salicyl" eluded them. Others also tried in vain[98] until Anschütz[99] succeeded in obtaining the product by *rapid* distillation, preferably under reduced pressure. This vindication of Couper aroused an interest in Anschütz, who was Kekulé's successor at Bonn and later to be his biographer. When, therefore, he came across Couper's other papers in preparing his biography of Kekulé[100] his latent interest was fired again, with the result that the whole significance of Couper's writings was drastically re-assessed and given its true value.[101]

Since then Atherton[102] has agreed with Anschütz in postulating a different structure for "trichlorophosphate of salicyl" from that which Couper had given it. The final word has yet to be spoken, perhaps, but as late as 1961 a further paper[103] admitted that "a conclusive decision between the two possible structures for Couper's compound is not yet possible", but at the same time advanced evidence for the *probability* of Couper's original suggestion, translated of course into modern terms. On the basis of reactivity, comparison with other compounds and infra-red spectra, the authors favoured a cyclic rather than an open-chain structure; more recently still (1966) they have definitely reverted to the alternative on the basis of chemical evidence and nuclear magnetic resonance spectra:[104]

Anschütz[99]
Atherton[102]
Pinkus et al.[104]

Couper[92]
Pinkus et al.[103]

It is interesting that the first compound for which anything approaching a modern constitutional formula should be given is still a matter for recent discussion.

It is now necessary to examine Couper's only other paper, in which he discusses more fully his theoretical ideas. This also appeared in several forms, and apparently in this order:

(a) A preliminary note, presented to the Academy by Dumas (14 June 1858).[105a]
(b) A full English paper.[105b]
(c) A full French paper, nearly identical with (b) but containing a slight change in symbolism and with a few extra paragraphs.[105c]

All these are entitled "On a New Chemical Theory".

In the preliminary note, Couper starts by distinguishing between "affinity of degree" and "elective affinity", the former being that possessed by an element capable of combining with another in several different proportions, and the latter being one element's affinity for several others.

Taking carbon as an example he illustrates the first case by the two oxides (which he writes as CO^2 and CO^4), and then quotes these two features of its elective affinity:

1. It combines with equal numbers of equivalents of hydrogen, of chlorine, of oxygen, of sulphur, etc., which can mutually replace one another so as to satisfy its combining power.
2. It enters into combination with itself. These two properties suffice, in my opinion, to explain all that is presented as characteristic by organic

chemistry. I believe that the second is pointed out here for the first time.[106]

Using a symbolism peculiar to himself, he reduces all organic chemistry to two types:

$$nC^2M^4 \quad \text{and} \quad nC^2M^4 - mM^2 \ (m<n)$$

or even, if n can be zero,

$$nC^2M^4 + mC^2M^2 \qquad (C^2 = 12)$$

Unfortunately he does not define n, m or M. But as Anschütz points out from the examples he gives of the first type (*e.g.* methyl and ethyl alcohols), M must represent the number of valencies available, not the number of univalent atoms linked to carbon, which is obviously less than M.[107] He also points out that the second expression must in that case apply to unsaturated compounds.

Couper proceeds to give the following formulae, where, it will be observed, C = 12 but O = 8, and our modern oxygen atom is replaced by the symbol O ... O:

$$C \begin{cases} O \ldots OH \\ H^3 \end{cases}$$

methyl alcohol

$$C \begin{cases} O \ldots OH \\ H^2 \end{cases}$$
$$\vdots$$
$$C \ldots H^3$$

ethyl alcohol

$$C \begin{cases} O \ldots OH \\ H^2 \end{cases}$$
$$\vdots$$
$$C \ldots H^2$$
$$\vdots$$
$$C \ldots H^3$$

propyl alcohol

$$C \begin{cases} O \ldots O \\ H^2 \quad H^2 \end{cases} C$$
$$\vdots \qquad \qquad \vdots$$
$$C \ldots H^3 \quad H^3 \ldots C$$

ether

$$C \begin{cases} O \ldots OH \\ O^2 \\ H \end{cases}$$

formic acid

$$C \begin{cases} O \ldots OH \\ O^2 \end{cases}$$
$$\vdots$$
$$C \ldots H^3$$

acetic acid

$$C \begin{cases} O \ldots OH \\ H^2 \end{cases}$$
.
.
.
$$C \begin{cases} H^2 \\ O \ldots OH \end{cases}$$
glycol

$$C \begin{cases} O \ldots OH \\ O^2 \end{cases}$$
.
.
.
$$C \begin{cases} O^2 \\ O \ldots OH \end{cases}$$
oxalic acid

or

$$C \begin{cases} O^2 \\ O^2 \end{cases}$$
.
.
.
$$C \begin{cases} O \ldots OH \\ O \ldots OH \end{cases}$$

Of these he observes that in propyl alcohol "the combining power of the atom of carbon that is situated in the middle is reduced to 2 for hydrogen, since it is combined chemically with each of the two other atoms of carbon"; and that in the acids the occurrence of *two* oxygen atoms attached directly to carbon is necessary to give the body the character of an acid, an effect that he thought was fundamentally electrochemical.

Finally he quotes his formulae for salicylic acid and "trichlorophosphate of salicyl" (see p. 73).

In this way Couper attempted to prepare the world for the more extended versions of his paper that were to follow soon. These versions are nearly identical; the French one has a few extra paragraphs, and the dotted lines of the formulae replaced by continuous lines. Except where stated otherwise, quotations will be from the English version.

The paper begins with a philosophical broadside against the Theory of Types. Claiming that "the end of chemistry *is its theory*", Couper flings a challenge to the "Typists" to show that their system is truly rational. On empirical grounds he is not seriously disposed to dispute it, although the non-conformity of peroxides is a difficulty, and the limitation to double decompositions is once again shown to be an over-simplification.

But it is chiefly on metaphysical grounds that the Type Theory is attacked. It must be rejected because its comparisons, though universally applicable, are on a fundamentally false basis, and because it explains nothing. Why is this? It occurs because the theory "begins with a generalization", that "old and vicious principle which has already retarded science for centuries". And even this generalization "lacks, moreover, the merit of being represented by a type having a known existence" (nO_H^H). He completes his case by a *reductio ad absurdum*, suggesting that the same principles would be involved in analysing the contents of a book in which "a certain word would serve as a type, and from which by substitution and double decomposition all the others are to be derived". Although one "would state certainly an empirical truth...", the method "would, judged by the light of common sense, be an absurdity. But a principle which common sense brands with absurdity, is philosophically false and a scientific blunder".

It is always refreshing when an established system is looked at from a new point of view, and very often disastrous. For the last decade at least chemistry – and above all chemistry in Paris – had been under the spell of Comte's positivistic philosophy (*cf.* p. 19). But Couper had the advantages as well as the disadvantages of a newcomer, and had imbibed a different outlook. The "common sense" of which he speaks is that of Hamilton, and his philosophy of science is derived from Kant. The right method is that of Descartes – analysis as far as one can go, and then an outward-working synthesis, unhindered by Gerhardt's preconceived agnosticism. Here is an echo of another element in Couper's background, a sturdy Scots religious faith, impatient with the self-blinded.

One could perhaps sum up Couper's objections to the Type Theory by saying it was based on a philosophically unsound and chemically meaningless premiss, achieved its stated objects only partially and then by false comparisons, failed to explain anything and (worst of all) tended to imply that nothing was explicable. All these ideas are present in Couper's diatribe.

Nor are matters mended by the doctrine of copulated compounds, says Couper, for this again implies that a whole is more than the sum of its parts. This too must be abandoned.

So Couper turns to his own theory which, he maintains, fulfils the necessary conditions of respecting the unity of chemistry, of comparing all known bodies, and of tracing the general principles common to all the elements.

As before he distinguishes the two kinds of affinity, and applies this to carbon. Again he shows that other elements cannot form chains, quotes his two types of organic compounds and gives a few examples. There are several "orders of complicity" in the elements, *i.e.* what we should now call the valency of the atom linked directly,

TABLE III

Compounds	Couper's formulae in Phil. Mag.[108]	Couper's formulae in Ann. chim. phys.[109]	Modern formulae
butylethyl ether	C....O....O....C ∴ H² H²∴ C....H² H³....C ⋮ C....H² ⋮ C....H³	C{ O—O / H² H² } C CH² H³C CH² CH³	CH₂–O–CH₂ CH₂ CH₃ CH₂ CH₃

formic acid	O....OH
	 C⋰..O²
	     H
		C{ O–OH / O² / H }	OH
			 O=C
			      H

glycol	O....OH C⋯ ⋮....H² ⋮....H² C.... O....OH	C{ O–OH / H² } { H² / O–OH } C	OH CH₂ CH₂ OH
tartaric acid	....O....OH C ⋮....O² ⋮....H C.... ⋮ O....OH ⋮....O....OH C.... ⋮ H ⋮....O² C ....O....OH	C{ O–OH / O² } C{ H / O–OH } C{ O–OH / H } C{ O² / O–OH }	OH C=O C(H)(OH) C(OH)(H) C(=O)(OH)

and singly, to carbon. Once more there are some rather odd electrochemical speculations.

Couper concludes with a large number of formulae, some of which it will be convenient to tabulate, somewhat in the manner of Anschütz. (See Table III.)

His efforts for certain other polyhydroxy compounds as glycerol, mucic and saccharic acids and glucose were less happy, but not seriously in error.

His most ambitious formula, however, was included in the French paper only. For cyanuric acid he wrote

$$\begin{array}{c} \text{H}—\text{O}—\text{O}—\text{Az}—\text{C}—\text{Az}—\text{O}—\text{OH} \\ \quad\quad\quad\quad\quad\quad | \quad\quad\quad | \\ \quad\quad\quad\quad\quad\quad | \quad\quad\quad\;\; \text{C} \\ \quad\quad\quad\quad\text{C}—\text{Az} \Big\{ \\ \quad\quad\quad\quad\quad\quad\quad\quad\; \text{O}—\text{OH} \end{array}$$

On the O = 16 scale this becomes

$$\begin{array}{c} \text{HO}—\text{Az}—\text{C}—\text{Az}—\text{OH} \\ \quad\quad | \quad\quad\quad\; | \\ \quad\quad \text{C}—\text{Az}—\text{C} \\ \quad\quad\quad\quad | \\ \quad\quad\quad\;\; \text{OH} \end{array}$$

The modern formula may be written as follows (with hydroxyl attached to carbon):

$$\begin{array}{c} \text{N} \\ \text{HO} \diagup \;\; \diagdown \text{OH} \\ \;\;\; \| \\ \text{N} \diagdown \diagup \text{N} \\ \text{OH} \end{array}$$

Thus, seven years before Kekulé's paper on benzene, Couper was proposing a cyclic structure for cyanuric acid.

Now there is so much similarity between this paper and that of Kekulé that a number of important questions arise. The first is how these papers differed, and this may briefly be summarized in the following way:

(1) *Differences of approach:* Kekulé approached his theory from

Types; Couper started (apparently) from philosophy. Kekulé had received a schooling from Liebig, Gerhardt and Williamson, while his Scottish colleague came fresh from only a year's apprenticeship in Wurtz's laboratory. Indeed it has been pointed out that the architectural background of the one and the linguistic training of the other can be discerned in these papers, though this seems rather far-fetched in view of Kekulé's reluctance to employ truly structural formulae. The essence of the matter is surely, as Anschütz realized, that Couper was able to approach things with his mind a good deal more free from preconceived ideas.

(2) *Differences of aim:* it follows that Kekulé was concerned to explain types in terms of valency, whereas Couper believed that types, being an unfortunate and arbitrary invention, needed not explaining but exploding.

(3) *Differences of formulation:* here we may note, first, that Couper was limited by his curious "double atom" of oxygen, and, secondly, that Kekulé was restricted by his failure to recognize what Couper so clearly enunciated, variable valency. Moreover, it is obvious from the modes of discussion that Couper's formulations had a constitutional significance that was intentional, while Kekulé's did not.

(4) *Differences of scope:* Kekulé's paper had a far wider sweep than Couper's in that it did not confine itself to carbon, but applied to other elements and to radicals. Couper, on the other hand, in his crusade to emancipate organic chemistry from the Theory of Types, was content – or compelled – to fight on a far narrower front. This is not to minimize the importance of what he did, however, for it was on carbon above all that attention had to be focussed at that time.

The final observation that has now to be made is that Couper's paper for all its merits was largely an isolated flash of genius, without obvious antecedents and unfortunately without much influence. It was Kekulé who was to be the prophet of valency, and Couper's place was for years accorded less recognition than even that of Frankland.

Undoubtedly the main reason for this was that Kekulé's paper appeared almost at the same time, and Kekulé quickly claimed priority.[110] The extent to which this claim was justified will be discussed later (see p. 124), but the important thing is that Couper suffered a total breakdown in health, and never published a reply. Consequently his ideas sank largely into the background until Anschütz re-discovered them.[111] But it is not true to say that Couper was totally without influence, and traces of his ideas may be discerned in three places.

First, an account of Kekulé's and Couper's theories was given by the latter's superior, Wurtz, in a review[112] that year. Preference was given to Kekulé's ideas as these claimed less, and were based more on experience and less on hypotheses. Yet the ideas of Couper must have impressed themselves upon Wurtz to warrant this public, though critical, recognition of his junior's work, and it is not far-fetched to suppose that Wurtz's later enthusiastic support of the valency theory owed something at least to Couper.

Secondly, a fellow-student of Couper in Wurtz's laboratory was A. M. Butlerov, who in 1859 published a paper claiming that some of Couper's ideas had been put forth by himself in verbal statements (but not in writing), though at the same time discussing them at length.[113] Unquestionably Couper's paper, in one way or another, made a deep impression on the Russian chemist who was to be responsible for launching the Theory of Structure.

Finally, when Couper had returned to Scotland he spent a few months at Edinburgh University under Lyon Playfair (who had also been in charge of Frankland in London). In the same university was the young Crum Brown whose later graphic formulae contributed so much to the success of the Structure Theory. The extent to which Couper may have influenced Crum Brown must always be conjectural.[114] But it is at least noteworthy that at the same Scottish university were two who played a significant part in fostering the fundamental theory of organic chemistry.

CHAPTER IV

Nomenclature

THE TERM "VALENCY" has several shades of meaning today. It has a numerical significance (implying the number of atoms of, e.g., hydrogen, combining with one atom of the element concerned). Secondly, it is used in the general sense of chemical linking of atoms ("the phenomenon of valency"). Finally the word has been particularized to mean the actual links, or bonds, themselves ("carbon has four valencies linked to chlorine"). The last of these is at present more colloquial and less generally accepted than the other two.

In this brief account of the early nomenclature, it will be convenient to base the discussion on these three aspects, considering first the names given to radicals or atoms with definitive values for their valencies; next, the evolution of the word "valency" for the general phenomenon; and finally the origin of the word "bond" in this connection.

1. NAMES USED IN GROUP CLASSIFICATION

We often speak now of a univalent or divalent atom (or radical). The earliest counterparts to these adjectives were "monobasic", "bibasic", *etc.*, for the terms arose in discussions of acids and salts, where the use of such words had been employed for some years in roughly their current sense. If the first real steps towards valency were taken by Williamson (see p. 54), he also must be held to be the originator of this use of these terms. He wrote of a "bibasic compound",[1] "the tribasic radical of chloroform",[2] *etc.* Kekulé used similar terms in 1854.

However, "polybasic" radicals were soon to be replaced by "polyatomic" ones, the terms being interchangeable in Kekulé's famous paper of 1857.[3] When shortly afterwards a distinction was drawn between "basicity" and "atomicity" (see p. 84), the former term ceased to have any connection with valency.

But this phraseology too was to be temporary. With the rise of the word "valency" and similar terms, the adjectives used to denote combining power became etymologically related to these terms. In German the situation was well put by Kekulé:

Man hat es nämlich zweckmässig gefunden, Adjectiva zu bilden, welche direct die Anzahl von Aequivalenten ausdrücken, die von einem Atom oder einem Radical repräsentirt werden; man spricht also von monovalent, bivalent, trivalent etc., oder man bedient sich, und sogar vorzugsweise, statt dieser lateinischen Worte der deutschen Ausdrücke: einwerthig, zweiwerthig u.s.w.[4]

That is,

It has been found useful to construct adjectives which indicate directly the number of equivalents that are represented by one atom or one radical; hence one speaks of "monovalent", "bivalent", "trivalent" *etc.*, or, instead of these Latin words, more particularly of the German expressions "einwerthig", "zweiwerthig" and so on.

Writing a few years later, Schorlemmer[5] indicated that German writing preferred the terms native to the language, though today both systems are used.

On the other hand, France and to some extent England favoured the Latin terms. This, as in the quotation above, was often with a fine disregard for etymology, Greek or Latin prefixes being attached to a Latin root, apparently on considerations of euphony alone. The list furnished (in English) by Hofmann in 1865 gave a different selection: "univalent, bivalent, trivalent and tetravalent".[6] In France this method of representation was rapidly accepted, and has been retained since. English writing, on the other hand, contained another system as well.

In 1864 appeared Odling's *Tables of Chemical Formulae*,[7] in which elements were classified according to their valencies as "monads", "dyads", "triads", "tetrads" and so on. The words themselves were not original, but they had not been used in this way before. They seem to have first appeared in Laurent's *Chemical Method*,[8] where, however, they had a meaning exactly opposite to that which Odling gave them. Laurent called oxygen a "monad" and hydrogen a "dyad" in water, these terms referring not to the valencies but number of atoms of each. In other words they applied to the atomicity in the modern rather than the older sense. This work was translated into English by Odling himself nine years previously, so this is presumably where he first met the words.

This phraseology became rapidly accepted in English chemical literature, doubtless part of the reason being Frankland's use of it in both his *Lecture Notes* of 1866[9] and scientific papers.[10] At this time Frankland exerted a strong influence on a certain section of British chemical thought. Although it has been suggested that this nomenclature did not last long, this is not true of England where, for instance, it appeared as late as 1932 in Kingzett's *Chemical Encyclopedia*.[11] But elsewhere the system never gained support and it remains as a minor monument to the great interest in valency shown in this country in the 1860s, where for a time English chemists had a particularly distinctive contribution to make.

2. ORIGINS OF THE WORD "VALENCY"

This word, which stands primarily for the general phenomenon of chemical binding or linking, has had a curious history. The first observation one must make is that the original phrases used to convey its present meaning (or something approaching it) had no etymological connection whatever with "valency". The simple elegance of Frankland's phrase "combining-power"[12] seems never to have been generally recognized, and Couper's "degree of affinity"[13] was doomed from the outset by the ambiguities and uncertainties residing in the concept of "affinity". Better able to command response was Hofmann's suggestion of "atom-fixing capabilities", delivered in a popular lecture at the Royal Institution.[14] This found an echo in Frankland's "atom-fixing power",[15] and possibly in the "atomic value" of Williamson.[16] Yet even in these was an etymological germ from which "valency" was to spring, though before this development is sketched it is necessary to discuss another line of thought altogether, which, had it been followed, would have obviated the necessity for any new terms.

Frankland had approached the concept of valency from a standpoint that was fundamentally dualistic and electrochemical. Being steeped in the radical tradition he was naturally interested, as was Berzelius before him, in the forces holding two parts of a molecule together. This idea of "force" or "power" was not unnaturally present in his mind when he sought for a phrase to describe what he had discovered. Other workers in England could hardly escape being slightly influenced by this. But meanwhile others had been groping to the same goal from the opposite direction, the Theory of Types, chief among them Kekulé. The chief objects of *their* study had happened to be inorganic and organic acids. Recognition of the "tribasic radical of chloroform" by Williamson[17] and of the "bibasic sulphur atom" by Kekulé[18] had led to the latter's paper of 1857 dealing *inter alia* with "the basicity of radicals".[19]

"Basicity", however, was never more than a transient antecedent to "valency". In the paper just mentioned the terms "monatomic" and "monobasic", "diatomic" and "dibasic", *etc.*, are used interchangeably. It was therefore but a step to identify the nouns "basicity" and "atomicity", and it appears that this may have been done in some places, for in 1860 a paper by Kekulé on bromosuccinic, tartaric and malic acids ends with the assertion that "the basicity of an acid is thus independent of its atomicity".[20] Later Wurtz made the same distinction.[21]

The term "atomicity" was employed by Kekulé in his *Lehrbuch* of 1861,[22] and elsewhere.[23] By the middle of the 1860s it seems to have been the commonest expression in use, and again English writers, including Williamson,[24] Frankland,[25] Brodie[26] and many others, were quick to adopt it. The earliest reference to the word in English that has been quoted in the *Oxford English Dictionary* is from a review of Kekulé's *Lehrbuch* in 1865, which incidentally states the development of this idea by Kekulé to be "one of the most salient features of the system of Kekulé's book".[27] But Williamson's paper pre-dates this reference by a year. On the Continent, Wurtz[28] and Erlenmeyer[29] offer other examples.

Today "atomicity" has a different, almost opposite, meaning, implying the number of atoms present in a molecule. The first reference to this modern use of the word that has been found is in a periodical mentioned later, the *English Mechanic* (see p. 103), where it is defined as "the number of atoms in the molecule".[30] By 1874 it was in general use[31] but by 1888 its equation with "valency" was definitely "out of fashion".[32] Meanwhile, it was coming under attack from several quarters, and several alternatives were being put forward.

As far back as 1858, Odling had suggested "equivalency" to denote the ability of one substance to be exchanged for another, "taking the atom of hydrogen as the unit of equivalency".[33] This gave a bismuth atom, for example, "a threefold equivalency".

Others had used similar terms, but the lack of atomic implications in a word that had had long empirical associations caused the invention of "the word atomicity... for the purpose of describing those properties of atoms which were described by the word 'equivalence' ".[34] Thus "equivalence" was not a great improvement on "atomicity". But the chief attack on the latter came from Hofmann in 1865:

We are in want of a good appellation to denote this atom-fixing power of the elements. The vague and rather barbarous expression, *atomicity*, has drifted into use for this purpose; the elements have been called *monatomic, diatomic, triatomic,* and *tetratomic,* accordingly as their respective molecule-forming minimum weights are capable of saturating 1, 2, 3 or 4 standard atoms. These expressions are faulty, because they are open to misinterpretation, as if intended to denote the atomic structure of the respective elementary molecules themselves; a situation of confusion, the possibility of which should always be sedulously avoided in scientific nomenclature.

We shall escape this by substituting the expression *quantivalence* for *atomicity*; and designating the elements *uni*valent, *bi*valent, *tri*valent and *tetra*valent, according to their respective atom-fixing values.[35]

Here then was yet another factor injected into a situation that was already complicated enough, but this one gained much support. Roscoe spoke of the importance of quantivalence in a review in 1869,[36] and used it in his *Elementary Chemistry* two years later.[37] It also appeared in J. P. Cooke's *The New Chemistry* of 1874,[38] and in books of the following decade by Stallo,[39] Wurtz[40] and others. The *Journal of the Chemical Society* employed it until about 1885.

Even in England, however, usage was by no means uniform. Frankland's *Lecture Notes* (1866) assert that "this combining power of the elementary atoms is usually termed their *atomicity* or *atom-fixing power*",[41] but in the 1870 edition he adds the word "equivalence" as a further synonym, though without reference to "quantivalence".[42] In 1877 he refers to "the doctrine of atomicity or equivalence".[43]

The final stage in the development of this linguistic chaos came at the end of the 1860s with the introduction of the word *valenz* (German) and its anglicized forms *valence* or *valency*. That this is a contraction of *quantivalence*, or its German equivalent, is fairly certain. Writing on 1 July 1868, Wichelhaus, in an important paper on phosphorus derivatives, announced:

"*valenz*" is used as a shorter word in place of "quantivalence" introduced with the same meaning by A. W. Hofmann.[44]

No doubt it was this quotation that induced the writers of two well-known works to attribute the introduction of this word to Wichelhaus,[45] the first of whom, indeed, refers specifically to this paper. However, shortly before this Wichelhaus had employed it in another paper (on a closely related subject), but without definition; on 23 March 1868 he wrote:

One is not therefore inclined to regard the total *valenzen* in ethyl phosphite of three oxygen atoms and three ethyl groups as being satisfied with one phosphorus atom, or, in other words, to suppose that the phosphorus atom here has a valency of nine (*neunwerthig*).[46]

Even this, however, is not the first time the word appeared in print. In the previous year, a short paper by Kekulé on mesitylene used graphic formulae, of which it was said:

When I have gone to the trouble of explaining my views on atomic constitution of chemical compounds, I have used for some years a method by which atoms of different *valenz* were represented by different sizes.[47]

Thus Kekulé definitely pre-dated Wichelhaus. But it is not necessary to suppose that either was aware of breaking specially new ground. Wichelhaus had been a friend of Kekulé since he came to work under him at Ghent four years previously, and it is certain that the new ideas were under constant discussion. The cumbrous "quantivalence" was ill-suited to colloquial speech, and nothing is more probable than that in the ebb and flow of conversation the first two syllables were dropped. Hence the appearance of the shortened term in a paper could occur without special notice. Possibly, however, after its appearance in two publications some comment was made and Wichelhaus felt a short explanation was desirable. If anything can be read into this it is surely the intensity of thought and preoccupation with major issues that could allow the introduction of a new term without an immediate definition of it for the benefit of those outside that active circle.

It is important to note that this word was not coined directly from the Latin *so far as its chemical uses were concerned*. As will be seen (p. 88) its appearance in England took place so soon after this that "valence" or "valency" must be regarded as a simple absorption into English of this same word. But the point is far from obvious from a perusal of standard works on etymology. Thus Bailey's *Etymological Dictionary* gives the origin as *valens*,[48] while the *Oxford English Dictionary* states *valentia* and *valere*,[49] the latter word being also quoted by Findlay.[50] The meanings of these are as follows:

valens: strong, powerful, energetic, able.
valere: primarily *to be strong or well*; in a secondary sense *to be worth*.
valentia: a mediaeval Latin term meaning *value*. (*E.g.*, "rex ... statuit ut qui ad valentiam xv. solidorum non habent in bonis ad hujusmodi contributionem nullatenus compellerentur".[51])

It is doubtless from these sources, and especially the third, that the word "valency" arose as used in a non-technical sense in certain English writings of the seventeenth century.[52] But as we have seen, its chemical applications arise from a shortening of "quantivalence", and not directly from the Latin at all. Retracing our steps, we can see that Hofmann's choice of this word was in all probability governed by its kinship to "equivalence", which, however, he rejected. This was an ordinary English word given a technical significance at the beginning of the last century, and this is undoubtedly derived from the above Latin roots + *aequus* (= equal). Therefore "valency", in a chemical sense, does originate from *valere* and its derivatives, but in a roundabout way.

An interesting feature of the whole case was that the last step in the evolution of the word came in a German paper, rather than in a language more rooted in Latin. Although today English and French usage is confined to "valency", the Germans almost from the beginning have employed, in addition to *Valenz*, the word *Wertigkeit*, whose literal meaning corresponds to "value" or "worth", and is thus the Teutonic parallel to both "equivalence" and "quantivalence". An early appearance of the word is in another article by Kekulé in Fehling's *Handwörterbuch* (1871): "Die Aciditat ist wie die Basicitat, ein specieller Fall der Werthigkeit oder der Aequivalenz."[53] Before this it had occurred in the title of a book by Geuther in 1869.[54]

The probable derivation may be summarized as shown on p. 88.

It remains to point out the varied reception the new word received in the English-speaking world. In the less sophisticated publications

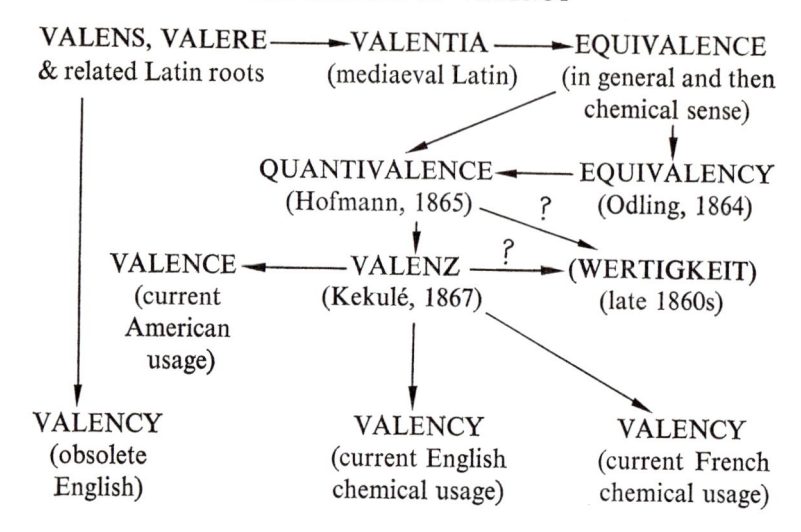

```
VALENS, VALERE ──────▶ VALENTIA ──────▶ EQUIVALENCE
& related Latin roots   (mediaeval Latin)   (in general and then
        │                      │                chemical sense)
        │               QUANTIVALENCE ◀────── EQUIVALENCY
        │               (Hofmann, 1865)    ?    (Odling, 1864)
        │                      │        ?
    VALENCE ◀─────── VALENZ ──────▶ (WERTIGKEIT)
    (current         (Kekulé, 1867)      (late 1860s)
    American              │
    usage)                │
        │                 │
    VALENCY           VALENCY           VALENCY
    (obsolete         (current English  (current French
    English)          chemical usage)   chemical usage)
```

its popularity was exactly matched by that of graphic notation (see p. 103), and the first reference that has been discovered in connection with the present work is in a pseudonymous letter (by "Urban") in the *English Mechanic*, where a correspondent, one G. E. Davies, is advised "to use the at present adopted term 'valency' instead of 'atomicity', and 'dyvalent' instead of 'dyatomic' ".[55] This reference (12 March 1869) is a little earlier than the first quoted in the *Oxford English Dictionary*,[56] which refers to the same periodical but a later issue (19 November). In this, "Sigma", later identified as J. T. Sprague,[57] writes as follows:

The molecule is "a body in which all the attractions or valencies are satisfied, leaving the combined atoms to act as a whole from one centre, so far as such forces as gravitation, cohesion, heat, *etc.*, are concerned".[58]

He had written a week earlier that

some chemists, for good reasons, attach distinct meanings to atomicity and valency, but the words are commonly used as synonymous.[59]

In the issues of this journal for 1869, enthusiastic discussion of valency included the question of terminology, and its merits over "atomicity".

In more exalted circumstances, at the 1875 meeting of the British Association, Harcourt remarked that "many substances are known whose existence is contrary to the theory of valency and saturation".[60] The following year saw the publication of the ninth edition of the *Encyclopaedia Britannica*, and here again "valency" appears as a word now generally accepted.[61] It is therefore all the more curious to note the omission of the word in Frankland's writings until well after these dates. However, his joint textbook with Japp (1884) stated that atomicity was sometimes called valency;[62] and in his autobiographical *Sketches*[63] he speaks of "valency" instead of "equivalence" which he had used in a parallel passage in 1877.[64]

It is interesting to observe that the transliteration of *valenz* to *valence* was never followed in America by changing the final vowel, and "valence" it remains today. In England, however, this must be accorded the status of a hypothetical, transient intermediate.[65] Does this suggest that perhaps after all the word "valency" had maintained a tenuous existence in this country all the time since the days of Jacobite drama? If so, it is possible to understand how it was able so quickly to assume a new and more profound meaning at the indirect instigation of a school of German chemical enthusiasts.

The subsequent fortunes of the word "valency" have been linked with the later developments in chemical theory. Recently, its use in a numerical sense has been impeded by the ever-widening range of phenomena that it has had to cover. Thus the distinction between covalency and electrovalency has led one writer to remark:

One should not try to establish a single all-embracing definition of the valency of an atom in a molecule or ion, but should recognize that the word "valency" can be used to signify either of two distinct, but equally useful, concepts.[66]

Another author, lamenting its ambiguities in co-ordination chemistry and elsewhere has observed: "The word valency is overworked."[67] It will be interesting to learn what alternatives future developments may offer.

3. THE WORD "BOND"

For some years after the birth of the theory of valency most chemists were reluctant to speak of "chemical bonds". There is little doubt about the reason for this hesitance. The use of the phrase committed the user to a more detailed, less vague picture of a molecule than many were at that time prepared to accept. For the same reason, graphic representations of molecules took time to gain general assent.

Kekulé, with characteristic restraint in this matter, preferred "affinity units",[68] and this became a common phrase in German literature (*Verwandschafteinheiten*).

It is to Edward Frankland more than anyone else that chemistry is indebted for the fertile conception of a "chemical" bond. His first formal statement on this matter appeared in 1866:

> By the term *bond*, I intend merely to give a more concrete expression to what has received various names from different chemists, such as an atomicity, an atomic power, and an equivalence. A monad is represented as an element having one bond, a dyad as an element possessing two bonds, *etc*. It is scarcely necessary to remark that by this term I do not intend to convey the idea of any material connection between the elements of a compound, the bonds actually holding the atoms of a chemical compound being, as regards their nature, much more like those which connect the members of our solar system.[69]

Now the last sentence of this appears almost *verbatim* after an account of the use of graphic formulae in the *Lecture Notes* published in the same year.[70] When the latter was issued in a new edition four years later, however, there is a significant change; the last phrases are recast thus:

> ... the bonds which actually hold the constituents together being, as regards their nature, entirely unknown.[71]

Again, the edited version of his paper that appeared in the *Experimental Researches* in 1877 contains a similar phraseology about the unknown nature of the bonds.[72] It would thus seem that Frankland was retreating slightly from a position of excessive literalism in the conception of a bond, and was anxious to avoid even a gravitational metaphor. But there was no withdrawal from a *recognition* of such bonds. In fact in the collected volume just mentioned at least three passages from earlier papers have been altered so as to involve more specifically this very idea. Thus the references to "one or both ethyl bonds",[73] "one oxygen bond",[74] and "exchange of three oxygen bonds for a like number of carbon bonds"[75] are all modifications of the originals.

Frankland's precise definitions of a bond vary slightly. In 1866 it was a "point of attachment ... by which it [an atom] can be united with any other element";[76] in 1870, however, he "named each unit of atom-fixing power a *bond* – a term which involves no hypothesis as to the nature of the connection".[77] It seems that about 1866 he began to be aware of the danger of too vivid a physical picture, and by four years later was anxious "to avoid any speculation as to the nature of the tie which enables an element thus to attach itself to one or more atoms of other elements".[78] Probably the reason was that with former adherents of the Type Theory, especially, his views were making little headway, and it was necessary to prune from them all unnecessary parts that would savour of dogmatism about things unknown or unknowable.

It is also possible that Frankland's addiction to the word "bond" rendered superfluous the word "valency", at least when this was used simply in the general modern sense of "chemical linkage". That colloquial English usage did permit the employment of "valency" as synonymous with "bond" is evident, *e.g.*, from another pseudonymous contribution in the *English Mechanic* for 1869, where "Urban" wrote of atoms being linked by their "valencies" and of "one valency satisfied".[79] This kind of thing may have been behind Frankland's reluctance to speak of "valency".[80]

Of the general refusal to admit the reality of bonds, or specific links between specific atoms, little need be said here, except that it lasted a long time, even with those who had done much to foster the idea of valency. Use of the word "bond" was specifically deprecated by Williamson in 1869,[81] for example. But by the 1870s few except Kolbe were prepared openly to oppose it. Since then it has become an indispensable concept in chemistry[82] – much more than a mere synonym to "valency". In our own day it is significant that the revolution in the teaching of elementary chemistry has produced, amongst other ideas for the *avant garde*, a new system of presentation known as CBA, the "Chemical Bond Approach".[83]

CHAPTER V

Valency Notation

THE SYMBOLIC REPRESENTATION of chemical ideas has always been of the greatest importance and is now a subject of international agreements on convention, comparable in some ways to those dealing with nomenclature.

Notation, like all forms of symbolism, is an attempt to express the unknown in terms of the known, the abstract in terms of the concrete. But it is also a kind of shorthand, and does not always set up a model, even on paper, with which we are invited to compare the "real" object. In the early stages of valency theory, the notation used to represent chemical bonds on paper at first had this second significance and was in fact little more than an economy in words; but later an increasingly vivid physical picture emerged, and with it a more realistic symbolism.

It will be convenient to discuss these developments by dealing with the various elements or symbols one at a time.

1. SUPERSCRIPT DASHES

These were introduced in 1855 by Odling and are still sometimes used. Dashes were employed to indicate the number of atoms of hydrogen to which one atom of an element is equivalent, *i.e.*, its valency. Thus he wrote,

$$H', Sn', Sn'', Bi''', etc.$$

The immediate object of this paper, "On the Constitution of Acids and Salts",[1] was to show that all salts were reducible to the water type, $\left. \begin{array}{c} H \\ H \end{array} \right\} O$.

In substituting a definite number of hydrogen atoms "the number of atoms introduced may vary very considerably, providing the total exponential value remains the same"; examples given included replacement of H'H'H' by K'K'K', $1\frac{1}{2}$Sn'', Sn''+K', Bi''', *etc.* His salts were then based upon

$\left.\begin{array}{l}H'\\H'\end{array}\right\} O''$ $\left.\begin{array}{l}H'H'\\H'H'\end{array}\right\} 2O''$ $\left.\begin{array}{l}H'H'H'\\H'H'H'\end{array}\right\} 3O''$ *etc.*

Kekulé's famous paper of 1857[2] refers to this publication by Odling, and uses multiple dashes for the polyvalent groups holding the molecules together, *e.g.* in

$S\overset{..}{O}_2, Cl_2$ $\left.\begin{array}{l}H\\S\overset{..}{O}_2\\H\end{array}\right\} \begin{array}{l}O\\ \\O\end{array}$ *etc.*

It suited his purpose better to place the symbol above rather than after that of the atom or group. In his *Lehrbuch*[3] he gave due place to Odling's contribution to notation, and used it frequently in the book. He continued to employ it in this way for several years in some of his papers, but so far as organic chemistry was concerned its value became more limited as structural and quasi-structural formulae became more fully written, and the overall valencies of various groups became so well known that little was to be gained by using Odling's suffixes.

Although Kekulé's works probably contributed more than most to the acceptance of this simple idea, he was not the first to use Odling's suffixes. Within a few months of Odling's paper, one appeared by Wurtz[4] in which tribasic nitrogen was represented as Az'''. He and other chemists used the suffixes extensively after this. They formed an integral part of Frankland's proposals for nomenclature and notation,[5] where they were applied especially to organic radicals as formyl (CH), Me''', methylene (CH$_2$), Me'', ethylene (C$_2$H$_4$), Et'' and so on. As is now usual, they were omitted in the cases of monovalent elements, oxygen and carbon. Frankland extended their use to mark multiple bonds, and placed them *before* an atom to imply the presence of "latent atomicities" or the difference between the maximum valency and that actually used in the substance concerned:

$''\left\{\begin{array}{l}CH_2\\CH_2\end{array}\right.$ (ethylene); 'Hg'$_2$Cl$_2$ (mercurous chloride)

Where the valency was greater than three, presumably for reasons of convenience, Frankland employed Roman numerals instead of dashes.

D*

A year before this paper appeared, Hofmann[6] had dispensed with dashes altogether in favour of Roman numeral superscripts even for valencies of 1, 2 and 3. This had been suggested on grounds of uniformity and convenience.

2. PARENTHESES

By these are meant symbols of the type ( ), rather than the brackets written as { }.

To deal with the difficult problems of nomenclature and notation the London Chemical Society set up a committee in 1864. In that year, Williamson was induced by this circumstance to read a paper to the Society, giving his own (unofficial) views. The paper,[7] which was generally approved,[8] contains no mention of valency, and apart from letters and numbers uses these parentheses as the only symbols. They are employed to enclose groups of atoms that are not separated from each other during the course of a chemical reaction, e.g., in:

$$Ag(NO_3) + KCl = AgCl + K(NO_3)$$

Now such a concept of a group of atoms remaining unchanged through a chemical reaction was familiar to the adherents of the theory of radicals. It is not, therefore, surprising to learn that the pioneer in the use of such symbols was Kolbe, although the system had received sporadic attention before, particularly when the parentheses enclosed a group of atoms that was being doubled or trebled, *etc.* Thus his formula for acetic acid,

$$HO.(C_2H_3)C_2,O_3 \qquad [C = 6; O = 8]$$

emphasized the persistence of the methyl group through a variety of reactions. Of the nineteen formulae for acetic acid given in Kekulé's *Lehrbuch,* only two apart from those of Kolbe contain these parentheses.[9]

On the whole, the adherents of the Type Theory found such symbols unnecessary, and when this view became broadened into valency, full use was made of the new structural formulae so in any case there was no great call for these signs. Even Frankland used them only rarely, except, like everyone else, for simple inorganic formulae as $Ca(NO_3)_2$. It was only later that *widespread* use was made of this kind of symbol in organic chemistry, when considerations of space made it sometimes necessary to contract the familiar structural formulae to compressed statements of the following kind (an early example, due to Kekulé):

$$CH_3-CH(OH)-CH_2-CH_2(OH),$$
"butylene glycol"[10]

Generally speaking, however, parentheses were not found in the formulae of the early period of valency theory, and, where they were, expressed the valency only in the indirect sense that they emphasized the overall value for a radical or group of atoms. Only Kolbe continued to use them for a number of years as a general practice, and he was one of the last leaders in chemical thought to surrender to these new ideas.

3. BRACKETS

Brackets have had a curious history in chemistry. Their first systematic employment seems to have been by Gerhardt in his Type Theory.[11] Here their function was to "link" on paper groups of atoms *without implying any actual physical arrangement in the molecule that could correspond with this method of writing.*

It was Kekulé who opened up the inner meaning, hitherto unrecognized, of these bracketed formulae, and as we have seen (p. 59) he showed that it was possible to set up mixed types in which a polyatomic radical held together the rest of the molecule. Thus in the formula for sulphuric acid the brackets took on a new significance. They could be regarded as implying a definite linkage between the atoms or groups they spanned. Thus[12]

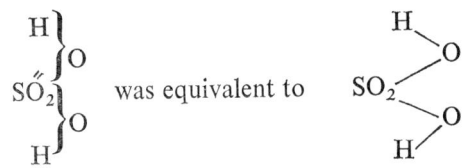

$$\left.\begin{matrix} H \\ S\overset{''}{O}_2 \\ H \end{matrix}\right\} \begin{matrix} O \\ \\ O \end{matrix} \quad \text{was equivalent to} \quad SO_2 \begin{matrix} H \diagdown \\ \diagup O \\ \diagdown O \\ H \diagup \end{matrix}$$

In Frankland's 1866 paper on notation[13] the bracket is exclusively used to indicate linking between groups written vertically over each other, as

$$\left\{\begin{matrix} CH_3 \\ CH_3 \end{matrix}\right. \qquad \left\{\begin{matrix} CH_3 \\ O \\ CH_3 \end{matrix}\right. \qquad ''\left\{\begin{matrix} CH_2 \\ CH_2 \end{matrix}\right.$$

These formulae, he tells us, were used in his lectures at the Royal College of Chemistry in the autumn of 1865. They also appear in many of his papers in the previous decade. Yet later in 1866 they were to be largely superseded by a better system.

One man, however, would have nothing to do with this kind of bracket. Williamson, in many ways surprisingly ahead of his time, refused to employ these relics of the Type Theory. In this he was consistently maintaining a position he had taken up ten years previously, when, as Kekulé remarked, "Williamson urged for clear formulae, without commas, Kolbe's 'buckles' or Gerhardt's brackets. This was excellent training for making the mind independent".[14] As we have seen (p. 94) the only extraneous symbols he would admit in 1864 were simple parentheses.

But Williamson was a lone voice, and if he abolished brackets Odling restored them.[15] Long after many chemists had accepted the new structural ideas of Williamson and Kekulé the Type Theory was still retaining a hold on others, and its bracketing system of notation was far from dead, as can be seen from the writings of Odling.

It is interesting, but probably coincidental, that the only time brackets of this kind are used today is to denote non-localized chemical attractions. Just as the "Typist" was reluctant to attribute structural significance to his formulae, so now we are not prepared to directionalize an ionic link or to localize a unit charge in a mesomeric system. For both these purposes a "curly bracket" is sometimes used, e.g.,

4. TOUCHING OR INTERSECTING CIRCLES

It is necessary now to follow a sequence of ideas that was to lead away from the main stream of thought and to have little ultimate influence. So far, the three items considered (Odling's superscripts, Williamson's parentheses and the brackets of the Type Theorists) all found a place in the formulae of the last half of the century, and in some sense have survived until today. But the symbols we are about to consider made only a transient appearance in the 1860s, and in retrospect seem an isolated phenomenon which left little impression on chemistry as a whole. Nevertheless while they were in use they undoubtedly contributed to the spread of the ideas of valency.

In 1861 Joseph Loschmidt published from Vienna a memoir entitled *Chemical Studies*,[16] the first part of which was a list of 368 formulae, ranging over the whole of organic chemistry. These formulae were written out graphically, *i.e.*, in such a manner as to show the way in which the components were joined, each atom being individually represented. Loschmidt's publication has the distinction of anticipating Kekulé's ring structure for benzene.

Four years later Loschmidt published a paper "On the magnitude of the molecule of air",[17] in the course of which he gave further details of his formulae. Assuming atoms to be "vanishingly small" in comparison with their sphere of influence, and that even in solids the space between them is large, he denoted the atoms at first as points or very small circles. The resultant formulae, like this for water,

tended to defeat the whole purpose of clarity, and so he went on to modify them. Circles were drawn to stand for the "equilibrium sphere", the radius representing the distance when attractive and repulsive forces were neutralized. Hence the circles now touched, and each valency bond (as we should now say) was symbolized by one contact of this kind. Thus water was now

Sometimes Loschmidt differentiated the atoms by concentric circles and by variations in size. Where two polyvalent (*mehrstelliges*) atoms were linked, their circles were made to intersect, rather than touch, and the overlapping area was marked by oblique strokes to denote the "number of places where one atom passes over the other". In other words, the intersecting circles stood for multiple bonds. This was a definite advantage over the systems used by Couper and Kekulé.

The following are a few more examples of Loschmidt's graphic formulae:

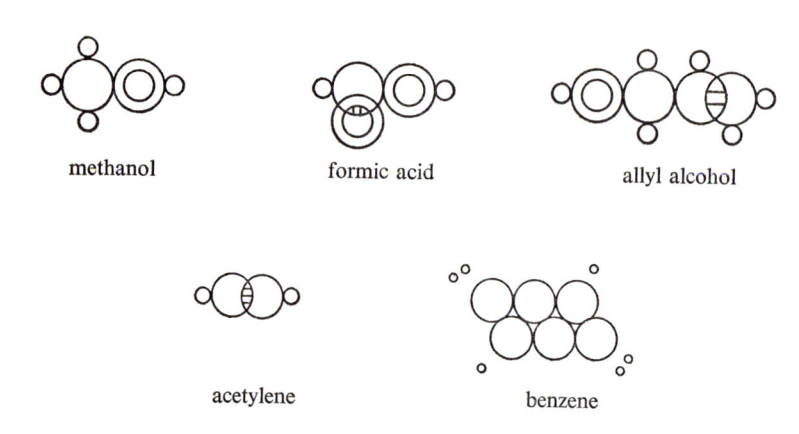

methanol   formic acid   allyl alcohol

acetylene   benzene

In the same year as the first of Loschmidt's two publications (1861), appeared volume i of Kekulé's *Lehrbuch der organischen Chemie*.[18] Written mainly in terms of typical formulae, it nevertheless had a footnote[19] in which the author introduced his own system of graphical formulae. The following are some examples:

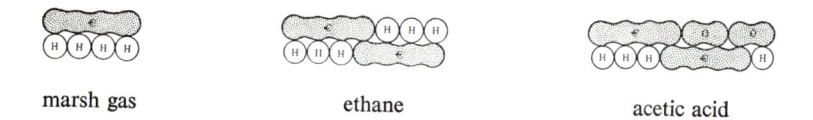

marsh gas   ethane   acetic acid

Each unit represents an atom, and the size of the unit depends on the combining power (valency) of the atom concerned. The symbols look as if they have arisen from the coalescence of several circles. As with Loschmidt, the linking of two atoms is indicated by touching of two symbolic units. But there are two important differences: (i) the linking is deemed only to occur between units placed above or below each other; there is no "horizontal" bonding, *e.g.*, between the hydrogen atoms in methane; (ii) multiple bonds are indicated by contact between adjacent units in two places.

Many years later, Anschütz was able to examine a notebook belonging to one of Kekulé's students at Ghent, M. Holzmann, containing notes from Kekulé's lectures delivered in 1857/8.[20] This notebook has similar formulae to those published in the *Lehrbuch*, e.g., these for marsh-gas:

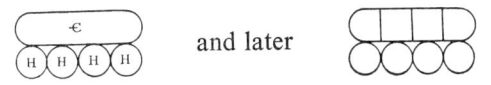

and later

Apart from this occurrence, this type of graphic formula appeared only once again in Kekulé's writings, and that was in his paper of 1865 on the constitution of aromatic substances.[21] Here he was primarily concerned to use these methods of formulation to serve as a basis for discussion of the aromatic compounds. But he does give several simpler aliphatic examples, as propyl and methyl-ethyl alcohols (propan-1-ol and propan-2-ol):

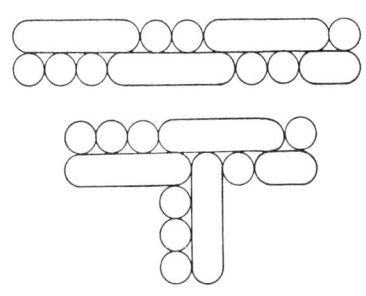

It will be observed here that the second formula cannot be interpreted quite so obviously as the others quoted before, for the "lower" carbon atom must be linked "horizontally" to its three hydrogen atoms. This possibility is inevitable with any chain of carbon atoms greater than two in length, because a 3-membered chain can obviously be bent thus:

$$\begin{array}{c} C-C \\ | \\ C \end{array}$$

Consequently, it would seem that this kind of symbolism was already doomed, for graphic formulae must aid clarity or their purpose is lost, as Loschmidt so clearly saw. It is therefore all the more surprising to read that Kekulé preferred this system to that of Loschmidt and to that of Crum Brown (see p. 101).[22] Indeed, a letter to Erlenmeyer (4 January 1862) contains the contemptuous phrase "Loschmidt's confusion-formulae".[23]

However, neither Loschmidt's nor Kekulé's system was to have a lasting influence. It is true that Wurtz,[24] Naquet,[25] Blomstrand[26] and others employed for a short time symbols like those of Kekulé, but the increasing use of the Crum Brown method inevitably caused them to fall from favour. With Loschmidt the chief difficulty was the inaccessibility of his writings, particularly the 1861 pamphlet. Anschütz had the greatest difficulty in locating it, though when he had done so he issued it as no. 190 in Ostwald's *"Klassiker"*.[27] Were it not for this, Crum Brown's formulae would have had to fight a harder battle for supremacy.

5. LINEAR REPRESENTATION OF BONDS

The use of a straight line to stand for a valency bond seems today so obvious as to render all other methods unnecessary, if not absurd. Yet it was not until the 1870s that the practice became common, and opposition to it remained almost until the death of Kolbe (1884). Partly, of course, this was because of genuine doubts as to the reality behind the symbol (see p. 47). A straight line is a much more tangible symbol than a pair of touching geometrical figures, a bracket or a set of dashes; it corresponds so closely to well-known physical objects like rods, tie-bars and so on. Hence once it was admitted as a symbol it was harder to maintain the intellectual outlook that there was possibly (or probably) nothing in the "real" molecule that did hold one pair of atoms together, even if one conceded any physical meaning to the word "molecule". Moreover, it was a complete break from the Type Theory in symbolism as well as in its essential idea. To use a bracket with a new meaning (valency) was a less severe innovation than addition of a new set of signs as well.

Credit for originating linear representation of bonds by lines has been usually given erroneously to Couper. The source of this mistake appears to be a paper by Anschütz where he says "Couper deserves the credit of having introduced into constitutional formulae the lines indicating union of atoms".[28] As has been pointed out by Wheeler[29] at the recent Kekulé/Couper Symposium, this view was stated several times.[30] In fact, however, Couper was anticipated by William Higgins in 1789.[31] In a work that was far ahead of his time,[32] Higgins wrote such formulae as:

LINEAR REPRESENTATION OF BONDS

$$\text{I——d} \qquad \text{S——d}$$

where I = hydrogen (inflammable air),
       d = oxygen (dephlogisticated air),
and S = sulphur.

These suggestions proved abortive, however, and although his priority remains, his influence seems negligible so far as graphic formulae are concerned.[33]

If Couper was not technically the first to use lines in formulae, he can claim the credit of introducing them to represent *valencies*, for his papers of 1858 show that he had grasped the essentials of these ideas in a way totally untrue of Higgins.

Examples of Couper's formulae will be found on pp. 74 and 78. Dotted lines were used at first, full lines slightly later. They were greatly superior to those of Kekulé of about the same time, but they did lack certain features that would have made them more useful, the chief of which was provision for indicating double bonds. The failure to write the hydrogen atoms separately, and the strange "double oxygen" atoms are less severe defects. But their ephemeral nature was not primarily due to any faults they had (which could have been easily rectified), but to Couper's breakdown in health and withdrawal from the chemical scene. Nevertheless, as has already been suggested, their influence probably persisted in the writings of Crum Brown, who for a few months, while he was preparing for his doctorate at the University of Edinburgh, was a colleague of Couper.

The thesis for Crum Brown's doctorate[34] contained for the first time examples of the graphical system of notation that was to have such a profound influence, and that is almost universally employed today. The formulae were first published in 1864[35] (not 1865 as Tilden says[36]) and were used by him several times after that.[37] Some examples are given on p. 102.

Now these formulae are an advance on any up to this time, in that they show each atom separately, and clearly indicate all single and multiple bonds. But such clarity was not without its dangers; the less ambiguous a method of representation, the more searching questions may be asked of it, and the more strictly must its conventions be laid down. With these symbols of Crum Brown the first difficulties were those of spatial relationships. How far, if at all, was it justifiable to infer a symmetrical, planar disposition of actual atomic valencies? At this stage, Crum Brown was wise enough to recognize that no such justification existed, at least at that time, and wrote that he "did not mean to indicate the physical, but merely the chemical position of the atoms". It is perhaps permissible to see in this philosophical distinction some of the influence of Couper, or at least of the Scottish universities generally where philosophy and the humanities had conceded less to science than elsewhere. At all events, a clearer explanation of the difference between "chemical" and "physical space" would have helped his case considerably.

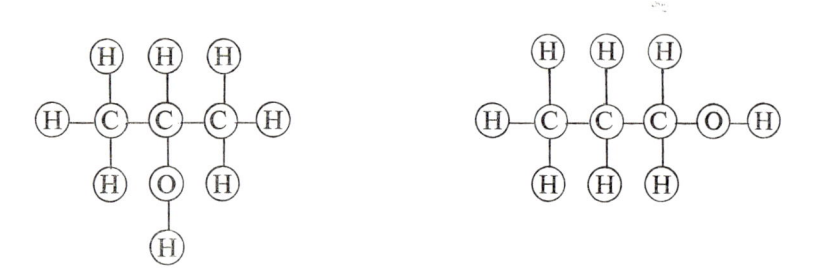

Perhaps this is part of the reason for the neglect of Crum Brown's influence on the Continent, especially in Germany. Other factors helped in this, however; in particular the Edinburgh publications seem not to have been very accessible in Germany. Thus Anschütz[38] believes Kekulé cannot have heard of Brown's 1865 paper or he would have taken it into consideration in the German version of his own paper that came out ten months later.[39] It is surprising that careful German historians like von Meyer[40] and Ladenburg[41] give Crum Brown not even a passing reference in their histories of chemistry.

How Kekulé dealt with the difficulties raised by Crum Brown's notation will be considered later. Meanwhile it is pleasant to record that in Britain, and especially England, it met with almost immediate success. One reason was doubtless that Crum Brown's formulae showed that only two saturated monohydric alcohols containing three carbon atoms could exist (propan-2-ol and propan-1-ol). It was not difficult to show the identity of alcohols corresponding to both these formulae:

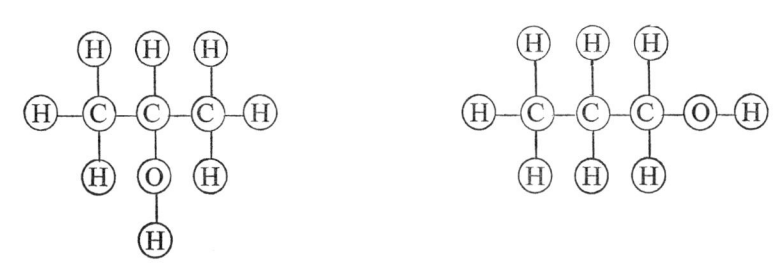

```
    (H) (H) (H)                  (H) (H) (H)
     |   |   |                    |   |   |
(H)–(C)–(C)–(C)–(H)          (H)–(C)–(C)–(C)–(O)–(H)
     |   |   |                    |   |   |
    (H) (O) (H)                  (H) (H) (H)
         |
        (H)
```

Whereas Kekulé's original graphic formulae implied three alcohols, such ambiguities did not arise with Crum Brown's system. This kind of achievement would naturally arouse admiration in any country but it is suggested that there was a special reason at this particular time for its success in England and lack of it in Germany and France.

Since the establishment of the Mechanics' Institutes in the 1820s, the demand for popular education in technical subjects had steadily grown in England until by 1850 there were about 700 Institutes and 107,000 members.[42] Establishment by the Government in 1854 of the "Science and Art Department", following the 1851 Great Exhibition's demonstration of the need for technical skills, had given further impetus to this tendency. Hence in the period under review (the 1860s) the demand arose for methods of teaching that would appeal to the unsophisticated, working-class man and youth.[43] Periodicals came into being whose purpose was to convey technical information informally and simply to a class of reader that the learned journals would leave quite untouched. Perhaps the most outstanding of these was an organ known as the *English Mechanic and World of Science*, whose popular appeal seems to have been considerable; it was founded in 1865, and issued weekly at 1*d*.

In France and Germany the situation appears to have been different. University education in chemistry seemed comparable to, if not better than, that in England.[44] But popular technical education lagged behind, owing doubtless to the different political circumstances of these countries. Chemical education of the masses was by no means so big an issue, at least until after the upheavals of the 1870 war. Then, of course, Germany took the lead.

If these views are correct, it is not hard to see that Crum Brown's formulae would be readily accepted in England as a simple means of understanding organic chemistry by those who lacked the benefit of a (probably Continental) university education.

The means by which they were to be first propagated outside the learned journals was a book published in 1866 by Frankland: *Lecture Notes for Chemical Students*. In the preface to this, Frankland states:

I have extensively adopted the graphic notation of Crum Brown which appears to me to possess several important advantages over that first proposed by Kekulé.[45]

In a letter to Crum Brown the same year he wrote:

There is a good deal of opposition to your formulae here, but I am convinced that they are destined to introduce much more precision into our notions of chemical compounds. The water-type, after doing good service, is quite worn out.[46]

Frankland introduces the symbols thus:

To give a concrete expression to these facts, the atom of hydrogen may be represented as having only one point of attachment or *bond* by which it can be united with any other element, zinc as having two such bonds, boron three and so on. Thus the atoms of these elements may be graphically represented in the following manner:

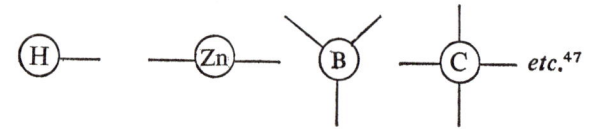

(H)—   —(Zn)—   (B)   —(C)— *etc.*[47]

He went a little further than Crum Brown in that he also symbolized "latent atomicities", corresponding closely to our unpaired electrons. Examples of his formulae are given on p. 105.

Meanwhile, a lecture delivered by Hofmann at the Royal Institution in 1865 employed as illustrations models made from croquet-balls and rods, diagrams of which became known as Hofmann's "glyptic formulae"[48] (see Plate 7).

These graphic formulae met with immediate success. Within a year they were being used quite widely, in most cases with the circles omitted. In the British Association Report for 1867 are found the formulae[49]

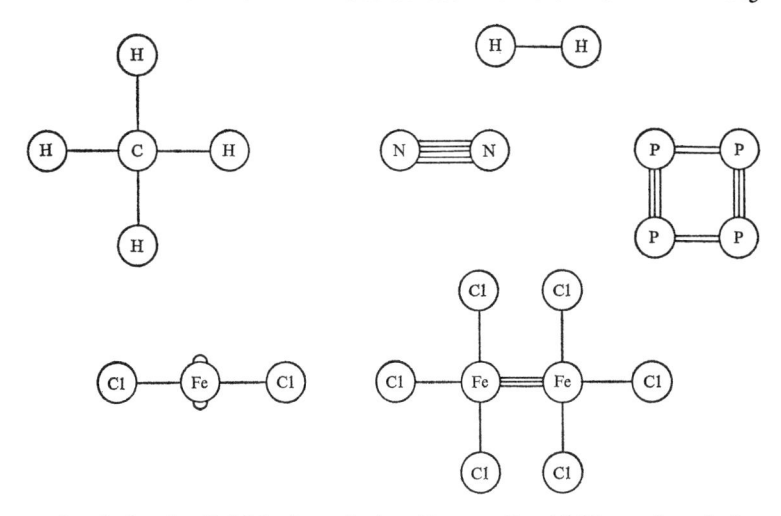

$$\begin{array}{cc} CH_2Cl & CH_3 \\ | & \text{and} \quad | \\ CH_2Cl & CHCl_2 \end{array}$$

But the old methods persisted in many quarters, and in 1868 Frankland was publicly regretting the lack of uniformity in notation. A textbook issued in 1870 used both Crum Brown's *and* Kekulé's notations, together with other devices such as strokes and commas; this is not without interest because J. P. Cooke, the author, was professor at Harvard, and so the book indicates that the new system was becoming used in America as well.[50] A fact in agreement with the thesis of the previous pages is the speed with which the Crum Brown/Frankland notation appeared in the pages of the *English Mechanic*, where it was firmly established by 1870. A rather curious discussion took place here over many weeks as to whether or not Frankland's *Lecture Notes* were the officially approved book for "Government classes"; the answer, rather inconclusive, appeared to be that they were not.[51]

By 1872, Frankland had brought out a revised edition of the organic part of his *Lecture Notes*, which he published as vol. ii. The same general symbolism is used, without the circles. Two other English textbooks published about this time, by Valentin (1872)[52] and Miller (1874),[53] also used Frankland's notation. The review of the former in *Nature* said that "it appears that of late years this system has gained much ground".[54] But the old ideas died hard, and Type formulae were sometimes employed side by side with graphic, as in the British Association Report for 1873 when we read of[55]

$$N \begin{cases} C_6H_5 \\ CH_3 \\ CH_3 \end{cases} \text{(H. E. Armstrong), and} \quad \begin{matrix} CH_2-CO \\ | \\ CH_2-CO \end{matrix} \Big\rangle O \text{ (W. H. Pike).}$$

A light-hearted comment on the place of this system of notation in English chemistry was the following:

> Though Frankland's notation commands admiration,
>   As something exceedingly clever,
> And Mr. Kay Shuttleworth praises its subtle worth,
>   I give it up sadly for ever:
> Its brackets and braces, and dashes and spaces,
>   And letters decreased and augmented
> Are grimly suggestive of lunes to make restive
>   A chemical printer demented.
> I've tried hard, but vainly, to realize plainly
>   Those bonds of atomic connexion
> Which Crum Brown's clear vision discerns with precision
>   Projecting in every direction.[56]

It is not without significance that the "Kay Shuttleworth" mentioned is Sir James Kay-Shuttleworth, a medical man of immense influence in the development of wider education, and first Secretary to the Committee of the Privy Council for Education (1839).[57]

Meanwhile, on the Continent the acceptance of graphic formulae was more slow and less certain. In 1865, Wilbrand used a system of notation that had little influence afterwards but was a genuine attempt to portray unsaturated and even cyclic compounds in terms of graphic formulae. Thus he wrote single, double and triple bonds as follows:[58]

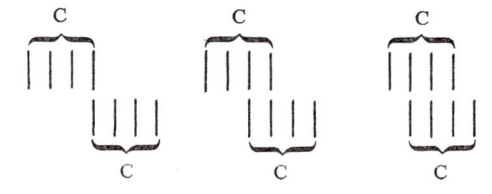

The chief influence for the use of graphic formulae was wielded by Kekulé. At first he had rejected those of Crum Brown as being unsuitable, and had invented his own notation (see p. 98). In 1867 he explained why he found the British system unacceptable. If all the bonds are confined to one plane it is impossible to form triple links, as in C≡N. He was therefore led to propose a tetrahedral model in which triple bonding could occur. But for practical purposes he employed symbols very similar to those of Crum Brown in the same paper.[59] Two years later he began to employ even simpler graphic formulae like the following:[60]

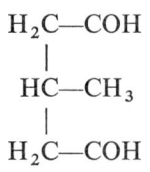

$$\begin{array}{c} H_2C\text{—}COH \\ | \\ HC\text{—}CH_3 \\ | \\ H_2C\text{—}COH \end{array}$$

Gradually the practice became common, and models like those brought from Edinburgh to Ghent by Dewar in 1867[61] doubtless helped to create the general impression that was favourable to the type of graphic formulae now in use.

CHAPTER VI

The Question of Priority

THERE NOW REMAINS one major topic for discussion relating to the birth of the theory of valency. This is a question that was hotly argued for many years after the events to which it relates, and is still to some extent a current issue. To which chemist can be fairly ascribed the origin of the idea of valency? Lest this should be regarded as an issue as dead as many others from the last century, it may be recalled that as late as 1930, H. E. Armstrong, reviewing Anschütz's two volumes on August Kekulé,[1] called for a re-examination of the whole matter, and the setting up of a historico-chemical "Court of Appeal" to deal with it.[2] Little was heard of this suggestion, though Anschütz expressed himself in agreement provided it dealt impartially, adding that "if Kekulé should be accused before it, I should be ready to defend him".[3] But the publication by Anschütz the previous year of Kekulé's *History of Valency*[4] did raise new issues, and these do not appear to have yet received adequate attention.

Now, if a controversy between the leading men of ability in a given field is to be maintained over a long period, it can only be for two reasons: there must be, on one side or both, errors of either fact or judgment. In a dispute over priority in formulating a new doctrine, be it scientific or theological, these may be stated thus. In matters of fact there must be ignorance of either date or contents of the documents concerned; or errors of judgment may exist whereby statements are taken to mean different things by different contenders. With regard to the latter, it is notoriously easy to misinterpret words and phrases used for the first time, or (worse still) with a new meaning.

It might be supposed, however, that a well-documented science like chemistry would be relatively free from disputes as to matters of literary fact. Even in the nineteenth century it was fairly easy to check dates and contents of papers, at least in the common journals. Yet it is strange to find that every one of the main contestants here suffered from the hazards of publication. Couper and Frankland both had the frustration of delayed publication; the former because of carelessness by his superior Wurtz,[5] the latter for a similar reason: "the paper was at once ordered to be published in their *Transactions* by the Council of the Royal Society, but was inadvertently laid aside in his private drawer by the Secretary (Professor Stokes), and its publication was, in consequence, delayed a twelvemonth".[6] Those not informed of these circumstances (as Kekulé) had therefore reason for an error of dating.

In another paper,[7] written jointly by Frankland and Kolbe, only the latter's name appeared, due to some editorial oversight, apparently.[8]

To these problems of publication must be added one more acute still, namely when to publish, or even whether to publish. Kekulé was dissuaded by his friends from early publication of his valency theories,[9] and was advised by the editor of *Annalen* (Volhard) not to publish his polemical *History of Valency Theory*[10] at all.[11] In some respects the birth of an idea is of greater significance than its first appearance in print, and the historian is wise to remember this.

Thus the whole question of priority was complicated by these difficulties. But of far greater consequence was the existence of misinterpretations of what had been published in papers that were perfectly well known and whose date was hardly a matter of dispute at all. Two schools of thought arose, in the traditions or the Radical and Type Theories, and "each seemed to labour under an absolute inability to place itself in the mental position of the other".[12] In this lies the chief explanation of the controversy that ensued.

When to this is added the confusion with regard to atomic weights that was rampant in the 1850s, it is not surprising that authors had difficulty in understanding each other. Frankland and Kolbe used one system; Couper employed another, while (at this time) Kekulé oscillated between two sets of values which he used alternately in a series of papers. This chaotic situation was in no small measure responsible for the misunderstandings that arose.

In order to survey this complex issue of priority, it is desirable to separate some of the aspects of valency that form the substance of the various claims. Neglecting for the moment the demolition of useless hypotheses (copulation, reality of types, *etc.*), we can see the following factors becoming recognized in the period under review:

(1) Elementary atoms have a definite saturation capacity.
(2) This will apply to certain atoms that hold together different parts of a molecule, and in particular two or more radicals.
(3) Application of this to carbon leads to the recognition of a four-fold saturation capacity here.
(4) The direct linking of carbon atoms together will give a radical whose new saturation capacity can be simply calculated, and whose constitution can be uniquely defined in terms of these conceptions.

More concisely this may be rendered:

(1) Saturation of elementary atoms.
(2) Coupling characteristics of polyvalent atoms.
(3) Tetravalency of carbon.
(4) Linking of carbon atoms.

Until all these matters had been clarified it was not possible to say that the conception of valency had emerged. Much remained to be done after this; it was only the beginning. But even these few seminal ideas owed their origin to no one man, and so we have conflicting claims to consider. This consideration will form the main part of the present chapter.

It will, however, be convenient to precede this with a brief reference to the claims made in general by Kekulé, who is certainly the central figure, though not the only one. And the chapter will conclude with an account of the influence of Frankland and Kolbe (amongst others), who, if not accorded priority, played such an important part that the failure of their claim raises almost as many questions as its attempted justification.

1. THE CLAIMS OF KEKULÉ

Kekulé's place in the history of chemistry is assured, above all for the part he played in the development of theories which culminated in the doctrine of valency. His application of this to the problem of benzene has left him an unshakeable reputation, and few would dispute the words of Baeyer that Kekulé "has the glory of having founded a unitary system of organic chemistry and having proclaimed it to the world with the enthusiasm of a prophet".[13]

But this, of course, is not to suppose that he was *necessarily* responsible for launching any of the main ideas outlined above as constituents of the well-rounded doctrine of atomic valency. However, his opinion, voiced in 1864, leaves little room for doubt as to his own estimate of the position: "Unless I am mistaken, it was I who introduced into chemistry the idea of the atomicity of the elements."[14] In similar vein, he entered into a friendly correspondence with Wurtz[15] on his contributions of 1854, which Wurtz had overlooked, and published a paper claiming priority over Couper in the matter of "basicity of atoms".[16] He admitted, however, that he owed much to earlier publications, and acknowledged his debt to Williamson and Odling,[17] and also to Wurtz.[18]

Shortly after the genesis of his ideas, Kekulé published the first instalments of his *Lehrbuch der organischen Chemie*,[19] and the first volume included a masterly historical introduction to his subject. In this historical section, Kekulé deals fully with many aspects of theoretical development, and again applauds the achievements of Williamson, Odling and Wurtz. But a curious feature is the total absence of any reference to the theoretical contributions of Frankland, and scant acknowledgment of some of the major achievements of those who stood in the same tradition. Bunsen's work on cacodyl gains no applause in this section, even though Kekulé was a colleague of Bunsen and his own student, Baeyer, was working on cacodyl at that time.[20] It is possible still to agree with Anschütz that "Kekulé shows a superlative knowledge of the development of his science",[21] and at the same time admit that he was curiously myopic in matters close to his own great achievements, where these involved a form of thought alien to his own.

After this period, Kekulé said little on his claims to priority for some twenty years. Then, in 1881, came a renewal of an attack by Kolbe,[22] to which Kekulé wrote a reply, intended for the *Annalen* as part of a paper on "Carboxytartronsäure".[23] Dissuaded by Volhard, then editor of the periodical, he withdrew from the paper this and another polemical essay "On the history of the benzene theory", allowing the rest to go forward. Anschütz published both these controversial parts in his biography of Kekulé in 1929.[24]

In this *History of Valency Theory*, Kekulé dealt very fully with claims made by Kolbe and also Frankland, basing his views, so he says, on the *Histories* of Ladenburg[25] and Kopp.[26] His conclusions are as plain as they are self-confident: "I extended the idea of valency or atomicity of the radicals formulated by Williamson to the atoms of the elements also."

It must be remembered that Japp, in his Kekulé Memorial Lecture[27] of 1898, would not have had access to this document. This must explain his assertion that though Kekulé claimed priority, he never expressed himself on Frankland's claims. In some ways Japp's conclusions might be thought to be vitiated by his ignorance of this source, but they display such a penetrating understanding of the issues involved that they seem in fact to be little affected by this omission. Armstrong goes so far as to claim that Japp is fairer to all concerned than Anschütz himself.[28]

Kekulé's final statement of his views as to the origin of the valency theory appears in 1890, when as principal speaker at a celebration held by the German Chemical Society to mark the 25th anniversary of his benzene formula, he said:

Fifty years ago, the stream of chemical progress had divided into two branches. . . . At length, as the two branches had again approached much nearer to one another, they were separated by a thick growth of misunderstandings, so that those who were sailing along on the one side neither saw those on the other, nor understood their speech. Suddenly a loud shout of triumph resounded from the host of the adherents of the type theory. The others had arrived, Frankland at their head. Both sides saw that they had been striving towards the same goal, although by different routes. They exchanged experiences; each side profited by the conquests of the other; and with united forces they sailed onward on the reunited stream. One or two held themselves apart and sulked; they thought that they alone held the true course,–the right fair-way,–but they followed the stream.

Our present opinions do not, as has frequently been asserted, stand on the ruins of earlier theories. None of the earlier theories has been recognized by later generations as entirely false; all when stripped of certain ill-proportioned, meaningless excrescences, could be utilized in the later structure, and form with it one harmonious whole.

Here and there a seed may have lain in the ground without germinating; but everything that grew came from seed that had been previously sown. My views also have grown out of those of my predecessors and are based on them. There is no such thing as absolute novelty in the matter.[29]

This mild and generous statement is in sharp contrast to his then unpublished *History of Valency Theory*. Polemics would have been inappropriate on such an occasion, but Kekulé's general air of benign urbanity owed its appearance to more than this. One suspects that Volhard did not have to persuade him very hard to withdraw the offending essays from *Annalen*, for his silence on any claim to priority in his last thirty years seems to suggest that the issue was now largely dead as far as he was concerned. Only the diatribe of a Kolbe would stir him to reply. Probably he felt that, Kolbe apart, most chemists had considered his claims vindicated, and quite possibly the effects of time had produced a more generous view of Frankland. Of this speech, Armstrong has said it was a "considered confession",[30] and called for a more liberal assessment of the work of Kolbe and Frankland than Anschütz, for instance, was prepared to make.

Yet this view of Kekulé's speech as the considered confession of a veteran chemist, mellowed by age, and now prepared to be generous to all is fundamentally too simple. That it was "considered" is of course obvious, and the least one would expect of Kekulé. But to imply that here was a man retracting views he had written down – if not published – seven years earlier is not to do justice to the facts. In the first place, the unpublished essay contains arguments of considerable cogency, and comes down unhesitatingly on the side of Kekulé's own priority. Nothing had happened since 1883 to alter that opinion. Secondly, examination of the text of the speech shows that there is only one reference to Frankland (that just quoted), and none at all to Kolbe or to Couper. If it is necessary to give credit to Frankland and Kolbe – and surely it is – we shall hardly find the incentive to do so from Kekulé's 1890 speech. And this is to imply no disrespect to him, but merely to deny ourselves the pleasure of reading into the speech an understanding of Frankland's significance that Kekulé gave no sign of possessing. But as a general statement of the many-sidedness of scientific facts, and a call for scientific toleration, the speech remains a masterpiece.

2. CONTESTED MATTERS OF PRIORITY

We shall now examine in detail some of the issues in dispute.

(i) *The Saturation Capacity of the Elements*

Writing in 1877 of his paper of 1852,[31] Frankland says this:

It was evident that the atoms of zinc, tin, arsenic, antimony, *etc.* had only room, so to speak, for the attachment of a fixed and definite number of the atoms of other elements, or, as I should now express it, of the bonds of other elements. This hypothesis... constitutes the basis of what has been called the doctrine of atomicity or equivalence of the elements; and it was, so far as I am aware, the first announcement of that doctrine.[32]

A variant of this passage occurs in his *Sketches* years later,[33] when the words following the semicolon are omitted; he probably realized then that the claim was in too sweeping a form.

He also speaks at the same time of "the analogies... upon which I have founded the doctrine of atomicity",[34] of the 1852 paper where "some theoretical considerations in which molecular symmetry and the law of atomicity of elements are pointed out",[35] and of "claiming the discovery of the law of atomicity in the study of the organo-metallic bodies".[36]

These claims were hotly disputed by Kekulé. He attacked them on two chief grounds in his *Geschichte*. First, he objected that Frankland was not propounding a theory of valency at all, but merely framing an empirical law, a conformity to a principle (*Gesetzmassigkeit*).[37] He refers here to the phrase "without offering any hypothesis regarding the cause of this symmetrical grouping of atoms";[38] possibly he was influenced by the use of the same word in a passage criticizing Frankland's paper in Kolbe's *Lehrbuch der organischen Chemie*,[39] though it is also his own translation of Frankland's "law or tendency".

It must be admitted that there is some substance in this objection. At no point in this paper does Frankland approach the idea of a polyvalent atom uniting several parts of a molecule. The overall "symmetry" is what impresses him, not the detailed architecture. His own distinctive conception of a "bond" is absent from the 1852 paper, and hardly enters into any considerations of priority here. Nevertheless Kekulé is surely wide of the mark in suggesting that "the law of saturation capacity... is independent of each hypothesis, even of the basic hypothesis of the atomic theory".[40] The doctrine of atoms permeates the theoretical part of the paper, and Kekulé's inability to see this springs from his second objection to Frankland's claim that here was the beginning of valency.

This second difficulty lay in the fact that Frankland wrote in terms of the old atomic weights ($C = 6$, $O = 8$, *etc.*), whereas Kekulé's contributions to valency were expressed in the new system ($C = 12$, $O = 16$, *etc.*). Why this should invalidate Frankland's claim is not clear at a first glance, since erroneous atomic weight values and incorrect formulae can easily be put right. But the fact remains that a correct system of atomic weights was a matter on which Kekulé felt very deeply, and the famous Karlsruhe Conference of 1860 was conceived by him for the purpose of tackling the problem at its fundamental level.[41] There were deeper reasons for this than is at first apparent; he was not merely concerned with achieving a very desirable uniformity. In his dispute on this aspect of Frankland's work, a clue may be detected for the reason behind his attitude. He wrote:

There is then the further consideration that most chemists in 1853 did not yet clearly distinguish the ideas of atom and equivalent, and that Frankland also did not make such a distinction, but spoke now of atoms, now of equivalents, whereas valency theory assumed a sharp difference between the two ideas.

Frankland argued with a false atomic weight for oxygen, and had to be able to argue also for a false atomic weight for antimony, *etc*. His law does not even deal with atoms. As far as bodies are concerned, it denotes that they shall have a definite saturation capacity and deals with that relative quantity we regard as a radical.... The law refers not to atoms but equivalents. It is right for equivalents but false for atoms, and therefore false in the form arrived at by Frankland.[42]

In other words Kekulé is castigating Frankland not for using a wrong set of atomic weights, *but for not using any at all*. Whatever might be the term by which they were known, the figures were *equivalents*, and the formulae were not atomic but equivalent formulae. Symbolic representations of the laws of stoichiometry, they owed no debt to Dalton. To that extent Kekulé was justified in speaking of a degree of independence of the atomic theory, though he did not give sufficient weight to Frankland's atomic phraseology. But why should this differentiation between atomic and equivalent weights be so essential? In a word, because of the relation:

$$\text{valency} = \frac{\text{atomic weight}}{\text{equivalent}}$$

Confusion in either numerator or denominator would lead to false values for the quotient. Moreover, failure to appreciate the distinction between atomic weight and equivalent meant inadequate if not impossible understanding of the meaning of the resultant figure.

Yet when this is admitted, it remains true that Frankland gave a clear enunciation of the saturation capacity of the elementary atoms, even though he used wrong formulae and gave no explicit table of values. The situation is in some ways parallel to that of Boyle, who clearly enunciated the idea of a chemical element but left it to his successors to provide a list of substances fulfilling his conditions. It is tempting – but unwise – to draw nationalistic conclusions from this coincidence. Perhaps the position may be summarized thus: Frankland's use of incorrect atomic weights prevented him (at that time) from being able to arrive at correct values for the valencies; but the employment of equivalents in no way implied that Frankland was incapable of appreciating the distinction between these and (unknown) atomic weights. He was conforming to common practice, as indeed was Kekulé himself in some of his own early papers (see p. 62). Recognition of a definite combining power does not presuppose infallible knowledge of the relative weights of different atoms.

A further point in this connection is that Kekulé was not alone in his attack. His student Baeyer commented:

I would here remind the reader of the series of compounds which Frankland gave in his investigation on zinc ethyl, in which he drew a parallel pattern between the compounds of the metals with oxygen and those with the alcohol radicals, while the latter are analogous to hydrogen and not to oxygen. The view which is thus gained is purely superficial; it vanishes immediately when the proper atomic weight for oxygen is adopted, and the formulae which contain an odd number of that element are doubled.[43]

This was adequately answered by Frankland's reminder that in fact in his 1852 paper he regarded "the oxygen, sulphur, or *chlorine* compounds of each metal as the true molecular types",[44] and oxygen was emphasized because "it formed the only pentadic inorganic compound of arsenic",[45] the conventional value ($O = 8$) being taken.

A third objection raised by Kekulé is an interesting one.[46] Frankland was limiting his generalizations too much. Instead of comparing, *e.g.*, the combining powers of oxygen, chlorine, *etc.*, or even the metals with each other, he was content to speak only of a constant saturation capacity of individual metals. His compounds were too complex for simple comparisons easily to emerge. To this there can be no objection, but merely the observation that this shortcoming was connected with the second. If Frankland entertained any doubts on the values of his atomic weights (and it is unlikely that he regarded them as final), little was to be gained by comparison of figures which might themselves be devoid of significance.

The final objection to Frankland's claim is one that Kekulé did not voice explicitly, but was probably at least as powerful as the others. Frankland, like Kolbe, was approaching the problem from a point of view utterly alien to Kekulé, who, with all his mastery of chemical history and literature, never seemed to be able to appreciate at its true value a view expressed by an adherent of the radical school. Even though Frankland claimed,[47] and Kekulé admitted[48] that the former had moved towards his own position, a psychological barrier remained between them. And when, later, Frankland was found on the side of those who opposed Kekulé's doctrine of unalterable valency (see p. 188), the barrier was in no way lessened.

The fairest comments that seem to have been made on this controversy are those of Japp:

A discovery, made by an adherent of the radical theory and correctly formulated by him in terms of the old equivalents, does not become the property of the first adherent of the type theory who happens to translate it into the new molecular weights.[49]

We may therefore conclude that Frankland deserves the full credit for first propounding the doctrine of "saturation capacity" of the elements, and in so far as this is at the basis of valency, Frankland must be held to be its founder. But for the extensions of these basic concepts we must look elsewhere, and at this point as at so many others in this field we are again confronted with the figure of August Kekulé.

(ii) *Coupling Characteristics of Polyvalent Atoms*

Frankland had shown that there is a definite limit to the number of atoms that can combine with, for example, one atom of arsenic. If this number is greater than one it can be said that the arsenic holds together the other atoms. But, as we have seen, this step was not taken by Frankland. For one thing, the uncertainty about atomic weights prevented dogmatism as to whether, for instance, oxygen *could* combine with more than one atom at a time. Hence it is elsewhere that we must look for the first enunciation of the ability of polyvalent atoms to "couple" together two or more parts of a molecule. Indeed, this itself depended upon the clear distinction between "polyvalent" and monovalent atoms, and there is little doubt that for this double accomplishment Williamson was responsible.

That it was he who first showed that oxygen was "bibasic" and gave birth to the theory of polybasic radicals is scarcely contestable. Kekulé himself gives the concluding section of his historical account of organic chemistry to the work of Williamson, with that of Odling and Berthollet receiving honourable mention also. And elsewhere he is generous in his acknowledgment of their services[50] together with those of Wurtz and others[51] (contributions that have been described in Chapter III).

Yet despite this, Kekulé still maintains that it was he who initiated the idea of valency. So presumably his predecessors had arrived at something less, and his view may be understood from the statement:

The essentially new feature of my view lies in the first place in this, that I extended the idea of valency or atomicity of the radicals formulated by Williamson to the atoms of the elements also.[52]

The question becomes this: did Williamson or Kekulé first look beyond the radicals to the atoms composing them, and conceive of polyvalent *atoms*? On this matter Divers wrote to Anschütz that: "Williamson founded 'valency' as far as the combining power of a binary radical is concerned. He said the oxygen atom 'holds together' the H and C_2H_5."[53] Now it has not been possible to trace the use of the phrase "holds together" by Williamson as applied to an oxygen atom.[54] Nevertheless a reading of his papers from 1850 to 1853 will make clear that the idea was definitely in his mind even if it did not find the specific expression quoted. In the absence of such a categorical statement it is perhaps necessary to award to Kekulé priority for enunciating the doctrine of combining powers of polyvalent atoms, and to Williamson priority in conceiving it.

Kekulé did not specifically assert priority over Williamson, but merely made the all-inclusive claim above, doubtless because Williamson himself made no pretensions to a special position. It was otherwise with Wurtz, however. In the first volume of his *Répertoire de Chimie pure* he gave a report on Kekulé's paper "Über die Constitution und die Metamorphosen der chemischen Verbindungen", and in a footnote on "the idea of basicity of the atoms" remarked:

This idea is not new in science. M. Williamson, M. Odling (*Quarterly Journal of the Chemical Society*, vol. VII, p. 1) and myself (*Ann. de Chimie et de Phys.*, 3rd series, vol. XLIV, p. 306 *et seq.*) have neatly expressed it several years ago.

I do not think I am deceiving myself in admitting that my latest work on the synthesis of polyatomic alcohols has given this idea the experimental confirmation that had been lacking. The existence of polyatomic radicals that I showed in organic chemistry gives in effect solid support to the idea of polyatomic *elements*. However, M. Kekulé himself recognized in his memoir that my work has served as the point of departure for the views which he has developed.[55]

To this statement Kekulé sent a prompt reply in the form of a letter to Wurtz which has survived and which Anschütz has reproduced in his biography of Kekulé.[56] In this he draws attention to his paper of 1854 on thiacetic acid,[57] at the same time repeating his acknowledgment of the work of Wurtz himself and also of Williamson, Odling and Gerhardt. Wurtz's reply has also been reproduced by Anschütz; it acknowledges his mistake and assures Kekulé that "if I had known the two passages of your memoir I should have refrained from writing the little note to which you alluded".[58] The following month a note appeared to Wurtz's "Mémoire sur les glycoles ou alcools diatomiques" as follows:

In giving recently (*Répertoire de chimie pure*, vol. I, p. 24) several historical indications on the theory of polyatomic elements and radicals, I have forgotten to mention, with the names of MM. Williamson and Odling, that of M. Kekulé. I ought to remark here that in his memoir on thiacetic acid (*Annalen der Chemie und Pharmacie*, vol. XC, pp. 314 and 315; 1854) this latter chemist has insisted on the bibasic nature of sulphur.[59]

It is a pity that all questions of priority could not be settled as easily, and as amicably.

(iii) *Tetravalency of Carbon*

Thus far we have attempted to assess the various claims to priority in the establishment of the doctrines of saturation capacities of elementary atoms and of the coupling ability of polyvalent atoms. The extension of these ideas to carbon compounds was of immense importance, and was inevitable once the other concepts had been grasped. Once again several voices were raised in claiming priority, and a bitter dispute arose between Kekulé on the one hand and Kolbe and Frankland on the other concerning the recognition of tetravalent carbon. Another contestant was Couper, whose illness prevented any advocacy of his own claims. Yet it is ironical to note that the true priority lay with none of these, but with William Odling, who apparently never laid claim to the honour and was therefore never engaged in the controversy.

In 1855 Odling delivered a lecture at the Royal Institution "On the constitution of hydrocarbons" in which he added to the types ClH, OH^2, NH^3 a fourth, CH^4 ("coal-gas"). For this and other hydrocarbons he regarded as superfluous "the conception of self-existent constituent compound radicals".[60]

This was the establishment of the "marsh-gas type". But the lecture was not published except in the Institution's *Journal*. Anschütz assures us that this was comparatively inaccessible to Kekulé, and that he could not have seen it.[61] There seems little doubt that this was so, and this must explain Kekulé's silence on the matter. But for the formal enunciation of this marsh-gas type, with four hydrogen atoms linked to one carbon atom, the credit must go to Odling.

It is now necessary to turn to the Kekulé–Kolbe controversy, with which the name of Frankland is also linked. The dispute began mildly enough with some statements by Frankland in the mid-1860s. In his later *Experimental Researches* he wrote of his claim to "the discovery of the law of atomicity in the study of the organo-metallic compounds, and, in conjunction with Kolbe, its application to the compounds of carbon".[62] But a paper from 1866 is more specific:

The rapid progress that has of late been made in the application of synthetical methods to the production of most of the other great organic families, has rendered it possible to apply, with some prospect of success, a mode of notation, the principles of which were suggested in a lecture which I delivered at the Royal Institution in the year 1858. The adoption of this type, however, was nothing more than the application to the compounds of the tetrad carbon atom, of the principles which I had previously employed in the notation of organo-metallic bodies containing metals of various degrees of atomicity.[63]

A year earlier, Frankland (with Duppa) made a similar reference, but this time to a paper of 1857 (not 1858 as above):

It has been proved by Kolbe and Frankland nearly twenty years ago that methyl is a constituent of acetic acid, and in the year 1857 these chemists were the first to propose the derivation of this and a large number of other compounds from the carbonic acid or tetratomic carbon type.[64]

To these claims Kekulé made no quick reply, probably because he considered the facts spoke for themselves, and he had an unshakeable confidence in his own right in the matter. But in 1881 Kolbe produced his own account of the situation, and Kekulé was stimulated to reply two years later. In addition to underlining Frankland's remarks on behalf of them both, Kolbe pressed his own claims on the basis of parts of his *Lehrbuch der organischen Chemie*, first issued in 1857. He further criticized some of Kekulé's own contributions.

Thus there arises a situation of some complexity. It will be convenient to record each controverted publication in the order in which it appeared: Table IV is a summary of the chief facts.

TABLE IV

	Date of writing	Date of publication	Author	Reference
1	26 Dec. 1856*	14 Feb. 1857†	Kekulé	*Annalen*, *101*, 200.
2	Dec. 1856‡	23 March 1857†	Frankland and Kolbe	*Annalen*, *101*, 257.
3	15 Aug. 1857*	30 Nov. 1857†	Kekulé	*Annalen*, *104*, 129.
4	?	Jan.–June 1858§	Kolbe	*Lehrbuch*, nos. 6 and 7.
5	16 Mar. 1858	19 May 1858†	Kekulé	*Annalen*, *106*, 209.
6	?	Jan.–June 1859§	Kolbe	*Lehrbuch*, nos. 8 and 9.
7	Lecture delivered on 28 May 1858†		Frankland	*Proc. Roy. Inst.*, *2*, 540.

Notes on sources of dates in this table:

* From Kekulés' "Geschichte der Valenztheorie", in Anschütz, *op. cit.*, vol. i, pp. 559–60.

† From periodicals mentioned; some of the publication dates in (*) are in error.

‡ From Frankland's *Experimental Researches* (see note 8), p. 148.

§ From Hinrichs' *Booklist*, quoted by Kekulé in "Geschichte der Valenztheorie" (see above).

If it can be established that the first paper above really contained an enunciation of tetravalent carbon atoms, then it clearly pre-dated all other attempts except Odling's. It was therefore in Kolbe's interest to show that it did not, but he was hard put to it for a convincing demonstration. However, he did assail it on three grounds. In the first place he tries to show that Kekulé was not applying anything to a "marsh-gas type" in view of his own denial (see p. 63) that he was speaking of Gerhardt's "chemical types" at all.[65] Capital is made of the fact that in his paper on copulated compounds (see p. 65), Kekulé relegates the new type to a footnote, and hesitates to add the "marsh-gas type" to Gerhardt's three types. But the point of this argument is lost when one realizes that "mechanical types" were far more to Kekulé's purpose, for all he was showing was a similarity of *arrangement*, not of chemical behaviour.

Kolbe's second objection infers from the wording of Kekulé's statement (see p. 65) that he was referring marsh-gas to the chloroform type, rather than vice versa.[66] Any validity in this contention is precluded by the actual arrangement of the table given by Kekulé, and by the observation that a "chloroform type" is nearly as good as a "marsh-gas type" if one is using it so as to imply tetravalent carbon.

Kolbe's last objection seems to have more substance than the others. It is that Kekulé's paper on fulminic acid is couched in terms of the old atomic weights ($C = 6$; $O = 8$), and deals, therefore, with the double atom C_2 only.[67] Kekulé replied that this was a paper not primarily dealing with valency considerations, but only expounding his views on mercuric fulminate; his discussion on thiacetic acid (see p. 55) employed the new system three years before, thus showing that he was not committed to double carbon atoms. This was probably true, but unlikely to convince a man like Kolbe. He could quite legitimately have asked why the 1857 paper was less relevant to valency than that of 1854; did not the latter purport also to be primarily an exposition of some new experimental work? One has the impression, on reading the two papers through, that this argument could be quite well maintained. In 1854 Kekulé was less, not more, likely to have been "intelligible to my contemporaries without commentary"[68] than three years later, at least in Germany, had he used the customary atomic weights. But ultimately the point is this: with Frankland (p. 114) and with Kolbe (p. 123) Kekulé was adamant in his insistence that their claims were invalidated by false atomic weights, and he had no right to expect any different treatment from them.

This is not to deny Kekulé the priority over Kolbe, for he went over to the new values later that year (in his paper of 1857, above), long before Kolbe did. But it is necessary to concede this point at least to Kolbe, that a spurious system of atomic weights is as damaging to one man's case as it is to that of any other.

We can now examine the opposite viewpoint maintained by

Kekulé in criticism of Kolbe's claim to priority. Again there were three main points. First, he joined Ladenburg[69] in an appraisal of formulae like the following, given in the joint paper with Frankland:

$$2HO.C_2O_4 \quad HO(C_2H_3)C_2O_3 \quad \left.\begin{array}{l}C_2H_3\\ H\end{array}\right\}C_2O_2 \quad \left.\begin{array}{l}C_2H_3\\ C_2H_3\end{array}\right\}C_2O_2$$

carbonic acid   acetic acid   acetaldehyde   acetone

The chief ground for criticism lay in the apparently arbitrary way that water molecules were removed or allowed to stay. The deeper reason for this, however wrong, was observed by Ladenburg, but as Kekulé remarks "the whole observation stands in no kind of relationship with the tetravalency of carbon".[70] Valuable as these contributions were for the development of organic chemistry, their bearings on valency were slight and indirect.

Kekulé's second objection was levelled at subsequent utterances by Kolbe, notably in his *Lehrbuch*, where much use is made of the term "carbonyl", which for Kolbe was a double atom of carbon C^2. Thus, "carbonyl, which is bound with four oxygen atoms in carbonic acid, has the same value in this compound as, say, four hydrogen atoms that are indeed necessary to saturate the same number of oxygen atoms".[71] Once again we are confronted with wrong atomic weights, and the objections made by Kolbe to Kekulé's work must be applied here also. As Kekulé remarks, the tetravalency of a carbon atom cannot be inferred from that of a carbonyl radical or from a double atom. "If a chemist today says it is possible to assume in the series of iron oxides one double atom consisting of two iron atoms, ferricum, he neither asserts that he has discovered that iron is hexavalent, nor imagines he has expressed himself in any way on the valency of iron atoms."[72]

Finally, Kekulé contends that his publications were always ahead of Kolbe's in time, and brings together most of the dates in Table IV.[73] But this comparison is only significant if one regards the contributions as of equal, or near-equal, value, and Kekulé has made it abundantly clear that he does not. One suspects that this juxtaposition of dates may hide a greater regard for Kolbe's work than Kekulé was prepared openly to admit. Whether this is so or not, two conclusions may be safely drawn. First, the closeness of the first two papers precludes any possibility of one having any influence on the other, and if Kekulé was technically first in proposing tetravalent carbon his rivals must have arrived at the same conclusion quite independently. Secondly, if both these papers are discounted on the grounds of their incorrect atomic weight system, Kekulé's paper of 1857 on polyatomic radicals is second only to that of Odling in proposing a single tetravalent carbon atom (C = 12); and of these two statements, that of Kekulé is unquestionably the clearer.

A brief mention must finally be made of Frankland's paper of 1858.[74] This is certainly a clearer enunciation of a "marsh-gas type" in terms of C_2 than anything before it; but it is both later in time than Kekulé's two statements, and of course still in the old terminology. Frankland's own commentary on it (see p. 120) does not quite amount to a claim to priority, so the matter can probably be left as an important contribution, but a representation of a considered opinion rather than the crystallizing of a new idea. It is interesting to note another of the small variants in otherwise identical passages in his *Experimental Researches* and *Sketches*, indicating a slight change of emphasis over the years between them. To his earlier account of the origin of the valency theory he adds many years later the words: "The application of my theory of valency to carbon compounds, however, belongs substantially to Kekulé, whose brilliant application of this theory to carbon compounds generally, constitutes one of the most important epochs in the history of chemical science."[75] This is tantamount to complete acknowledgment of Kekulé's priority.

One other claimant to priority in the matter of tetravalent carbon could have been A. S. Couper (see p. 71). So, at least, Kekulé appears to have thought, for Couper's papers were rapidly followed by a disclaimer from Kekulé "stating the originality of the views expounded by me in my previous memoirs".[76] The very short period of time between Couper's note "On a new chemical theory"[77] and Kekulé's paper "On the constitution and metamorphoses of chemical compounds"[78] (respectively 14 June and 19 May 1858) almost inevitably called for a statement. So far as the tetravalency of carbon is concerned, however, the issue was actually dead already, in view of Kekulé's earlier papers and (especially) Odling's lecture at the Royal Institution. But the concept of self-linking carbon atoms was in a different class, and to that issue we shall now turn.

(iv) *Linking of Carbon Atoms*

There are only two serious contestants for priority in postulating chains of linked carbon atoms, Couper and Kekulé. Also, no doubt arises about their advocacy of this; both presented this as a new idea with perfect clarity and conviction (*cf.* pp. 74 and 69).

In Kekulé's Note on this point, he observes a number of similarities between his own and Couper's phraseology but emphasizes that his contribution was published a month earlier, and written two months before that. He admits that his formula for the number of hydrogen atoms combined with n carbon atoms is less general than Couper's – but also less vague. He adds "if M. Couper thinks he has discovered the cause of this difference in the basicity in the existence of a special kind of affinity, the affinity of degree, I am the first to recognize that I have no right to contest this priority with him".[79] But of course Couper claimed no such thing.

There can be thus no dispute on the technical priority of Kekulé in this fundamental doctrine of organic chemistry. Couper's failure to publish after this time and retirement from the chemical scene meant that his name sank into near-oblivion for half a century, so the issue between him and Kekulé was forgotten and the latter was universally regarded as the discoverer of the carbon chain. Nevertheless justice demands that Couper be accorded an equal share of the credit for this conception, if not the technical priority. The latter, indeed, might have been his also had not Wurtz mismanaged the situation (see p. 109). But there can be surely no question that the two men arrived at substantially the same conclusions in complete independence and total ignorance of each other's activity.

3. THE INFLUENCE EXERTED BY THE FOUNDERS OF VALENCY THEORY

Technical priority in a given field is no guarantee of permanent, or even transient, influence. When even a formal priority is difficult to establish, as in this case, it is quite impossible to make any *a priori* deductions from the events surrounding the birth of a theory as to how, and by whom, that theory will exert its influence in years ahead. The present objective will be, therefore, to attempt to assess the actual part played in the early growth of valency theory by Williamson, Odling, Kolbe, Frankland, Couper and Kekulé. Their main contributions have already been described, and the priority of these has been discussed; but how far each played a part in moulding the thoughts of his contemporaries and successors is separate from matters of priority, although related to them.

(i) *Williamson*

Williamson played a large part in the spread of chemical theories, both at University College, London, and as President of the Chemical Society. But the virtual absence of original work from his laboratory for the last twenty years of his active life must have diminished his influence with some, and it is to the 1850s that we must look for his main effect. So far as valency is concerned, Odling and Kekulé are greatly in Williamson's debt. Kekulé's acknowledgment of this has been referred to already (p. 61) and was repeated many times when speaking of the origins of the theory of valency. Had Williamson achieved no more than this he would have accomplished much.

His ideas on "polyatomic" radicals and condensed types found an echo in his friend and pupil Odling. Indeed the two men had much in common, due partly no doubt to their training under Gerhardt.

Williamson's influence outside the field of valency was very great. His ether theory was "almost universally adopted" by 1853,[80] and shortly after that the leading concepts of his view on salts were adopted "by several eminent chemists in Great Britain and France".[81] Acceptance of these ideas was an almost essential prerequisite to any understanding of the doctrines of valency.

(ii) *Odling*

Remembered chiefly today for an erroneous formula for bleaching-powder, Odling played a considerable part in shaping the chemical thought of his day. In teaching at the Royal Institution (1868–73), and at Oxford (until 1912), in his Presidency of the (Royal) Institute of Chemistry (1883–8), and as Secretary of the Chemical Society (1856–69) Odling was continually engaged in the propagation of the new chemical ideas. Concerning his influence at the Chemical Society Tilden has written: "It is mainly due to his activity and clear vision that, in spite of the general sluggishness of chemical opinion at that time, the reforms introduced by Gerhardt, and especially by

Cannizzaro, were accepted by his contemporaries."[82] Against this background his influence on the growth of valency theory must be judged. Two things stand out particularly. First was the marked impression he made upon Kekulé during the latter's stay in London; again the debt is often acknowledged. Secondly, his system of strokes to indicate numerical values for valencies was quickly accepted and often used. Any useful contribution to a symbolism is always of value in propagating the idea symbolized.

(iii) *Kolbe*

In his own day, Kolbe was probably the most controversial figure in the chemical scene. That he exerted a great influence by his views is undoubtedly true, but the part he played in the growth of valency theory is more debatable.

Unable, or unwilling, to see the force of physical arguments in chemical discussions, he adhered to the old atomic weights until 1870. Contemptuous of the aridity of much of later type theory, he nevertheless shared Gerhardt's structural agnosticism, and particularly opposed Williamson and Kekulé who must have seemed to him to have combined the worst features of chemical thought of the day: an approach that was strongly coloured by "typical" ideas, and at least the beginnings of definite conceptions of structure. Even Frankland, to whom he most closely approached, was unable at first to convince him of his own view of "saturation-capacity"[83] or to persuade him to employ the new Crum Brown structural formulae.[84]

Thus Kolbe was out of sympathy with the doctrine of valency *in the form and language in which it was currently expressed*. His general influence on its development must therefore be ranked as small. But, as has been shown (p. 32), he came very near to the idea of tetravalent carbon, and there can be no doubt that about this time he and Frankland exerted an influence on each other that certainly affected Frankland's thought for a long time to come. It is perhaps here that Kolbe's direct influence in the matter of valency was most directly felt.

Yet more can be said than this. His fine output of experimental work provided an increasing volume of new data which his opponents were able to interpret in terms of the theory of valency. Through his pre-eminence as a teacher, his textbooks and other publications, and (from 1870) his editorship of the *Journal für praktische Chemie* he contributed to the accumulation of much "raw material" of chemistry, so necessary for the vigorous growth of the subject.

He had never hesitated to criticize his opponents (*cf.* p. 121), and after 1870 became increasingly polemical in his writings. Possibly this was in part an unconscious reaction to his disappointment at the obvious success of the views he deprecated, and at the knowledge that in theoretical matters he was being left behind. According to Armstrong, who knew him well and held him in very high esteem, "Kolbe, in his last years so fixed his mind upon certain grievances as to be little short of a monomaniac".[85] The same writer considers that Kolbe was profoundly affected by the war of 1870 and that this was partly responsible for his obsessions. However this may be, his opposition to Kekulé and his school had two useful results. First, it caused Kekulé to write his *Geschichte der Valenztheorie*[86] which, though not published till later, has thrown some interesting light on his reactions to the controversy. Secondly, it helped those under attack to examine their own position more thoroughly. As Meyer has written:

It is well to remember that the critical acumen which was brought to bear upon the occasional errors of chemical investigation by Berzelius and Liebig, and at a later date by Kolbe, had a consolidating and not a disintegrating effect, even in those cases where the critic's argument had a strongly polemical, and – to the subject of the attack – a strongly personal flavour.[87]

Ladenburg's conclusion on Kolbe's influence on the theory of valency states the case with characteristic fairness:

His participation in the development of the notions as to the atomicity (valency) of the elements and radicals is not of importance, because, as I believe, he did not distinguish between molecule, atom and equivalent, and also ... because he had not then grasped the idea of the part played by the polyatomic radicals in holding the molecule together.[88]

(iv) *Frankland*

The fate of Frankland's views of the doctrine of valency constitutes one of the strangest features of chemical history in the last century or so. Clear though his exposition of them was, the impression created by chemical literature of that time and later is that they were almost completely overlooked. That this first impression is not quite correct we shall endeavour to show; but it is strongly reinforced by the omission of Frankland's name from Wurtz's *Histoire des Doctrines Chimiques*,[89] the historical section in Kekulé's *Lehrbuch der organischen Chemie*,[90] and other writings. Even more pointed is an omission of Frankland's name at the 1865 meeting of the British Association. Frankland himself was present at the meeting, and read a paper on the constitution of members of the acetic, lactic and acrylic acid series. But W. A. Miller, the chairman, spoke in these terms:

Owing to the labours of many distinguished men, amongst whom the names of Williamson, Kekulé, Odling, Cannizzaro and Wurtz are prominent, a classification of the elements into families has been made; and that this classification rests upon what is known as the *atomicity* of the elements.[91]

Admittedly Frankland's explicit views on classification of the elements had yet to be published,[92] but his part in the more fundamental task of establishing saturation capacities would surely have earned him a reference here if that part had been adequately recognized. The evidence is that it was not.

There seems little doubt that Frankland's small impact upon his contemporaries was partly because he spoke and thought in the forms of the dualistic Radical Theory, whereas they were mainly adherents of the Type Theory. There were other reasons as well, but this is the most obvious. Nowhere was the ostracism of Frankland's views more evident in the 1850s than in Kekulé, whose omission of Frankland from his *Lehrbuch* was but one of many instances of a strange failure to acknowledge his work. H. E. Armstrong had known both men well; he wrote thus:

The problem first solved by Frankland was in the air–chemists everywhere had it in mind, especially in France. Kekulé was in London in 1854 and consorted with Williamson, who like himself was under the enthralling influence of Gerhardt. Had he consorted with Frankland, a man infinitely in advance of himself and most other chemists as a worker, his attitude would not have been so independent. I have not been able to discover that Kekulé made the least attempt to exchange views with Frankland, having joined another camp. We are in face of a psychological puzzle.[93]

The two men had certainly met, in London[94] and for an evening at Wurtz's house in Paris,[95] for example, but there seems to be no record of any chemical intercourse between them.

To explain Frankland's apparent lack of influence by his difference of chemical tradition from Kekulé and the others does not seem sufficient. He had, it is true, received no schooling with Gerhardt in Paris, and he was steeped in the Radical tradition both from his work with Bunsen at Marburg and his friendship with Kolbe. But in the paper of 1852[96] he is moving from an exclusively dualistic position towards that of the Typists, and it is hard to believe that any of these was so rigidly inflexible that a total capitulation would be stipulated as a condition for acceptance of a theory. Rather would a variety of factors seem to have been at work, and it is suggested that this disparity in chemical background was only one of them.

When Frankland's ideas were first expressed they failed to make any profound impact upon Kekulé partly because they were tinged with dualism and partly because of the incorrect atomic weight system to which Kekulé objected so strongly in 1883.[97] Moreover it is likely that Frankland's theoretical contributions were to some extent overshadowed by his practical achievements in the organo-metallic compounds. His fame as the discoverer of "methyl" could well have diverted attention from his pronouncements added right at the end of a long paper.

Now Kekulé's ideas were germinating while he was working in London from December 1853 to the autumn of 1855. During this time Frankland was occupying the Chair of Chemistry at the newly formed Owen's College, Manchester, and the responsibilities of this would have meant absence from London for long periods at a time. Consequently his chances of meeting Kekulé would have been lessened, and a further factor thus arises to explain the lack of contact between the two men. When Frankland returned to London (1857) for the rest of his chemical career, Kekulé had departed for Heidelberg.

In 1866 Frankland published the first edition of his *Lecture Notes*. In this he committed himself to the doctrine of varying valency,[98] and represented molecules with structural formulae in the notation of Crum Brown. Neither of these features of the book was likely to commend itself to Kekulé (*cf*. pp. 175 and 99), and it is likely that it continued to discredit its author in his eyes. Shortly after this Frankland ceased much of his organic work, and by 1870 had withdrawn from most of the discussion that arose from the discovery of valency. According to Meyer,

INFLUENCE OF THE FOUNDERS OF VALENCY THEORY 131

this in all probability accounts for his service in developing such an important doctrine having been forgotten by many chemists, and precisely by those who have taken the most active share in the above discussions.[99]

These considerations probably apply to some extent to other chemists as well as Kekulé. The apparent failure by Odling and Williamson, for example, to recognize their countryman's work may be ascribed to much the same reasons. That the three were on terms of cordiality with each other is not surprising; it is on record that in 1861 they spent part of a yachting holiday at the Isle of Wight together.[100] But they too found the same extraordinary hold upon themselves by the Type Theory that Kekulé had experienced.

When all this is said, however, the distinction must be made between open acknowledgment of a theory and the slow unconscious yielding to its influence that may proceed quite independently. In the case of Frankland's idea of saturation capacity, it seems clear that its effect must have been felt in the opposite school of thought long before it was recognized. Thus there is no doubt that the papers of Frankland were circulating in Kekulé's laboratory at Heidelberg. Baeyer's work on cacodyl was done here, and in his paper on this work[101] he makes reference to Frankland's discovery of zinc ethyl; Kekulé himself[102] invites the reader of his 1857 paper "On the so-called copulated compounds" to compare his own work on thiacetic acid[103] with that of Frankland on zinc ethyl. In this paper Frankland showed how zinc ethyl confirms his own ideas on "the moleculo-symmetric form of the organo-metallic compounds".[104] Thus we may be fairly sure that Kekulé's school were coming under the impact of the teaching of Frankland. Japp believes "they were more indebted to it than they were at that time conscious of".[105]

Gradually it came to be recognized that chemical theory did derive in part from the impetus given it by Frankland. Wurtz, for example, amended his ideas by 1880 and acknowledged that Frankland had been the first to suggest the saturation capacity of elementary atoms.[106] Shortly afterwards, Ladenburg (himself a former student of Kekulé), wrote:

In the way in which the development actually took place, the influence of Kolbe, and more particularly that of Frankland, upon the supporters of the Gerhardt–Williamson school (Wurtz, Kekulé and Odling) can hardly fail to be recognized. Both schools were required, in order to raise the significance of the formulae to what it subsequently became.[107]

Frankland's influence, however, was not to be confined to this rather vague and general permeation of current thinking by members of the Kekulé school. Kolbe, in particular, found himself at first strongly opposed to the concept of a limit to an element's saturation, at least as explained by Frankland, and made this clear in his *Lehrbuch*.[108] Basically, as Frankland saw,[109] he was unable to agree to an electronegative element like oxygen being replaced by an electropositive radical like ethyl. After some correspondence, however, they came to a measure of agreement and published a joint manifesto:

We are of the opinion that in the oxides of the metals (and of course this applies also to the sulphides, chlorides, *etc*.) single and sometimes even all oxygen atoms can be replaced by the same number of atoms of a positive element or radical–hydrogen or methyl for example,–and perhaps also by the oxyradicals of the acids; and that in consequence of these singular substitutions, new conjugated compounds are produced, which constitute the oxides of independent conjugated radicals...[110]

Unfortunately that was as far as Kolbe was prepared to go, but it is a measure of Frankland's chemical insight, as well as his powers of persuasion, that even this was conceded. Again to quote Ladenburg:

It was not an easy matter for Kolbe to follow Frankland in his most recent developments; to assume that the affinity of the elements is always satisfied by the same number of atoms without regard to their chemical character amounted to giving up the electro-chemical theory altogether, and to admitting that the electro-chemical nature of the elements was without influence upon the formation of compounds.[111]

This statement brought a characteristically vitriolic reply from Kolbe:

This is one of the many erroneous assertions in Ladenburg's writing, which by a superficial study of sources and the influence of preconceived ideas give little favourable character to the book.[112]

Nevertheless it is difficult to dispute it.

Frankland's influence in the development of the theory of valency may be seen, finally, in his textbooks, particularly the *Lecture Notes* referred to already.[113] This book enjoyed great success and a second edition was issued (in two volumes) in 1870 and 1872.[114] That it was received so well seems in large measure due to the employment

INFLUENCE OF THE FOUNDERS OF VALENCY THEORY 133

of Crum Brown's symbolism, and other books followed also using "Frankland's notation" (see p. 105). The concepts of valency were absorbed with astonishing rapidity in Britain, particularly in the "semi-popular" scientific world, and as we have seen (p. 104) this was probably due to the clarity with which Frankland presented them in his books.

(v) *Couper*

Of the direct influence of Couper little definite can be said. It has been suggested (p. 80) that his colleagues Butlerov and Crum Brown may have owed him some of their structural ideas. But generally he was completely ignored, or the significance of his work was just not understood. A case of the latter was the British Association Report for 1859, where he was mentioned, but only in connection with his wrong atomic weight for oxygen.[115] Apart from Kekulé's Note[116] there was little to keep his name before the chemical public. Only former colleagues from Paris (*e.g.* Butlerov) referred to his work to any extent.[117]

This situation has continued until today with few exceptions. Wurtz made some amends for his earlier treatment of Couper in his *History of Chemical Theory*:

It should not be forgotten that Couper has developed analogous ideas without previous knowledge of the propositions enunciated by Kekulé, which have since exerted so great an influence on the development of organic chemistry.[118]

Ladenburg has written that

these two papers of Kekulé and Couper constitute the foundations of our views respecting the structure of compounds. As a consequence of them, organic chemistry took an altogether new direction, and they may be regarded as the most important advances of our science, on the speculative side, in recent times.[119]

In 1909 Anschütz did much to rescue Couper's work from the obscurity into which it had so undeservedly fallen. However, to celebrate the centenary of the advent of the Structure Theory, the London Chemical Society decided to designate the proceedings as "The Kekulé Symposium on Theoretical Organic Chemistry"[120] and Couper received no more than a passing reference. The American Chemical Society gave him fuller recognition, however.[121]

(vi) Kekulé

From what has been said in the last paragraph it will be plain that Kekulé's reputation today stands at a high level. All that needs to be added is that from the beginning his influence was considerable. Due in part to his ability and enthusiasm as a teacher, in part to the enshrining of his ideas in his *Lehrbuch*,[122] it was probably above all a consequence of his leadership of a succession of active research schools, all producing men of vigour and ability infected with the ideas (and idealism) of their teacher.

Yet this is not to say that Kekulé's doctrine of valency was immediately accepted. Eventually it gained general approval, largely as Kekulé's own applications of it in the fields of benzene chemistry and of the dicarboxylic acids produced such impressive results. But much remained to be done in clearing up many uncertainties in chemistry, and nothing in the literature suggests as dramatic a response as Japp seems to imply in remarking that "Kekulé's structural formulae cleared away at one stroke the entire brood of pseudo-constitutional formulae".[123] Ideally this is true, but in practice it was not. A valuable comment has been given by Armstrong: "Thinking back, my impression is that his immediate influence was in no way so great as it is represented to be, more particularly by Japp in his Chemical Society Memorial Lecture";[124] the same author thinks that the benzene theory received no special attention until the early 1870s. With valency a similar time-lapse seems to have been generally true, although it is significant that when the origin of valency is under discussion the name most frequently encountered is that of Kekulé. But to the question "who discovered valency?" no simple, direct answer can be given.

PART TWO

Early Applications of Valency

CHAPTER VII

Valency and the Classification of the Elements

THE THEORY OF VALENCY, once established, began to make an impression on several branches of chemistry, and in so doing became more firmly embedded as a foundation doctrine of the science. Not merely was its usefulness made clear by these applications to inorganic and organic chemistry, but the theory itself became modified and extended. Its value was first apparent in giving a new impetus to the classification of the elements, in making possible the Theory of Structure in organic chemistry, and in the development of the latter into stereochemistry. These matters are the subjects of the present chapter and the two following.

Classification of the elements had long been a matter of interest, but progress in evolving a satisfactory system had been delayed by lack of suitable data. Berzelius had pioneered a classification based on the electrochemical series, but it had been a continuous array of elements from oxygen at one end to potassium at the other, without subdivisions and with little regard for characteristics other than those strictly electrochemical.[1] Nevertheless, although its author deemed it "more correct than any other in giving an idea of chemistry",[2] not much was heard of this in the 1850s and 1860s, partly because of the eclipse of the electrochemical theory itself.

The possible bearing of atomic weights upon this problem had been hinted at by several writers in the early part of the century, including Döbereiner,[3] Dumas[4] and others. But these views were fragmentary and partly vitiated by the uncertainty then attaching to atomic weight values. When the latter difficulty became resolved, more serious efforts were made by de Chancourtois,[5] Odling[6] and Newlands[7] to place elements into groups on the basis of their atomic weights. These contributions were paving the way towards reception of the Periodic Law a few years later.[8]

In these preliminary efforts to reach a satisfactory classification, considerations of valency played almost no part. At the same time, however, another line of thought involved valency in a direct way, and used this concept as the basis for a different scheme of classification of the elements.

1. CLASSIFICATIONS BASED UPON VALENCIES

In the middle 1860s there is evidence of a small group in London by whom valency was viewed as a sound basis for classification of the elements. The first indication of this was Odling's division of elements into those of even valency ("artiads") and those of odd valency ("perissads").[9] As a basis of classification, of course, this was of very limited value, but it did emphasize the frequent variation in valency by increments of two units. His nomenclature was used by a number of his contemporaries.

In the same year, Williamson put forward a suggestion for the "classification of the elements in relation to their atomicities".[10] He attempted to group the elements in families according to their valency, though he recognized that the latter could vary as Odling had pointed out.

At this time Odling and Williamson were teaching at St Bartholemew's Hospital and University College respectively. Also in London, W. A. Miller was occupying the Chair of Chemistry at King's College, and was clearly aware of his colleagues' views. The following year he reported to the British Association that:[11]

Owing to the labours of many distinguished men, amongst whom the names of Williamson, Kekulé, Odling, Cannizzaro and Wurtz are the most prominent, a classification of the elements into families has been made; and that this classification rests upon what is known as the *atomicity* of the elements.

Two years after this, Miller included a classification of the elements on this basis in his *Elements of Chemistry*.[12] They were grouped in accordance with their lowest valencies; thus sulphur, selenium and tellurium under an "atomicity" of 2, while nitrogen, phosphorus, arsenic, antimony and bismuth were classed under 3.

Meanwhile, Frankland had left Manchester for London, and in the Chairs of Chemistry at St Bartholomew's Hospital and the Royal Institution was becoming more a recognized leader. Possibly reflecting the influence of Odling and Williamson, he published in 1866 two Tables of the elements, classified into groups according to their valencies.[13] These were repeated later,[14] with minor amendments (like the transference of vanadium from the hexad to the pentad group, doubtless reflecting the researches of Roscoe[15]). In Frankland's system, the highest valency then known for each element determined its place. Within most of the groups were several "sections", each containing elements of similar chemical character. Thus the "monads" contained:

1st Section: hydrogen;
2nd Section: fluorine, chlorine, bromine, iodine;
3rd Section: caesium, rubidium, potassium, sodium, lithium;
4th Section: silver.

It is possible that this work of Frankland may have had some influence upon Miller. Although his British Association address[16] omitted all reference to Frankland, the fifth edition of *Elements of Chemistry*,[17] published after his death in 1870, used the latter's notation.

These suggestions by the London chemists were in themselves of limited value only, in that they missed the fact of a periodicity with increasing atomic weight. Yet they threw into focus the importance of valency in a way altogether absent in the writings of those who used atomic weights as the sole criterion for classification. A fusion of the two views was needed. In fact, however, it does not seem that the systems based upon valency were of great influence upon the development of the Periodic Law. The part played by valency in this was recognized *after* it had been established from a study of atomic weight relationships.

2. PERIODIC VARIATIONS IN VALENCY

The rise of the Periodic Law is associated with the names of Lothar Meyer and Mendeléef.[18] The former chemist approached it in his *Modernen Theorie*[19] in 1864, when he observed that, when elements are arranged in groups with increasing atomic weights, the differences between the values for the latter increase also, being approximately the same for all groups. Now this arrangement of Lothar Meyer was on the basis of valencies, and is thus a link between the classifications based exclusively on these and the later systems of Lothar Meyer and those of Mendeléef. The artificiality of such an arrangement was plain from the inclusion of monovalent thallium with the alkali metals, and an insistence upon valency as the chief criterion was bound to lead to anomalies like this. Probably for this reason, the next approaches were made from atomic weights.

In 1869-70 appeared the great contributions of Lothar Meyer[20] and Mendeléef[21] in which the Periodic Law was enunciated and enshrined in the first published versions of the Periodic Table. Based essentially upon atomic weights, the Law dwelt upon valency only in passing, but these incidental references are of great interest.

A brief note from Mendeléef gave his famous Table, and included the observation that "The arrangement according to atomic weights corresponds to the valency of the elements, and up to a certain point the difference in chemical behaviour, *e.g.*, H, Li, Be, B, C, N, O, F".[22] A more detailed statement of the same year showed Mendeléef's attitude more clearly.

He went to some trouble to show the inadequacy of all systems of classification apart from those based upon atomic weights. Included in the unsatisfactory approaches were those based upon the electrochemical series and upon valency. His basic objection to the latter was the fact that valency was not constant for every element, and the difficulty of determining it. After a lengthy discussion he concluded:

There is thus no single general principle having the status of a critique that could support a judgment of the relative properties of the elements, in order to classify them into a more or less rigid system.[23]

Nevertheless, he recognized that there *was* a dependence between valency and atomic weight; in 1871 he wrote:

In that the periodic law brings out the dependence of these two magnitudes atomic weight and valency on one another, at the same time there is the possibility of determining one by the other, namely valency by atomic weight; therefore, if the law of valency determines the forms of chemical compounds, the periodic law will also determine the same; the latter, however, goes somewhat further because it determines at the same time those forms of oxygen (*Sauerstoffformen*) which the law of valency leaves out of consideration.[24]

To these statements all that need be added is the famous observation by Lothar Meyer:[25]

The saturation capacity of atoms rises and falls regularly and evenly in both intervals:

1-valent	2-v.	3-v.	4-v.	3-v.	2-v.	1-v.
Li	Be	B	C	N	O	F
Na	Mg	Al	Si	P	S	Cl

Thus out of a study of the periodic dependence of general chemical behaviour on atomic weights there emerged a new set of valency relationships that, for the first two short periods at least, revealed an underlying simplicity that was to prompt still more fundamental questions. For some years the Periodic Law gave a valuable check on valencies suggested on other grounds and, in stimulating a revival of inorganic chemistry (together with other factors), led to a multitude of new discoveries that were radically to alter the whole conception of valency. And it was through this means also that Lewis and others were brought to the electronic theory of valency (see p. 270).

CHAPTER VIII

Valency and the Theory of Structure

To suppose that in a molecule all the atoms have definite relationships with other atoms by means of fixed valency connections is to assume the Theory of Structure. The recognition of "constitutional" or "structural" formulae followed the acceptance of the theory of valency, and would indeed be impossible without it. It is therefore not surprising that within a few years of the papers by Couper and Kekulé in 1858, the Theory of Structure was gaining wide recognition, and was bringing an order into organic chemistry even greater than that produced by the Periodic Law in the inorganic field.

At first, however, its reception was slow for a variety of reasons; once established, it could not accomplish much until the additional problem of equivalence or non-equivalence of valencies had been settled; finally its development, chequered for a time by new difficulties, led ultimately to the solution of the latter in the recognition of "mobile valencies" and of stereochemistry. The last of these matters will be dealt with in the next chapter, the remainder being considered in the following sections of the present chapter.

1. FACTORS OPPOSING THE STRUCTURE THEORY

Examination of the chemical literature for the 1860s and even later reveals the surprising fact that the theory of structure was used only sparingly in some periodicals and hardly at all in others. Against the theory powerful factors were at work, the strongest of these being the Theory of Types.

Far from being dead, the Type Theory was enjoying an extension of its existence, partly, no doubt, because of its success in the previous decade, culminating in its fusion with radicals in valency. The strangest feature of all is the continued employment of Typical formulae by Kekulé, the man who more than any other had made them logically unnecessary. He was using them at least up till 1869. Yet he was not entirely consistent, and once observed:

FACTORS OPPOSING THE STRUCTURE THEORY

The notation of our formulae is a matter of convention. Most chemists are agreed in using typical notation, except a few who, nevertheless, prefer to represent the same ideas by a different form. When new facts are discovered, it is sufficient to apply to them the principles of the current notation in order to have formulae that summarize these facts. But it can happen that a new fact is found outside the limit of the cases foreseen by the theoretical principle that was the basis of that notation: it is then that it cannot be applied in a natural way to the facts.[1]

This explains part of the difficulty. Kekulé used Typical formulae because they often "fitted the facts", but where this was not the case (unsaturated and aromatic compounds especially) others should be employed. And it is fair to add that a better system of notation did not arise in general use until the late 1860s, when Crum Brown's formulae began to gain acceptance. In other words, Kekulé's employment of Typical formulae does not necessarily commit him to approval of that part of the Theory of Types that denied the possibility of knowing structural arrangements in a molecule.

There was, however, more to it than this. It is clear that Kekulé was reluctant to cast away the last shreds of structural agnosticism, as the following statement in his *Lehrbuch* (1861) demonstrates:

Rational formulae are only reaction formulae and are not structural formulae. They are nothing but expressions for the reactions of substances and for the comparison of various substances with one another; by no means can they express constitution, *i.e.*, the relative positions of atoms in the corresponding compounds.[2]

On the face of it, this looks like a complete repudiation of the basic tenets of Structure Theory. As Kekulé's influence was so important it will be desirable to examine this further, and to quote the long paragraph which follows the above quotation (in very small print and quite easily missed by a casual reader):

This deserves to be very specially stressed, because many chemists, strange to say, are yet of the opinion that from the study of chemical change the constitution of compounds can be deduced with certainty, and therefore the position (*Lagerung*) of the atoms can be expressed in chemical formulae. That the latter is not possible requires no special proof; it is self-evident that one cannot represent the positions of atoms in space–even if one had discovered them–by placing letters side by side in the plane of the paper, and that at least one requires a perspective drawing or a model. However, it is likewise clear that through the study of metamorphoses one cannot ascertain the positions of the atoms in compounds that are not reacting (*bestehenden*), because the manner in which the atoms of the substance undergo change and destruction cannot possibly prove how they are arranged in an unreacting compound that remains unattacked. At all events it must now indeed be held as a task of natural science to ascertain the constitution of matter, and therefore, if we can, the position of the atoms; but this cannot be attained by the study of chemical changes, but only through the comparative study of physical properties of unreacting compounds. Thus it will perhaps be possible to establish the constitutional formulae for compounds, which must naturally be unchangeable for one and the same substance. However, even if this is successful, different rational formulae (decomposition formulae) are still always permissible, because a molecule that is produced by atoms positioned in definite ways can be split in different ways under different conditions, and therefore can yield fragments of different sizes and compositions.[3]

Two extremely important points emerge from this. The first is that Kekulé drew a distinction between formulae expressing structure and those expressing reactivity. *He is not denying that structural formulae can be determined*, although that is the impression received from reading the main paragraph only (the first quotation above). He does, however, suggest that chemical reactions cannot convey structural information and that is the only deficiency in his viewpoint; this cautious attitude to arguments based upon reactivity is, of course, a legacy from Gerhardt. His respect for physical evidence probably owed much to Kopp under whom he had studied at Giessen. This may be also seen in a footnote to his 1858 paper "On the constitution and metamorphoses of chemical compounds".[4]

The second fact emerging is that he repudiated structural formulae *if these were interpreted in terms of spatial arrangement* of the atoms present. It is difficult for us to appreciate a position where complete uncertainty about steric matters led to complete indefiniteness on structural formulae. All that could be said was a general limitation of the kind that Kekulé gave. But this was no denial of the Theory of Structure; only a designation of its limits.

It seems, then, that Kekulé's opposition to the Theory of Structure was confined to the frequent use of out-of-date, but not erroneous, formulae; to an insistence on the validity of conclusions drawn from physical data only; and to a refusal to anticipate the conclusions of stereochemistry. The last of these was wholly admirable, the first harmless but the second highly restricting.

Kekulé was not the only chemist whose attitude did not help the growth of structural theory, however. Butlerov himself used for a time a Typical notation in his work.[5] In 1863 Playfair[6] tried to use "the strong and sound cement" of Gerhardt and his followers to build up "the remains of the venerable dualistic edifice", and others also held tightly to the Typical viewpoint. Odling opposed the new notation, and probably the new ideas behind it, up to 1870 (see p. 55). But opposition was also to come from another quarter.

Throughout the nineteenth century there had been an undercurrent of rebellion against the atomic theory, probably originating in Davy (p. 7). Occasionally this movement had been perceptible, but in the 1860s and 1870s atomic scepticism became a matter of urgent debate.[7]

In 1866, Brodie refuted the existence of atoms in a lecture[8] which gave the chemical world "a sensible shock", for up to this time "during all the discussions on 'atomicity', hardly a doubt has been raised as to the actual existence of atoms".[9] This brought forth vigorous replies from many, including Kekulé, who though admitting a philosophical difficulty in accepting the existence of atoms, found them "absolutely necessary in chemistry".[10]

Williamson replied the following year,[11] and returned to the subject in 1873.[12] In the latter statement he pointed out the consistency and utility of the atomic theory, and showed its great value for structural chemistry; for example it provides the simplest explanation of isomerism.

In 1877 the discussion centred on the use of an atomic system of notation; Wurtz[13] was in favour of this, but Berthelot[14] opposed it, regarding the atomic hypothesis as ill-founded. Indeed, Berthelot remained pointedly apart from all structural discussion, and was one of the most distinguished chemists of the century who did nothing whatever to foster the growth of the new doctrines of valency and of structure. Undoubtedly, his opinions, and those of others like him, reacted unfavourably on structural chemistry.

The third factor militating against an immediate acceptance of the Theory of Structure was the opposition of Kolbe.

The position taken by this controversial figure is very difficult to define. Few had done as much as he, in the years preceding the general recognition of valency, to advance the knowledge of chemical constitution. Yet he consistently ranged himself against the advocates of the Theory of Structure. Without being diverted into details, we may note three probable reasons for this opposition, which expressed itself particularly in his own *Journal für praktische Chemie*.

Firstly, he was an implacable opponent of all aspects of the Type Theory, and it was in the form of this theory that much of the early structural work was framed. Secondly, he held tenaciously to the view that all extensions of constitutional ideas to the actual spatial arrangement were unwarranted. Hence literal interpretations (like that of Naquet, p. 174) were suspect from the beginning, and Kolbe's opposition was aroused. A final reason may be that represented by Farrar and Farrar:[15]

Like Berthelot, he was an agnostic, both in religion and chemistry, and held firmly to the belief that the arrangement of atoms in a molecule was unknowable; chemical structures were "Phantasiebilder" and an attempt to introduce spiritualism into science.

It seems undeniable that Kolbe's basic opposition was philosophical, if not psychological, and more of this will be seen in connection with stereochemistry.

2. THE FOUNDING OF THE THEORY OF STRUCTURE

Much recent controversy has taken place on the matter of priority in the origin of the Theory of Structure. These disputes have centred around the name of Alexander Butlerov, and started with rather sweeping claims made on his behalf at a meeting of the U.S.S.R. Academy of Sciences in June 1951. We may quote:[16]

The second half of the nineteenth century was marked by the greatest event in the history of organic chemistry – the creation of the chemical structural theory by the Russian scientific genius Aleksandrov Mikhailovich Butlerov.

The case for Butlerov has been stated in several recent publications.[17] On the other side there are statements like the following:[18]

Kekulé and Couper ... are responsible almost entirely for the genesis of the Structure Theory. The claims of Butlerov, advanced by Russian writers, depend more on repetition than on documentary evidence.

Now any discussion as to the origins of the Structure Theory must involve clear conceptions of what is meant by this phrase. Unfortunately this clarity has not always been present, and this fact, coupled with philosophical and political causes, has produced a situation of the utmost confusion. In view of the direct relevance of valency to the origin of the Theory, it seems desirable to attempt some reassessment of the position, though no effort will be made to evaluate the contributions of Butlerov in a wider chemical context.

It seems that the phrase "origin of the Structure Theory" has been used in at least three distinct senses. It will be both convenient and essential to clarity to deal separately with each of these.

First, these words are employed in the sense of "introduction of the phrase 'chemical structure' ". On the priority in this matter there can be little doubt. Butlerov had used the words *Molecularstructur* and *Structur*, but on 19 September 1861, at a conference at Speyer, he introduced the phrase and idea of "Chemical Structure":[19]

> Starting from the assumption that each chemical atom possesses only a definite and limited amount of chemical force (affinity) with which it takes part in forming a compound, I might call this chemical arrangement, or the type and manner of the mutual binding of the atoms in a compound substance, by the name of "chemical structure".

Butlerov's priority in this respect is not at issue, and hardly could be. There are those who appear virtually to limit his contribution to one of mere words, but none who deny this much. It may, however, be remarked that a parallel development was the use of the phrase "constitutional formula" by Erlenmeyer when engaged in similar work in the 1860s.[20]

The phrase "origin of the Structure Theory" may also be taken to mean the introduction of the structural *concept* into chemistry, and here there is much more room for dispute. In particular, the names of Couper and Kekulé have been frequently advocated in the West as the true founders of the Theory. Butlerov's role is then reduced, at worst, to that of namer of the Theory, or, at best, as its most ardent advocate and consistent user. That both Kekulé and Couper were known to Butlerov is certain, and it is highly probable that their influence was far-reaching; he had stayed at both Heidelberg and Paris in 1857. But whether they rather than Butlerov were responsible for introducing the idea of chemical structure depends upon our further analysis of the meaning of this concept.

It is possible to mean by this phrase the idea of tetravalent chain-forming carbon atoms. In that case, Couper and Kekulé claim undoubted priority over Butlerov, their contributions having been made in 1858 (*cf.* p. 125). Yet even such a valuable theory as this would have been incapable *by itself* of being the basis of progress in organic chemical science. There always has to be present an additional assumption, usually employed today without conscious recognition, but nevertheless made: it is that the arrangement of atoms in a molecule affords – at least in principle – a *sufficient* basis for derivation of all chemical properties. These, in other words, are based not upon "structure" *and* some other factor (vital force, *etc.*), but upon "structure" alone. Hence a molecule will be uniquely defined by its structure. If, therefore, we can establish that this additional idea was absent in Kekulé but present for the first time in Butlerov, we can agree with G. V. Bykov who has made essentially this point in a recent book,[21] and in a reply[22] to the authors of the second quotation[23] in this section:

If one considers that the statement of dependence of the chemical properties of an organic molecule on its chemical structure, together with all consequences derived therefrom, is the basis of classical structural theory, then the *rise of this theory* is connected with Butlerov's name.

The characteristic reticence of Kekulé on structural matters makes an understanding of his real beliefs very difficult to assess. From his published works, however, it appears that he regarded structural considerations as a valuable if not an indispensable tool, but one whose objective reality was an open question. One may compare his views on atoms themselves (p. 145). As has been very aptly remarked[24] his conservatism here shows a remarkable parallel to that of Copernicus. Nevertheless, Kepler's dictum that "he failed to see how rich he was" is hardly applicable to Kekulé, for, unlike the astronomer, he was not held fast by the chains of an outmoded dogmatism. His conservatism was negative rather than positive, and his caution sprang more from an experience of the transience of some scientific doctrines than from a deep-rooted tradition. The quotation from his *Lehrbuch* on p. 143–4 makes plain his position; he did not deny that structural knowledge was ascertainable, but did not believe that it could be obtained from current chemical methods; therefore *at that time* certainty had not been reached.

The position of Couper is quite different. He had freed himself from the Type Theory and started his arguments from valencies; the trend of his philosophical attitude was towards an open-minded and provisional acknowledgment of chemical structure. In particular, the way in which he discussed his formulae implies that he regarded them as essentially constitutional. His use of the double oxygen unit was a drawback, but not a serious one; the reason given

THE FOUNDING OF THE THEORY OF STRUCTURE 149

by Ladenburg[25] was a quasi-structural one: "in order that he may not be obliged to assume in the formation of salts the replacement of the hydrogen by metal (and therefore reduction of the oxide). According to him O = 8 is bivalent, but one valence must always be satisfied by oxygen". It would seem that the one barrier to Couper's recognition as founder of the Structure Theory (in the sense of the present discussion) is simply that the views he almost certainly held were never expressed explicitly enough.

In Butlerov, however, appears a clear statement of the uniqueness of a structural formula:

Only one rational formula is possible for each compound, and when the general laws governing the dependence of chemical properties on chemical structure have been derived, this formula will express all of these properties.[26]

This may be compared with Kekulé's own views (p. 143); although the term "rational formula" was used in different senses, the gain in clarity obtained by Butlerov is surely obvious.

Now two recent statements have placed much emphasis on Butlerov's own claims, concluding that:

Though the name chemical structure has first been introduced by Butlerov the basic principles have been given by Kekulé and Couper. This view was also held by Butlerov, who only claimed recognition for his consistent application of the new ideas.[27]

The second of these publications[28] suggests that if the point above has not been adequately made, references to Butlerov's paper[29] of 1859 should be convincing. He writes:

The tetratomicity of the carbon molecule mentioned by Couper has already been assumed by Kekulé; to this latter chemist belongs the priority also of the view of affinity units remaining free when a part of this has been consumed in building a compound molecule.[29]

This, however, seems to miss the point completely, for there is no question of priority in the introduction of the ideas referred to here. Structure Theory must involve more than these, and so this claim has no direct bearing on the question under dispute. In any case this was in 1859, two years before the introduction of the doctrine.

Later, Butlerov had occasion to be more specific. After others had used the structural ideas to predict successfully isomers of, for

F

example, aliphatic hydrocarbons, Butlerov felt it necessary to write thus:[30]

I must expressly mention that in the chemical literature of recent years such ideas as those I have already published are presented as novel and original (without citation). . . . I am now obliged to assert that to me belongs an important part of the priority for the complete and consequent development of this principle. This assertion of mine, as I hope, will be found to be solidly based on the making of a closer acquaintance with my publications that have appeared since 1861.

To this came a reply from Lothar Meyer,[31] who felt it was a personal attack on him. Meyer considered the writing of structural formulae was "the obvious consequence" of the two propositions of tetravalent carbon and its formation into chains:

However, Kekulé combined these propositions long before Prof. Butlerov, who learned of the chain-formation from Kekulé and suggested calling it structure. The establishment of these formulae, carried through by a series of combinations, permutations and the calculus of variables required neither unusual genius nor great gifts of discovery.

As Leicester[32] has pointed out, this involves the twin errors of treating Butlerov's contributions as "a mere manipulation of symbols" and of denying credit to one whose ideas now happen to be universally obvious. Butlerov's reply included the following:

I have always been, and still am, a long way from misunderstanding the brilliant services of Kekulé in the theoretical and practical branches of our science. Like Herr Meyer, I hold that the idea of chemical structure is a consequence of the recognition of the valency of the elements and especially the proposition first recognized and expressed by Kekulé that carbon is tetravalent. I also believe that the idea of chain-formation by atoms cannot be neglected. I believe that the views expressed by Couper (unfortunately in brief only) are almost identical with those generally accepted and that the formulae given by Couper are actually rational formulae in the present sense of the word, *i.e.* constitutional formulae and formulae of chemical structure. . . . Apart from the recognition of the new propositions we still find for several years more much which either had become superfluous or did not agree with these propositions. This includes the use of types, both compound and mixed, the old method of determining formulae only by reactions so that several rational formulae could be given for one and the same substance, *etc.* Similarly, assumptions which did not entirely agree with the new principles were not entirely absent from the writings of Kekulé even *after* I had expressed my views on chemical structure and made them the basis of the chief principles of my theoretical speculation.[33]

In other words Butlerov was claiming for himself a degree of detachment from ideas inimical to Structure Theory not shown by others at the time. He acknowledges their part in laying the foundations, but makes clear his own claim to have gone substantially beyond these.

Thus, in so far as arguments based upon Butlerov's own claims can have any validity, there is a definite body of evidence in support of his priority in founding the concept of the Structure Theory. That the opposite impression has become so widespread is probably due to confusion between the premises of Structure Theory and the doctrine itself, and also to the following statement included by Butlerov[34] in his 1861 paper where the theory was introduced:

I am far from thinking that I am proposing here a new theory; I believe rather that I have expressed an idea of the kind that belongs to very many chemists. I must even remark that a similar idea, though not clearly enough grasped or expressed, lies at the basis of the views and formulae of Couper.... I will only pronounce that it is true that the idea of atomicity wholly free from typical views should be used in all cases as the basis for the observation of chemical constitution, and that the same seems to place in our hands a means for remedying the undesirable position of chemistry.

And this appears to be the situation. The ideas to which Butlerov gave expression were to some extent common property, and his distinction was formally to introduce them, and thereby to gain for himself a just reputation to be the founder of the Structure Theory.

The third sense in which the last phrase could be used is that of one who consistently applied the new concept. It is undeniable that this was also true of Butlerov, but as time progressed others began to apply it with equal success, often without full realization that they were using his hypotheses. Particularly valuable were the efforts of Erlenmeyer whose ideas of chemical constitution were in some ways in advance of those of Butlerov.[35] The man whose significance has been consistently under-rated in this connection, however, was Crum Brown, in certain respects having a far clearer understanding than Butlerov possessed in the early 1860s. The latter was hampered by the spurious idea of primary and secondary valencies, and his early structural theory was built upon this fallacy. It was Crum Brown who first gave a logical refutation of this (p. 154), and his system of notation was far superior to the *quasi*-Typical formulae at first used by the Russian chemist. Unfortunately some of his papers were published only in the *Transactions of the Royal Society of Edinburgh*, a journal that may have been rather inaccessible on the Continent, and Brown's structural ideas appear not to have been as widely recognized as they deserved.

It seems, therefore, that Butlerov's application of the Theory of Structure was not unparalleled elsewhere. His claim to be the founder of the theory rests upon his extension of the ideas of Kekulé and Couper into the doctrine that he first formulated in September 1861.

3. THE EQUIVALENCE OF VALENCIES

With establishment of the structural doctrine, the first and most immediate problem awaiting solution was that of the equivalence of valencies. Until it could be established that all valencies in carbon were equivalent to each other, for example, all arguments about structures of carbon compounds would rest upon an uncertain foundation.

When this matter was first raised, opinion was generally in favour of a disparity between at least some of the valencies of carbon. This arose from the early confusion between the gas obtained by the electrolysis of acetic acid or from the action of methyl iodide on zinc ("methyl"), and that formed from potassium and ethyl cyanide or from zinc, ethyl iodide and water ("hydride of ethyl"). These gases were assumed to be different by Frankland on the basis of an apparent difference in their products with chlorine,[36] an error which was to act as a serious brake on structural chemistry for some years.

To explain this supposed "isomerism" Butlerov introduced the idea of "primary" and "secondary" affinities:

It is necessary to distinguish the quantity of affinity from its intensity, *viz.*, the smaller or greater energy with which it tends to become active. This intensity varies as the body reacts with one or another substance, and according to the conditions under which this action takes place. It seems also that if part of this quantity of affinity of polyatomic elements, especially carbon, has been combined, all the other conditions remaining equal, however, the part remaining free shows another degree of intensity. Thus it is probably not the same, for instance, if the diatomic molecule *A* combines with first a monatomic substance *B* and then with an atom of a monatomic substance *C*, or when it happens the other way round, first with *C* and then with *B*. Thus the four hydrogen atoms which are held under the influence of the four units of affinity in marsh-gas cannot be substituted with equal ease. In this manner we are to led distinguish *primary, secondary* affinities, *etc*.[37]

In any consideration of the subject of valency, this distinction is of the greatest importance, especially as it was later shown to be fallacious. The following year, Butlerov elaborated his theme:

The simplest and most ordinary compounds of the polyatomic elements, the isomers of the hydrocarbons, *etc.*, lead us to this assumption. In fact there exist at least two isomeric bodies of the formula C_2H_6, methyl and hydride of ethyl; now since hydrogen in all its compounds appears with only one affinity, the affinity of the carbon must necessarily be regarded as the cause of the linking of the two carbon atoms here. As a natural chemical explanation of the isomerism of both the known bodies, I believe it must be assumed that the units of affinity joining the carbon atoms are not the same in the two cases.[38]

He did not believe that there was necessarily an "absolute difference in the free affinities", but that in reaction they sometimes played a different part. He applied this to possible isomerism of the chloro-methanes; his conclusions[39] may be summarized thus:

TABLE V

Types of different affinity (each letter = different type)	No. of isomers of CH_3Cl	CH_2Cl_2	$CHCl_3$
abcd	4	6	4
aabc	3	4	3
aabb	2	3	2
aaab	2	2	2

In "hydride of ethyl" he concluded the carbon atoms were linked as in ethylene, while in Frankland's "methyl" these must be the same as those joining the iodine to carbon in methyl iodide.[40]

Kekulé also pointed out in his *Lehrbuch* the separate existence of "methyl" and "hydride of ethyl", writing them as $\begin{cases} CH_3 \\ CH_3 \end{cases}$ and $C_2H_5.H$ respectively.[41] He did not, however, point out that this implied differences in valency, and was characteristically silent on the possible cause for the anomaly. It is possible that in the following quotation from his paper on itaconic and pyrotartaric acids he is implying a belief in differences of affinities:[42]

The hydrogen of an organic molecule which is only indirectly related to the carbon is separated and replaced more easily than other atoms of hydrogen combined directly with carbon. For the rest, the nature of this hydrogen combined indirectly, or rather *the nature of the place it occupies, depends on the nature of the elements which surround it*. It is easily replaced by metals when it is placed near two atoms of oxygen; it does not have this property when there is only a single atom of oxygen in the vicinity.

At about the same time, Erlenmeyer was also persuaded of the existence of different kinds of affinity; discussing isomers he wrote:[43]

The possibility of their existence can be explained, according to our view, only on the supposition that the different affinities of polyatomic elements are not all endowed with an equal tendency to combine with a given element.

In 1864, however, two papers had been read which made this speculation useless, and within a few years it had ceased. Crum Brown, at the Royal Society of Edinburgh, produced a fresh approach to the whole problem of isomerism by using his new graphical formulae. This caused him to doubt whether "methyl" and "hydride of ethyl" were "metameric" (isomeric), though at that stage he was not prepared positively to assert their identity. He was, however, able to reveal an inconsistency in Butlerov's doctrine of primary and secondary valencies:[44]

By carrying this argument a little further, and making use of no additional assumption, we arrive at an absurdity: thus, the carbon radical of the acetic series is the same as that of oxyacetic (glycollic) acid, that again is the same as that of oxalic acid, therefore as that of oxalic nitrile or cyanogen gas; but in cyanogen gas we have the two carbon atoms united by two *primary* affinities; but we have before proved, that in the acetic series they are united by a primary affinity of the one, and a secondary affinity of the other. It is obvious, then, that at least one of our assumptions is false.

The argument can be summarized thus:

$$PRIMARY \quad \begin{matrix} CN \\ | \\ CN \end{matrix} \longrightarrow \begin{matrix} COOH \\ | \\ COOH \end{matrix} \longrightarrow \begin{matrix} CH_2OH \\ | \\ COOH \end{matrix} \longrightarrow \begin{matrix} CH_3 \\ | \\ COOH \end{matrix}$$

$$SECONDARY \quad \begin{matrix} CH_3 \\ | \\ I \end{matrix} \longrightarrow \begin{matrix} CH_3 \\ | \\ CN \end{matrix} \longrightarrow \begin{matrix} CH_3 \\ | \\ COOH \end{matrix}$$

and these are identical.

Thus, assuming the carbon–carbon bond remained unchanged throughout, he showed that by one route it was primary, and by the other secondary.

Although, two years later, Frankland deemed it "far from improbable" that a distinction might exist "between the values of the different bonds of an element – of carbon for instance",[45] he expressed a minority view. Nevertheless, the problem of isomeric hydrocarbons C_2H_6 remained until solved in 1864 by Schorlemmer, who had been an assistant at Manchester during Frankland's tenure of the Chair of Chemistry, and was now working there under Roscoe. By means of carefully conducted experiments he was able to show that "methyl" and "hydride of ethyl" gave rise to the *same* product with chlorine, ethyl chloride.[46] This became generally recognized as conclusive evidence, though for some unaccountable reason Frankland as late as 1877 was apparently unconvinced, and suggested a repetition of his own experiments which were then "nearly the only evidence in support of the non-identity of the two series of hydrocarbons".[47] However, in 1872 Schorlemmer was able to report to the Chemical Society that after his discovery "a few chemists only have made use of the above hypothesis in order to explain certain cases of isomerism".[48]

With this difficulty removed, the equivalence of the four valencies in carbon became recognized. Other facts, of course, later substantiated this, as the non-existence of isomers of the halogeno–methanes predicted by Butlerov if the valencies were not equivalent.[49] An interesting piece of physical evidence came from the thermochemical researches of Thomsen, which showed that the progression from methane to its mono-, di-, tri- and tetramethyl derivatives was accompanied by a regular increase in the heat of combustion, thereby implying equality of the four valencies of carbon.[50]

Carbon, of course, was the most important example of a polyvalent element for the progress of structural chemistry. But others also came under discussion, amongst them nitrogen. The problem of its variable valency, to which we shall return, raised the further question of the equivalence of its valencies in its higher state of combination. Thus Butlerov pointed out that the non-existence of NCl_5 and NH_5 suggested a non-equivalence of the affinities of the nitrogen.[51] It was Hofmann, however, who showed conclusively that all four alkyl groups in the quaternary ammonium salts possess a fundamental kind of equivalence. Thus he showed that the product from triethylamine and amyl iodide on conversion to the "oxide" (quaternary ammonium hydroxide), followed by heating, decomposes into ethylene, water and amyldiethylamine. This suggests an equivalence of the four alkyl groups in the iodide. The reactions may be represented in modern symbols thus:

$$C_2H_5-\underset{\underset{C_2H_5}{|}}{\overset{\overset{C_2H_5}{|}}{N}} \xrightarrow{+C_5H_{11}I} C_2H_5-\underset{\underset{C_2H_5}{|}}{\overset{\overset{C_2H_5}{|}}{N}}{}^{\oplus}\text{-}C_5H_{11} \xrightarrow{OH^-} C_2H_5-\underset{}{\overset{\overset{C_2H_5}{|}}{N}}-C_5H_{11} \\ +C_2H_4+H_2O$$

In Hofmann's own words:

The preceding considerations clearly show that, whatever the actual disposition of the molecules in ammonium or its congeners may be, the atoms rearrange themselves whenever the fourth equivalent of hydrogen, or of its substitute, joins the compound.[52]

Following this discovery there came a succession of others relating to elements possibly showing more than one state of combination. With these the theory of structure led to a more profound investigation into the question of variable valency, and so far as inorganic chemistry was concerned, the two became inseparably linked. This was to a smaller extent also true of parts of the organic branch, and these developments of Structure Theory will be encountered again in connection with the even larger issue of variable valency (Chapter X).

4. THE CONCEPT OF TAUTOMERISM

The greatest difficulty encountered by the Structure Theory, steric considerations excepted, was in the phenomenon now called "tautomerism". The history of this has been recently reviewed,[53] and it is only necessary to call attention to the part it played in the development of valency theory.

In the early 1860s, acetoacetic ester was assigned an enolic formula by Geuther:[54]

$$CH_3.C(OH):CH.COOC_2H_5$$

but was regarded as ketonic by Frankland and Duppa:[55]

$$CH_3.CO.CH_2.COOC_2H_5$$

However, as is now well known, this compound gives reactions of both enols and ketones.[56]

Butlerov obtained two olefins from the action of sulphuric acid on tert.-butyl alcohol, and this led him to suppose that an equilibrium might exist between the two molecules: "the molecule will always behave in two or more isomeric forms".[57] This brilliant extension of his Structure Theory was followed by a discovery from another leader in this field, Erlenmeyer, who obtained from compounds of the vinyl halide type hydrolysis products isomeric, but not identical, with the corresponding vinyl alcohols. He inferred from this an isomeric change at the moment of reaction.[58] A slightly different kind of situation arose in the case of isatin examined by Baeyer.[59] Only one isatin was known, although two methyl derivatives were obtained. Again, hydrogen cyanide was known to give rise to two series of derivatives; Butlerov had suggested isomeric change took place here also,[60] and Crum Brown suggested the molecule HCN behaved ambiguously.[61]

In 1885, Laar put forward a rather different view.[62] Since all the pairs of isomers differ in the position of hydrogen he suggested that a mobile hydrogen atom oscillated between the two positions, echoing Kekulé's theory of benzene structure. Unlike the view of Butlerov, Laar's hypothesis involved one rather than two molecular species and is an interesting anticipation of resonance, though not, of course, to compounds where the latter could occur. The word "tautomerism" was introduced by Laar.

A reconciliation of the two views was achieved by Hantzsch and Hermann[63] who pointed out that the essential difference between them hinged upon whether or not both isomers were isolable. The final vindication of these ideas came with isolation of keto and enol forms of acetoacetic ester.[64]

What is the relevance of all this for valency? The study of tautomerism played an essential part in the development of Structure Theory, and without a recognition of either oscillating atoms or isomeric change the theory would soon have foundered in face of the difficulties in the chemistry of acetoacetic ester and other substances. Tautomerism, therefore, was a valuable subsidiary hypothesis (or pair of hypotheses) that gave vigour to the Structure Theory and thus indirectly supported the views of valency on which this was based.[65]

Incorporation of these facts into a larger scheme did not occur until the twentieth century, when tautomerism was conceived as part of a vast system of ionizations and isomerizations, and the underlying reasons for its existence became clear. Until then, however, its dependence upon fine structural considerations, and therefore on valency, did not emerge.

CHAPTER IX

Valency and Stereochemistry

WE ARE CONCERNED here with that extension of valency theory in which the bonds are conceived as being in some cases directed in space. The development of the science of stereochemistry has been traced elsewhere,[1] but the purpose of the present chapter is to indicate the regions in which the idea of spatially directed valency became particularly valuable, and to trace the course of its progress in the early years of the theory of valency.

Just as the initial growth of valency arose from the study of organic chemistry, so its extension to three-dimensional space also began here, and until the end of the century stereochemistry was centred on the concept of the "tetrahedral carbon atom". As is well known, this hypothesis was independently advanced in 1874 by van't Hoff[2] and le Bel,[3] primarily as an explanation of optical isomerism. Before this, however, ideas of stereochemistry had been in existence for some time. Arising first from rather vague speculations, these rudimentary views became tinged with structural notions before they acquired the definiteness given them by van't Hoff and le Bel, and as the century closed they were finding application in chemical reactivity. All this happened before the revolution culminating in Werner had become very evident, and long in advance of any electronic interpretation. Several phases may therefore be detected, although they overlapped to a considerable extent.

1. RECOGNITION AND APPLICATION OF DIRECTED VALENCIES

Possibly the earliest appearance of a stereochemical idea appears in 1808 in some speculations of Wollaston:

I am further inclined to think that, when our views are sufficiently extended, to enable us to reason with precision concerning the proportions of elementary atoms, we shall find the arithmetical relation alone will not be sufficient to explain their mutual action, and that we shall be obliged to acquire a geometrical conception of their relative arrangement in all the three dimensions of solid extension.[4]

In similar vein, Gmelin, forty years later, wrote:

Nearly all chemists adopt the Atomic Theory. They determine the relative weights of the atoms, and their relative distances from one another.... Why, then, should we not likewise throw out suggestions with regard to their relative positions?[5]

Long after this, Wislicenus uttered the words:

The facts compel us to explain the difference between isomeric molecules possessing the same structural formula by the different arrangements of their atoms in space.[6]

One year later, in 1874, the tetrahedral theory was launched. It was, however, far from being the first reference to such a geometrical arrangement. For some time past there had been a series of tentative proposals that a carbon atom might have its valencies so disposed, and these came from approaches to two different problems.

The first moves began in Paris. In 1860, Pasteur, who had discovered much about the optical behaviour of the tartrates and devised three methods of resolving racemic mixtures,[7] delivered two lectures[8] to the Société Chimique de Paris in which recognition was accorded to the importance of molecular dissymmetry; here he stated that:

When the atoms of organic compounds are asymmetrically arranged, the molecular asymmetry is betrayed by the crystalline form exhibiting non-superimposable hemihedrism.[9]

Regarding the tartaric acids he speculated thus:

Are the atoms of the dextro-acid grouped on the spiral of a right-handed helix, or are they situated at the apices of an irregular tetrahedron, or are they disposed according to some other asymmetric arrangement? We do not know. But of this there is no doubt, that the atoms possess an asymmetric arrangement like that of an object and its mirror image.[10]

Important though this was, it did not involve recognition of a tetrahedral disposition of *valencies*, but of *atoms*. In 1860 the idea of a valency bond was not sufficiently developed for the former. But Pasteur's views along with those already quoted promoted the idea

1. *Edward Frankland*

2. *August Kekulé*

3. *Archibald Scott Couper*

4. *Alexander Crum Brown*

5. *Alexander William Williamson*

6. *William Odling*

of a three-dimensional molecule, and injected the new concept of tetrahedral groupings.

There is no doubt that the influence of Pasteur was keenly felt by both le Bel and van't Hoff during their work with Wurtz in Paris. It is in the paper of le Bel,[11] however, that this was most perceptible. Starting from the views of Pasteur on molecular dissymmetry, he argued in general terms that a molecule with four different groups attached to one carbon atom would be asymmetric and optically active, but if any two (or more) of these four became identical the asymmetry and optical activity would at once vanish.

This paper was clearly in the Pasteur tradition, with experimental facts interpreted spatially, and further speculation kept to a minimum. The argument is abstract and general, there are no diagrams and the tetrahedral hypothesis seems to be presented with some reserve. He was, for example, not prepared to draw from it the conclusion that in ethylene all six atoms would be coplanar.[12]

The paper by van't Hoff presents a rather different aspect. Its author assures us that he and le Bel when in Paris together "never exchanged a word about the tetrahedron there, though perhaps both of us already cherished the idea in secret". He adds that the idea occurred to him the previous year, in Utrecht, when reading the paper by Wislicenus[13] on lactic acid, and continues:

On the whole, le Bel's paper and mine are in accord; still the conceptions are not quite the same. Historically the difference lies in this, that le Bel's starting point was the researches of Pasteur, mine those of Kekulé.... My conception is, as Baeyer pointed out at the Kekulé festival, a continuation of Kekulé's law of the quadrivalence of carbon, with the added hypothesis that the four valencies are directed towards the corners of a tetrahedron, at the centre of which is the carbon atom.[14]

Certainly the tetrahedral disposition of valencies is more prominent here than in le Bel's paper, and is illustrated with great clarity. After commenting on the inability of "present constitutional formulae" to explain certain cases of isomerism, he concluded:[15]

Theory is brought into accord with facts if we consider the affinities of the carbon atom directed towards the corners of a tetrahedron of which the carbon atom itself occupies the centre.

This explains the non-isomerism of compounds of the types $CH_2R'_2$, $CH_2R'R''$ and CHR'_2R'', as well as the existence of two isomers when a carbon atom bears four different substituents. By assuming that double bonds can be represented by two tetrahedra with a common edge, he was able to explain the isomerism of maleic and fumaric acids, and similar pairs of geometrical isomers (see Plate 10).

Now on van't Hoff's own showing, this theory owes much to Kekulé. What then is the extent of the debt? The admission by van't Hoff rather implies that his rôle was to unite the ideas of valency (from Kekulé) and tetrahedra (from Pasteur). It is probable, however, that at least the germ of the tetrahedral hypothesis also can be traced to Kekulé. Now it has been already mentioned that Kekulé[16] had introduced a three-dimensional atomic model, an improvement on the planar one used by Dewar (see p. 107). It is likely that this tetrahedral model may have exerted an unconscious influence on van't Hoff who, before his stay in Paris, had spent a year with Kekulé at Bonn. Anschütz, shortly afterwards an assistant at the same laboratory, refers to this model as follows:

In his Bonn lectures he always used it for the explanations of the binding-relationships of the carbon atoms in saturated and unsaturated hydrocarbons, isomerism of paraffins and of paraffinic alcohols, that between acetaldehyde and ethylene oxide, benzene, and so on.
It may even have been thus that the young van't Hoff received, from the teacher he admired, the first impression that led him to the development of the theory of asymmetric, tetrahedral carbon; historically, more than this cannot be established.[17]

Kekulé's atomic model was originated to overcome the inability of planar arrangements to indicate triple bonding, and it seems hard to understand why a model so literally conceived should not have led its designer – or his students – to van't Hoff's conclusions several years earlier. A reason for this will be suggested in the next section, but meanwhile its probable influence on van't Hoff is noted.

The immediate applications of this theory were in the realm of optical isomerism, and here it was at once successful. Its application to geometrical isomerism followed rapidly and the course of these events has been recently reviewed.[18] Following the application to maleic and fumaric acids and the crotonic and chlorocrotonic acids by van't Hoff,[19] the view received powerful support from Wislicenus, who amongst other achievements showed that maleic and fumaric acids were respectively *cis*- and *trans*-isomers:[20]

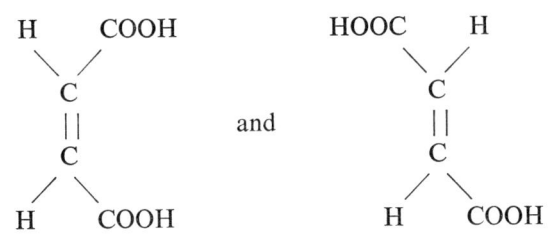

$$\begin{array}{ccc} \text{H} \quad \text{COOH} & & \text{HOOC} \quad \text{H} \\ \diagdown \diagup & & \diagdown \diagup \\ \text{C} & & \text{C} \\ || & \text{and} & || \\ \text{C} & & \text{C} \\ \diagup \diagdown & & \diagup \diagdown \\ \text{H} \quad \text{COOH} & & \text{H} \quad \text{COOH} \end{array}$$

Extension of these ideas to nitrogen was slow, doubtless because the low energy barrier between the two stereoisomers of Nabc prevented their separation – and thus the recognition of the pyramidal nature of trivalent nitrogen. An interesting but premature suggestion of a nitrogen atom oscillating above and below a plane containing three substituents was made by Walter[21] in 1873; this was part of a proposal to explain the varying valency of phosphorus in dynamic terms, and also involved a precognition of the trigonal–bipyramidal arrangement of the pentachloride. But the first direct evidence for directed valencies in nitrogen came with the isolation of two[22] and then three[23] dioximes of benzil. In 1890, Hantzsch and Werner proposed *cis/trans* isomerism involving the nitrogen might be responsible,[24] a suggestion made more probable by the proof by Goldschmidt[25] of the structural identity of the two benzaldoximes discovered by Beckmann.[26] The case of the diazo-compounds was considered by Hantzsch in 1894.[27] Optical resolution of a quaternary ammonium compound was effected in 1899 by Pope and Peachey of allylbenzylmethylphenylammonium iodide.[28]

The modern formulae following illustrate the main points of the previous paragraph:

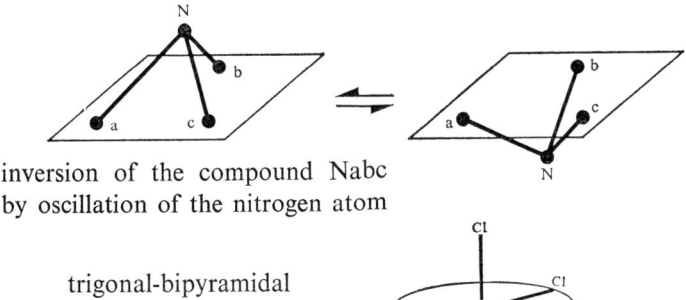

inversion of the compound Nabc by oscillation of the nitrogen atom

trigonal-bipyramidal structure of phosphorus pentachloride

the three dioximes of benzil

two benzaldoximes   isomeric diazocompounds

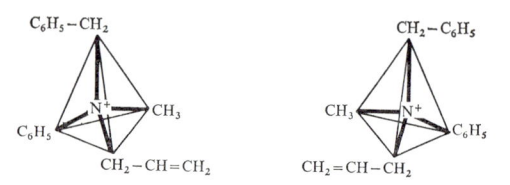

two forms (enantiomorphs) of allylbenzylmethylphenyl-ammonium iodide (the iodide ion not being shown)

The other main application of the concept of directed valencies came in connection with reactivities. The famous "strain theory" of Baeyer will be discussed later; it seems to have owed at least as much to Kekulé as to van't Hoff.[29] As an explanation of the heightened reactivity of olefins and of the stability of 5- and 6-membered rings, it enjoyed success for some years. One of Baeyer's students, Victor Meyer, accounted for the low reactivities of certain *ortho*-disubstituted benzoic acids in terms of steric hindrance.[30] This again drew attention to the importance of a knowledge of positions of atoms in space.

With all this work, however, the reception accorded to the new views was by no means universally cordial. Scepticism was followed

2. OPPOSITION TO DIRECTED VALENCIES

In the early days of the Structure Theory considerable confusion existed between the structural and steric significance of the formulae in use. It thus became almost a commonplace for authors to deny that their structural formulae implied anything of the spatial dispositions of the atoms. This was no doubt behind Crum Brown's distinction between "physical position" and "chemical position";[31] Frankland's *Lecture Notes* warn the reader that "The graphic formulae are intended to represent neither the shape of the molecules nor the relative positions of the constituent atoms",[32] and in the Preface he suggests that knowledge of atomic constitution is worth the risk of such spatial misinterpretations.[33] Many other examples of this kind of warning could be given.[34]

While it is easy to understand a natural desire not to go beyond the limits of legitimate theory, the reluctance of chemists to accept directed valencies after van't Hoff's paper is hard to explain. Yet opposition there certainly was, and, more important than that, a general aloofness and reluctance to apply the tetrahedral hypothesis. This is evident from the quite slow progress made in the last quarter of the century. Some of this was, of course, a result of experimental inaccuracy but other factors may be discerned in definite opposition to the van't Hoff theory.

The first of these to appear seems to have been a manifestation of the common antipathy to the old *Naturphilosophie* which was thought to have tainted the new views. This was probably behind Kolbe's strong reactions to not only the Structure Theory but also this extension of it to three dimensions. His objection is one of his best-known pieces of invective:

A certain Dr J. H. van't Hoff, of the veterinary school of Utrecht, finds, so it seems, no taste for exact chemical investigation. He has thought it more convenient to mount Pegasus (obviously loaned by the veterinary school) and to proclaim in his *La Chimie de l'Éspace* how, during his bold flight to the top of the chemical Parnassus, the atoms appeared to have grouped themselves through universal space.[35]

Fear of a revival of this philosophy may have been quite common among some of the German chemists, particularly among those who had been students of Kolbe, upon whom he seems to have exerted a strong influence. Amongst these was Claus, whose objections enshrine probably the most deep-seated of all reasons: fear of a conflict between chemical and physical laws. This second factor has been well described in a recent review of the tetrahedral hypothesis by Sementsov:

The physics of that period asserted that the direction of an attractive force is determined by the positions of the attracted bodies, and cannot have a direction independent of the position of the attracted body.[36]

In accordance with this, Claus opposed not only the tetrahedral disposition of valencies but "the assumption that the valency of a polyvalent atom of such a kind should be regarded as a single force divided *a priori* in the atoms present".[37]

A denial of valencies altogether was the logical outcome of Claus's speculations, and this indeed happened (see p. 201). Others expressed similar sentiments, *e.g.*:

We cannot assume that valencies act across empty space, free of atoms; it is only possible on paper or in a model where there are lines or wires but not forces.[38]

Probably this respect for physics accentuated Kekulé's caution about structural matters generally to such a level that he was unable to use his own model to explain the very cases of geometrical isomerism that were the subjects of his own experiments. Nothing else seems adequate to explain his life-long bewilderment about the cases of maleic and fumaric acids (*cf.* p. 227).

A third factor militating against the ready acceptance of van't Hoff's theory was its treatment of multiple bonds. Lossen could not see how the tetrahedral model could explain the structure of ethylene, and concluded that "Polyvalent atoms cannot in general be represented as material points" but must have different kinds of area over their surfaces.[39] He had earlier challenged van't Hoff's representation of acetylene in terms of tetrahedra on the grounds, apparently, that it was a cumbrous way of saying the molecule was linear.[40] On the subject of multiple bonds le Bel also had difficulty in accepting the tetrahedral theory; he did not admit the planarity of ethylene until 1882,[41] and ten years later changed his mind again.[42]

Another factor was crystallographic in nature. This seems to have been the ostensible reason for le Bel's attitude, for he had found that carbon tetrabromide and tetraiodide formed biaxial rather than cubic crystals,[43] and from this concluded the carbon was at the apex of a square-based pyramid, on which basis he attempted the fruitless resolution of citraconic and mesaconic acids;[44] the tetrahedral hypothesis would have told him this was impossible (*cf.* p. 161). In 1926 le Bel's pyramidal theory was revived by Weissenberg in an attempt to explain the crystalline form of certain derivatives of pentaerythritol.[45]

These then were the main or subsidiary causes for the hostility to the tetrahedral model. To overcome the difficulties, especially that of conflict with physical laws, other alternatives were suggested. Instead of interpreting molecular dissymmetry in terms of directed valencies, some suggested the cause lay in the shape of the atoms themselves. Amongst these are included the ideas of Wunderlich,[46] Knoevenagel,[47] Knorr,[48] Auwers[49] and others. The conception of zones was suggested by le Bel (each atom having a spherical zone of repulsion),[50] and Werner (see p. 217).

Nevertheless, the tetrahedral model gradually gained ground, and (as Sidgwick says[51]) was finally vindicated by the resolution by Mills and Warren[52] of a molecule whose optical activity was inexplicable on any other theory advanced (including Werner's); only a tetrahedral atom (nitrogen in this case) was capable of explaining the facts:

PART THREE

Variations in Valency

CHAPTER X

The Problem of Variable Valency

THE NEXT THREE CHAPTERS deal with problems encountered almost as soon as the idea of valency was becoming generally accepted. In the last years of the nineteenth century, and the first years of this, the principal difficulty in the field of valency was its extension to *all* types of compound. With the vast majority of aliphatic and alicyclic compounds no problem arose in deriving formulae from the premiss of tetravalent carbon, monovalent hydrogen, *etc.* But the theory ran into serious difficulties with cases where the conventional values would not fit the facts. Basically the issue became whether or not an atom could have variable valency. In the case of the inorganic compounds with which this chapter is concerned the issue was quite clear cut, and disputants knew precisely the nature of the controversy: fixed *v.* variable valency. Organic chemistry, otherwise a unified triumph for valency, was yet troubled with two problems whose nature eluded a simple analysis. The phenomenon of unsaturation could be defined in terms of a variable valency or it could be stated in a different way which would involve new issues altogether, particularly those now part of stereochemistry. Finally there existed another group of substances, known as "aromatic", no adequate explanation of which existed until the advent of wave mechanics. These two groups are considered in the following two chapters.

Now the problems raised by these classes of compounds were of great interest in themselves. But their importance lay beyond any intrinsic fascination they might exert. It was very largely through a study of these matters that new conceptions of the nature of valency arose, and, as is often in science, prolonged examination of essentially familiar data led ultimately to a modification of original ideas and the recognition of new facts and the opening up of vistas of unimagined depth and variety.

The issue of variation in valency is nearly as old as the concept itself. By the middle of the 1860s it was the subject of lively interchanges, and it remained an issue for many years after that. Chief amongst the upholders of a fixed unvarying valency was Kekulé, supported at least for a time by Carey Foster, Wichelhaus and others. Opposed to this view were many others, amongst them Naquet (an early protagonist), Wanklyn (whose proliferation of valency states was the least inhibited) and Michaelis (whose thrusts seem to have been the most penetrating). On the simple issue of whether an atom could act with more than one valency (saturation capacity), opinion eventually decided that it could, though Kekulé remained unconvinced all his life. But on the subsidiary questions that arose from the discussion a great deal more was to be heard before any measure of unanimity was reached.

From a viewpoint inevitably coloured by the findings of modern chemistry it is extremely difficult to see how the doctrine of unvarying valency could ever have been held, still less maintained over a long period. In the 1860s, however, inorganic chemistry was not filled with the variety of compounds whose existence today quantum theory and its derivative hypotheses make at least credible. Sulphur, nitrogen and phosphorus provided the chief examples of elements for which several valency states appeared likely, but the emphasis on organic chemistry which we have already noticed would focus attention far more upon the fixity of the valencies of the elements involved there. Furthermore most chemists were still thinking in forms not far removed from those of the Type Theory, and in this fact we have probably the chief explanation for the persistence of the concept of constant valency. That this is so may be inferred from the general persistence of the theory (see p. 142), and also from a few comments made near the time. In 1869, for example, Buff wrote that a knowledge of combining laws led to "the law of absolute atomicity which may be regarded as a residue of the rigid types of a view that once prevailed".[1] Ten years later Schorlemmer admitted "it is easy to see that the theory of types must have led to the conception of a constant atom-fixing power".[2]

In fact one can urge that, had the electrochemical theory possessed more influence, it would have fostered the idea of a variable valency. The ease with which its exponents added or removed water, for example, outside a copula would have suggested a less rigid view of combining power. Possibly Kekulé saw this, and regarded variation of valency as an evil legacy from dualism.

The last half of the nineteenth century saw a large growth in the number of inorganic compounds known, and this began to be appreciable in the 1860s. In particular the chemistry of the transition metals came under closer scrutiny, and developments in this field brought to light many new compounds containing vanadium, niobium, tantalum and other metals, all of which show to a marked degree a multiplicity of valency states. The need to explain the existence of these substances intensified the disputes and gave added point to them. More familiar cases such as the oxides of carbon and nitrogen, the hydrated salts and some other anomalies were already waiting for explanation.

Before discussing the course of events in detail, it will be well to state briefly the origin of the view of fixed valency, especially as some confusion arose in the early years as to the source of this idea.

In 1862 Erlenmeyer was reviewing Erdmann's book *Ueber das Studium der Chemie* in instalments of his *Zeitschrift für Chemie*. In the course of his remarks he said:

According to the atomic theory, to put it shortly, all atoms possess an unalterable weight. If now we make the further assumptions,
(1) that in each atom the number of affinities contained is also equally unalterable (G. C. Foster),
(2) that not only the affinities of unlike but also those of like atoms can join with each other in compounds (Kekulé),
we can explain by these the law of multiple proportions for all bodies.[3]

The reference to Foster is presumably to a statement of his in Wurtz's *Répertoire de chimie pure*, made in a review of the famous paper of Griess on "New compounds produced by the substitution of nitrogen for hydrogen".[4]

Deprecating the concept implied in the last clause of the title, Foster said in an extensive footnote: "The idea that atomicity, or capacity for combination, is a property of elementary atoms as fixed and as unalterable as their weights seems to us the only one that, in the actual state of science, can serve as a basis for a general theory of chemical combinations."[5]

Foster's name features in a paper by Wurtz a little later. Speaking of atomicity, he asked permission "to develop the sense that I attach to this definition differently from that given by other chemists, notably M. Foster and M. Kekulé".[6]

Nevertheless, Foster did not seem to regard himself as the originator of the idea, for in a letter to Kekulé he suggested that for Erlenmeyer to mention his name in the review cited above "was very insane of him".[7] Anschütz considers that Foster knew quite well that the view of fixed valency began with Kekulé, and this appears likely.[8] But there does not seem to be a *direct* statement by Kekulé before 1864 to confirm such an assertation, though fixity of valency is implicit in all that he wrote. However, in 1864 he expressed himself very directly in *Comptes rendus*. Earlier that year Naquet had issued a statement distinguishing "substitution value" from "saturation maximum", and proposing that sulphur, selenium and tellurium should be regarded as tetravalent. As a tentative model for an atom he suggested the following:

If one supposes, for example, the atoms are endowed with certain hooked extensions for the purpose of linking up with similar hooked extensions of other atoms in order to produce combinations, it is evident that the number of hooks carried by one atom of a body will represent the atomicity of the latter absolutely. If the further supposition is now made that the hooked extensions are apt to unite only indistinctly with the similar extensions of all bodies, it is conceivable that the atomicity shown, or the relation of a radical, can sometimes stay below its absolute or true atomicity.[9]

Kekulé's reply to this was immediate and direct. It put his readers in no doubt of his position and left an impression on chemical thought for many years, even though its reception was by no means favourable. For this reason quotation of its opening paragraphs will be appropriate:

Several memoirs recently published, among them a Note presented by M. Naquet to the Academy in one of its later sessions, tend, it seems to me, to throw a certain confusion on the theory of the atomicity of the elements. I think it my duty to intervene in the debate, so much the more as it was I, unless I am mistaken, who introduced into chemistry the idea of the atomicity of the elements. I shall therefore expound as concisely as possible several fundamental ideas of this theory, and especially those which seem to me to be appropriate for a clarification of the points under dispute.

It has been known for a long time that elementary bodies combine according to the laws of constant and multiple proportions. The first of these laws finds a perfect explanation in Dalton's atomic theory; the second is explained only generally and rather vaguely by that theory. What Dalton's theory only explains is why atoms of different elements combine in certain ratios with each other. I think I have explained the whole body of these facts by what I have called the *atomicity of the elements*.

The theory of atomicity is therefore a modification which I had believed could be contributed to Dalton's atomic theory, and, as I regard it, can be understood so that atomicity is a fundamental property of the atom, which should be as constant and invariable as the atomic weight itself.

To wish to admit that atomicity could vary, and that one and the same body could function now with one atomicity and now with another, is to use the word in a totally different sense from that which I gave when I proposed it; this is to confuse the notions of atomicity and equivalence. No one doubts any more that one and the same body, the same element, could be able to function with different equivalents. The equivalence can vary, but not the atomicity. Variations in the equivalent, on the other hand, should be explained by the atomicity.[10]

Kekulé continued with an attempted *reductio ad absurdum* of his opponents' arguments. If iodine in iodine trichloride is assumed to be "triatomic", this means that phosphorus in its tri-iodide must have an atomicity of 9; moreover the similarity between the two halogens implies an identity of valency, and so chlorine must also be taken as triatomic, which gives iodine in ICl_3 a value of 9, and so on. The rest of the paper is concerned in advocating molecular compounds, and this will be returned to later. Meanwhile it is worthy of note that the next five years saw a succession of papers combating Kekulé's views, one of the first being a reply by Naquet.

Dealing chiefly with the issue of "molecular compounds", it also vigorously rebutted the argument just outlined about the valencies in iodine trichloride. He made the perfectly fair comment that Kekulé had forgotten "that elements can enter into the compounds they form with diverse substitution values",[11] and that *if* there is variation in valency the case is self-consistent, and arguments like these are without point. Such a presentation "Kekulé's opponents might justly regard as a travesty of their views".[12]

There is a serious defect in the view expressed here by Kekulé that does not appear to have received attention before, probably because no major review of his work has appeared since Anschütz's biography of 1929, when material was first published that bears strongly on this. In the "Geschichte der Valenztheorie",[13] appearing then for the first time, the burden of his criticism of Kolbe and Frankland was that they had confused the difference between molecules and equivalents, and even between atoms and equivalents.

Consequently they had not recognized the true meaning of the expression,
$$\text{valency} = \frac{\text{atomic weight}}{\text{equivalent}}$$
Yet in the paper just quoted he stated that "the equivalent can vary, but not the atomicity". Now if this expression has the importance attached to it that Kekulé implies, all should have seen that (given constant atomic weight) the "atomicity" *must* vary with the equivalent. The point is, of course, that the equation above cannot be universally applied, but in that case some of his strictures on Frankland and Kolbe lose their force. It was impossible to have it both ways.

It was with this memoir of Kekulé's that the issue became generally discussed and assumed a prominence that it is hard to overestimate. As time progressed the terms of the problem became more complex. Initially the question was to explain the existence of NH_3 and NH_4Cl, of PCl_3 and PCl_5, and similar pairs of compounds. And to these were soon added cases like VCl_3 and VCl_4, which, though at first sight parallel examples, could not be discussed in anything like the same terms. Later still emerged large numbers of ammines and hydrates upon which classical valency, as we may call it, was powerless to give a convincing statement. But all of these – and many others – may be condensed into the basic question: *how may one explain the existence of more than one inorganic compound between two elements or radicals?*

Answers to this question may be classified into the following groups:

(1) Theories involving fixed valencies only.
(2) Molecular compound theory.
(3) Admission of variable valency.
(4) Conception of auxiliary valencies.

Each of these will be considered in turn.

1. THEORIES INVOLVING FIXED VALENCIES ONLY

The point of view expressed by Foster and Kekulé was not without adherents in the decade following 1864. Further, many reacted against a tendency to allocate a new valency value each time a compound arose that was not immediately explicable in terms of the old one, even though these people would not go as far as Kekulé and assert any variation to be absolutely impossible. On the other hand the idea of "molecular compounds" did not commend itself to them as a universal answer to the problem.

Fortunately for these chemists it was possible in some cases both to preserve an accepted figure for the valency of an element and at the same time to keep clear of "molecular compounds". For inorganic compounds three ways of doing this were possible, using the hypotheses of polymerization, chain formation and valency pairing.

(i) *Polymeric Theory*

If the empirical formula of a compound implied that an atom had a lower valency than usual, it might be possible for the anomaly to disappear if several units corresponding to this empirical formula were joined together, using the "missing" valencies to do this. The resultant "polymer" would have to be justified on other grounds, of course, and of these the most probable would be vapour density, if the substance was sufficiently volatile.

Some of the early examples were unable to survive this test. Thus in 1871 Sestini suggested that ammonia and nitric oxide might be written as follows[14] in order to preserve the pentavalence of nitrogen:

$$H_3N=NH_3 \quad \text{and} \quad O-N\equiv N-O$$

Monovalent oxygen does not seem to have troubled him in the latter case, but he was aware that such doubling was inconsistent with vapour density data. By a curious distinction between "physical" and "chemical molecules" he satisfied himself on the matter, but few others were convinced.

An interesting example where vapour densities were an issue again was mercurous chloride. The familiar dimeric formula Hg_2Cl_2 had been proposed before the fixity of valency became a controversial topic; it was advocated by Kekulé on its analogy with ethylene chloride, at the Karlsruhe conference of 1860.[15] This being accepted, it could be easily explained in terms of divalent mercury atoms:

$$Cl-Hg-Hg-Cl$$

Vapour density measurements, however, showed this dimerization was incorrect.[16] Kekulé supposed that it decomposed into 2HgCl immediately on vaporization,[17] but Williamson admitted that he was worried by the behaviour of both this and nitric oxide "which are to me simply unintelligible". On the occasion of this remark, however, he was able to add that Odling had just discovered that gold leaf held in the vapour became amalgamated.[18] Thus instead of supposing a dissociation of the type

$$Hg_2Cl_2 = 2HgCl,$$

one could infer

$$Hg_2Cl_2 = HgCl_2 + Hg,$$

so avoiding the concept of monovalent mercury.

In some cases, however, the compound was not volatile enough for vapour density measurements. Thorpe, anxious to preserve the triatomic nature of vanadium in Roscoe's newly discovered compounds VOCl and VOCl$_2$,[19] doubled their formulae thus:

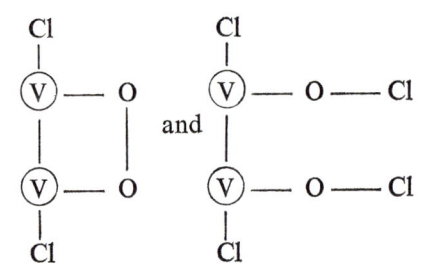

$$\begin{array}{cc}
\begin{array}{c} Cl \\ | \\ V - O \\ | \\ V - O \\ | \\ Cl \end{array}
& \text{and} \quad
\begin{array}{c} Cl \\ | \\ V - O - Cl \\ | \\ V - O - Cl \\ | \\ Cl \end{array}
\end{array}$$

remarking, "so far as I am aware there is nothing to disprove such a method of representation; it has at least the merit of preserving the triatomic nature of vanadium in their compounds, and shows in a simple manner relation to vanadyl trichloride".[20]

A curious example of this approach is the case of ferrous chloride. At an early stage the formula of this was doubled,[21] thereby overcoming the discrepancy between the valency of iron in ferrous and ferric chlorides. Now a precedent for a doubling of the formula had existed in this higher chloride; Deville and Troost had shown that two volumes of ferric chloride vapour contained six volumes of chlorine. Hence one molecule of ferric chloride, in the vapour state, contained six atoms of chlorine, and must therefore be taken as Fe$_2$Cl$_6$.[22] A similar result was obtained by Scheurer-Kestner on

THEORIES INVOLVING FIXED VALENCIES ONLY 179

basic ferric acetates.[23] The conclusion as to the valency of iron reached by Wurtz was that the double atom Fe_2 must be hexavalent.[24] Friedel was unable to see how the overall valency could have this value, and suggested that each iron atom was tetravalent, as in iron pyrites, FeS_2.[25] But if this were accepted, one now had ferrous iron trivalent and ferric tetravalent, and the original difficulty of differing valencies remained. If, however, a double bond was conceded between the iron atoms in the lower chloride, a uniform tetravalency would be established. By this time, however, so many hypotheses had entered in that any degree of certainty was small. Thus Frankland's *Lecture Notes* of 1870 contains formulae corresponding to all possibilities, which may be reproduced in modern notation thus:[26]

$$\text{Cl-Fe-Cl} \quad \begin{array}{c} \text{Cl} \ \ \text{Cl} \\ | \ \ \ | \\ \text{Fe=Fe} \\ | \ \ \ | \\ \text{Cl} \ \ \text{Cl} \end{array} \quad \begin{array}{c} \text{Cl} \ \ \text{Cl} \\ | \ \ \ | \\ \text{Cl-Fe} \equiv \text{Fe-Cl} \\ | \ \ \ | \\ \text{Cl} \ \ \text{Cl} \end{array} \quad \begin{array}{c} \text{Cl} \ \ \text{Cl} \\ | \ \ \ | \\ \text{Cl-Fe-Fe-Cl} \\ | \ \ \ | \\ \text{Cl} \ \ \text{Cl} \end{array}$$

But in the earlier edition only the first and third were recorded;[27] at this time variable valency was less of an issue. Later still the monomeric nature of ferrous chloride was shown by cryoscopic methods[28] and by vapour density measurements at 1400°.[29]

In general, then, it may be said that polymeric formulae were not successful in justifying constancy of valency; and it may be added that where vapour density observations did necessitate them, as with acetic acid,[30] aluminium chloride[31] and others, the problems of valency that they raised were greater than those which the early proponents of polymeric formulae had attempted to solve.

(ii) *Chain Theory*

It will be noticed that the polymeric formulae were derived by joining two atoms of the same kind together. They were in a sense a special case of a larger group of formulae in which atoms were linked in long chains rather than in clusters involving some of them with unacceptably high valencies. There was, however, no question in the following examples of arbitrary doubling of empirical formulae; where the latter were doubled (as in H_2O_2 and S_2Cl_2) there was plenty of evidence supporting this. The following examples

(a) *Phosphorus*

Compounds of phosphorus were extensively studied in the last half of the previous century, largely owing to the use of the powerful reagents phosphorus pentachloride and pentoxide. These were readily accessible and opened up several new fields of discovery in organic and inorganic chemistry. The pentahalides were hailed by Kekulé as molecular compounds (p. 190), but it was not necessary to suppose this was true of some of the oxygen compounds where chain formulae could be constructed so as to preserve a nominal valency of three.

In 1868, Wichelhaus had established to his own satisfaction the trivalency of phosphorus in ethyl phosphite (or "phosphorous acid ethyl ether"),[32] and in the "superchloride", which he regarded as a molecular compound.[33] It therefore followed that the oxychloride, $POCl_3$, must be written not as

$$O=P\begin{array}{c}Cl\\ -Cl\\ Cl\end{array} \quad \text{but as} \quad P\begin{array}{c}O-Cl\\ -Cl\\ Cl\end{array}$$

Although his initial axiom was rejected by some (p. 190), others found these convincing to a degree, but lacking in independent experimental proof. One of these was Thorpe, who, as has been mentioned, was anxious to preserve the trivalency of vanadium and had therefore had recourse to dimeric formulae for two of the new oxychlorides (p. 178). He wrote the well-known $VOCl_3$ in a manner similar to that for its phosphorus analogue:[34]

$$\begin{array}{c}Cl\\ |\\ \text{\textcircled{V}}-O-Cl\\ |\\ Cl\end{array}$$

In 1875 he attempted to obtain independent physical evidence for or against this view of phosphorus oxychloride. His arguments rested on Kopp's generalization[35] that the specific volume of oxygen in a compound depends upon its state of combination. The increment in the total specific volume for doubly bound oxygen (as in acetyl) was 12.2, but for singly bound oxygen (as in alcohol) it was 7.8. A similar situation arose with sulphur, so Thorpe applied the idea to $POCl_3$ and also $PSCl_3$, and found the specific volumes for oxygen and sulphur in these "almost identical with the values given by Kopp for these elements when without the radical" (*i.e.*, when singly linked). Consequently he considered that the formulae $PCl_2(OCl)$ and $PCl_2(SCl)$ were established.

The pentoxide of phosphorus could also be written in a chain formula, thus,

$$O=P-O-O-O-P=O$$

But such formulae provoked some scepticism and the inevitable query as to why the chain should be limited to this length, and why P_2O_6, P_2O_7 and still higher oxides should not exist.[36] However, this particular issue was shortly to die as evidence accumulated rapidly against the molecular compound theory for phosphorus compounds, and against the element's unvarying trivalency (p. 191).

With the recognition of pentavalent phosphorus in the 1870s there still remained the question as to whether it existed in compounds capable of representation in two ways, as phosphorus oxychloride. This last remained an open question for many years, but in other cases this was not always so. For example there was the case of triphenylphosphine oxide, $(C_6H_5)_3PO$, which could be formulated in two ways corresponding to the two projected structures for the oxychloride:

$$O=P\begin{smallmatrix}\diagup C_6H_5\\-C_6H_5\\\diagdown C_6H_5\end{smallmatrix} \quad \text{or} \quad P\begin{smallmatrix}\diagup O-C_6H_5\\-C_6H_5\\\diagdown C_6H_5\end{smallmatrix}$$

(I)   (II)

Now, in 1885 Michaelis and la Coste treated phenol with diphenylchlorophosphine producing a compound that had to have structure II:[37]

G

$$\begin{array}{c}\text{Cl} + \text{H}\!-\!\text{O}\!-\!\text{C}_6\text{H}_5 \\ \text{P}\!-\!\text{C}_6\text{H}_5 \\ \text{C}_6\text{H}_5\end{array} \longrightarrow \begin{array}{c}\text{O}\!-\!\text{C}_6\text{H}_5 \\ \text{P}\!-\!\text{C}_6\text{H}_5 \\ \text{C}_6\text{H}_5\end{array} + \text{HCl}$$

The product, a liquid, was different from the triphenylphosphine oxide that Michaelis et al. had made earlier,[38] for this was a solid, which must be isomeric and thus possess structure I.[39] In this way, spurious chain structures were gradually removed from chemistry.

(b) *Nitrogen*

Although similar to phosphorus in many ways, nitrogen did not cause the same kind of difficulties. It is easy to see that this must have been so, since nitrogen is unable to be pentacovalent and form compounds analogous to PCl_5, $POCl_3$, P_2O_5, *etc.* On the other hand its ammonium compounds were *par excellence* typical members of the group that came to be termed "molecular compounds".

Yet chain formulae were allocated to some nitrogen compounds, notably to those now known as azo-compounds and diazonium salts. Kekulé was chiefly responsible for the establishment of these formulae in 1866; he wrote azobenzene and "*diazobenzolbromid*" (benzene diazonium bromide) as, respectively,[40]

$$(C_6H_5)\!-\!N\!=\!N\!-\!(C_6H_5)$$
$$\text{and } (C_6H_5)\!-\!N\!=\!N\!-\!Br$$

In these cases subsequent events have been less hostile to these formulae than to those of the phosphorus derivatives. The first is now accepted as it stands, and the second is applied not to the diazonium salt but to the isomeric diazobromide. For the diazonium salts Blomstrand proposed a different formula[41] which is no longer a continuous chain:

$$e.g., C_6H_5\!-\!N\!-\!Br$$
$$|||$$
$$N$$

approaching closely to our conception with ionic bromide. Kekulé could not accept Blomstrand's suggestion because of its assumption of pentavalent nitrogen. Equally, he was compelled to write the symmetrical structure for the azoxy compounds[42] instead of the unsymmetrical one proposed later, involving again a disparity in valencies of the nitrogen:[43]

e.g., $C_6H_5-\underset{\diagup\diagdown}{\underset{O}{N-N}}-C_6H_5$ instead of $C_6H_5-N=\overset{\overset{O}{\|}}{N}-C_6H_5$.

(c) *Sulphur*

The birth of the theory of valency was closely associated with the assumption by Kekulé that sulphur was "bibasic". This was assumed in his paper in 1854 on thiacetic acid,[44] when compounds under discussion were referred in effect to a hydrogen sulphide type.

In the 1857 paper on "The so-called copulated compounds",[45] a great deal of the argument rests upon the concept of a divalent group, as SO_2, capable of holding together two monovalent radicals. This idea is elaborated further in the 1858 paper "On the constitution and metamorphoses of chemical compounds", and the sulphuryl group itself is analysed: "The radical of sulphuric acid SO_2 contains 3 atoms, each of which is diatomic", and which he wrote then as[46]

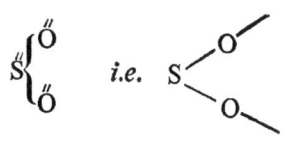

$$S\begin{Bmatrix}\overset{\|}{O}\\\underset{\|}{O}\end{Bmatrix} \quad i.e. \quad S\begin{matrix}\diagup O \diagdown \\ \diagdown O \diagup\end{matrix}$$

Thus committed to a diatomic sulphur atom, and bound by his own doctrine of unchangeable valency, Kekulé was forced to write all sulphur compounds with divalent sulphur, and this he did for the rest of his life. At the end of it he seems to have been the only chemist doing so. However, at first he had some support, particularly at Bonn, and formulae like the following were accepted for a time:

Cl—O—S—O—Cl   H—O—O—S—O—O—H
(Kekulé[47])   (Kekulé[48])

Cl—S—O—O—OH C_6H_5—S—O—O—O—H
(Müller[49])   (Barbaglia and Kekulé[50])

K—O—O—O—S—S—S—O—O—O—K
(Spring[51])

Now, these are, or are supposed to be, structural formulae. As such they must be confirmed or refuted by arguments based, not now upon vapour densities, but upon more directly chemical reasoning. Thus Blomstrand argued against the chain formula for sulphuric acid in 1869,[52] and others did the same. But nowhere has a more sustained and trenchant attack been found than in the writings of Michaelis.

Having made for the first time sulphur tetrachloride (from chlorination of the monochloride at low temperatures), and shown it to be a genuine chemical compound, Michaelis used the occasion to advantage, pressing home the variability of the valency of sulphur.[53] He had just shown it could be four, but Kekulé assumed it to be two, even in sulphuric and benzenesulphonic acids. Now if this is true, two kinds of arrangements are possible, *viz.*, with the carbon joined to oxygen (as in C_6H_5–O–S–O–OH) or to sulphur (*i.e.*, C_6H_5–S–O–O–O–H). One like the former had been suggested by Barbaglia,[54] on the basis of the action of phosphorus pentachloride on potassium benzenesulphonate, but had been queried by him later since benzyl disulphide (C_6H_5–S–S–C_6H_5) was oxidized by nitric acid to benzenesulphonic acid, thereby implying a C–S link.[55] Michaelis eliminated such formulae as the first one above by citing the reduction of benzenesulphonic acid to thiophenol instead of phenol.[56]

He then proceeded to show the untenability of the second linear formula. If it is accepted, we are committed to writing sulphuric acid as H–O–S–O–O–O–H, and therefore sulphuryl chloride appears as Cl–S–O–O–Cl. But such a representation is against the facts. Compounds having a group –O–Cl are unstable (hypochlorites, *etc.*), and sulphuryl chloride is not. The same considerations rule out Cl–O–S–O–Cl. Moreover, sulphuryl chloride is made by direct combination of sulphur dioxide and chlorine in sunlight, similar conditions being needed for phosgene from carbon monoxide:

$$SO_2 + Cl_2 = SO_2Cl_2$$
$$CO + Cl_2 = COCl_2$$

Now if sulphur is divalent the molecule SO_2 is of the kind O—O and would therefore have to break up under these mild conditions, which is surely unlikely. It is easier to regard it as unsaturated in the same sense as carbon monoxide, and to allow sulphur a hexavalent state in SO_2Cl_2.

Four years later Michael and Adair produced additional evidence for the hexavalency of sulphur, and against current formulations in terms of chains of atoms. This was in connection with the sulphones. If Ar and Ar' represent two different aromatic hydrocarbon radicals, the two sulphones predicted on the chain theory would be thus:

$$\text{Ar—S—O—O—Ar'} \quad \text{and} \quad \text{Ar'—S—O—O—Ar}$$

A symmetrical formula (Ar–O–S–O–Ar') would presumably be excluded by the method of synthesis (ArSO$_3$H + HAr → ArSO$_2$Ar + H$_2$O).

Michael and Adair attempted syntheses of several mixed sulphones by the action of a sulphonic acid on a hydrocarbon in the presence of phosphorus pentoxide. They found that one product separable from the mixture from the action of benzenesulphonic acid on naphthalene was identical with that from benzene and naphthalene-β-sulphonic acid.[57] Again, identical products were formed from benzene with *p*-toluenesulphonic acid and from toluene and benzenesulphonic acid.[58] They concluded that, although molecular compounds were a remote possibility, all the evidence suggested that the sulphones were based upon hexavalent sulphur, and in no sense could they be allowed a chain structure. With this, of course, present theories agree:

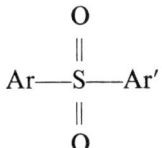

$$\text{Ar—}\underset{\underset{O}{\|}}{\overset{\overset{O}{\|}}{S}}\text{—Ar'}$$

(d) *Oxygen*

Although the controversy raged for a long time over variation of the valency of oxygen, it never assumed the importance of phosphorus or sulphur. Chains of oxygen atoms certainly were employed in formulae, but in order to avoid admitting an increase in the valency of other elements rather than oxygen itself. One exception exists that is of some interest. Despite the advocacy of tetravalent oxygen by some, Wurtz[59] in 1880 wrote hydrogen peroxide as H–O–O–H.

His reason seemed to be that there were two hydroxyl groups present, but others found this unconvincing, and the compound will be met again as a "molecular compound" and as a case of tetravalent oxygen (pp. 195, 199).

(e) *Chlorine*

The possibility of higher valency states for the halogens did not seem to give much trouble in the decade or two following 1860. The four oxyacids of chlorine (all of which were known at this time) were represented by (literally) croquet-ball models by Hofmann in 1865,[60] with a linear arrangement throughout, in modern formulae thus:

$$\text{H—O—Cl} \quad \text{H—O—O—Cl} \quad \text{H—O—O—O—Cl}$$
and
$$\text{H—O—O—O—O—Cl}$$

Wurtz, in 1864, recognized the difficulty of determining the valency of an element combined with several polyatomic atoms, and tentatively wrote "perchloric anhydride" as[61]

$$\overset{\prime}{\text{Cl}}-\overset{\prime\prime}{\text{O}}-\overset{\prime\prime}{\text{O}}-\overset{\prime\prime}{\text{O}}-\overset{\prime\prime}{\text{O}}-\overset{\prime\prime}{\text{O}}-\overset{\prime\prime}{\text{O}}-\overset{\prime\prime}{\text{O}}-\overset{\prime}{\text{Cl}}$$

Wurtz later said[62] that these formulae received support from the ideas of Graebe and Liebermann on the constitution of quinones, where two oxygen atoms were assumed to be linked together.[63]

Errors of this kind mattered much less than those where experimental proof could be brought to bear on them. Those who held them generally did so tentatively until a better hypothesis should be forthcoming.

(f) *Metals*

A final example of the application of these chain formulae must be those cases where metals seemed to show an anomalous valency, sometimes to a startling degree. Where small molecules such as water, ammonia and hydrogen chloride add to a salt, the products are usually inexplicable in terms of the valency theories generally received at this time. Compounds such as ammines, hydrates and complex chlorides were known in this period, and some attempts were made to formulate them, particularly by Wurtz and Blomstrand. The only way to avoid increasing the valency seemed to be to link up the adding molecules with each other to form a chain. Some of these retained favour until the end of the century, *e.g.*, Blomstrand's formula for H_2PtCl_6. But for many the vague concept of "molecular compounds" seemed preferable, and little more could be done until entirely new conceptions of valency had arisen.

THEORIES INVOLVING FIXED VALENCIES ONLY 187

The following are some examples from these two authors:

Cl.NH³
|
Ag.NH³

(Wurtz[64])

Cl.NH³
|
NH³
|
NH³
|
½Ca—NH³

(Wurtz[65])

$$\begin{array}{c} O \\ \diagdown \\ O \end{array} S \begin{array}{c} O \\ \diagup \\ O \end{array} Cu \begin{array}{c} OH_2 \\ | \\ OH_2 \\ | \\ OH_2 \\ | \\ OH_2 \\ | \\ OH_2 \end{array}$$

(Wurtz[66])

Pt—NH³—NH³—Cl

(Blomstrand[67])

$$\begin{array}{c} Cl \\ \diagdown \\ Cl \end{array} Pt \begin{array}{c} Cl—Cl—K \\ \diagup \\ Cl—Cl—K \end{array}$$

(Blomstrand[68])

Unfortunately, the precedent of organic chemistry with its chains of carbon atoms, that was doubtless the basis of these ideas, was not usually applicable in the inorganic field. And it is ironic to note that where such linking is valid, in mercurous chloride, it had to be abandoned on vapour density grounds (p. 177); and in another place where it was applicable (hydroxylamine, H_2N-O-H), it was unrecognized.[69]

(iii) *Theory of Paired Valencies*

One fact that emerged clearly during the early years of controversy between the adherents of fixed and of variable valencies was that the apparent difference in values was often 2 or a multiple of it. Indeed, Odling[70] had classified elements into *artiads* if they had even valencies (2, 4, 6, *etc.*), and *perissads* if these were odd (1, 3, 5, *etc.*). Further, Cannizzaro was the first to point out the analogy between olefins, which, for example, add two bromine atoms, and inorganic substances which behave in the same way.[71] In 1861 Butlerov wrote that "the amount of free affinity always represents an even number" for any atom, so that while an atom with one free affinity had to combine with another, an atom with 0, 2, 4, *etc.*, free affinities did not have to do so.[72]

These theories all seemed to be tending towards the position taken up by Frankland in 1866. Having quoted some of the cases of apparently variable valency, he then propounds what he later[73] called the "law of variation in the atomicity of elements":

These remarkable facts can be explained by a very simple and obvious assumption, *viz.*, that one or more pairs of *bonds belonging to one and the same atom of an element can unite, and, having saturated each other, become, as it were, latent*. Thus the pentad nitrogen becomes a triad when one pair of its bonds becomes latent, and a monad when two pairs, by combination with each other, are, in like manner, rendered latent; and the hexad sulphur becomes, by a similar process, successively a tetrad in triethylsulphine-iodide [($S^{iv}Et_3I$)] and a dyad in sulphuretted hydrogen [($S''H_2$)].[74]

Frankland was thus led to distinguish *absolute atomicity*, the maximum number of bonds of an element, from *latent atomicity*, the number united together, and from *active atomicity*, the number used to link with other elements.

A nearly identical statement occurs in both editions of his *Lecture Notes*,[75] and these publications also include examples, in his own distinctive notation, of relevant formulae. The following[76] are those given (*cf.* p. 179) for ferrous and ferric chlorides:

the marks on the first iron atom symbolizing the pairs of "latent atomicities". The resemblance to modern formulae with unshared electron pairs is striking.

If this idea is accepted, as Frankland pointed out, most substances can be represented in simple formulae, and without recourse to "molecular" hypotheses. It was doubtless the failure of nitric oxide and monomeric mercurous chloride to conform to this principle that so perturbed Williamson (p. 178). Only the hydrates, ammines and analogous compounds remained obviously outside the system (apart from one or two odd anomalies). A very similar idea was expressed by Wright in 1872. He explained the usual difference in valency states of two units by saying that one or more pairs of affinities satisfy each other.[77]

However, Frankland's teaching on this point does not seem to have made the impression that might have been expected, particularly on the Continent. Basically there were two reasons. First, it came just at a time when new developments were taking place in the field of transition metal chemistry, and within a few years many compounds were discovered in which the metal varied its valency by *one* unit at a time. Vanadium,[78] tungsten,[79] tantalum,[80] niobium,[81] and many other transition metals were being examined with a thoroughness that was quite new in this area. And not only was this unitary variation in valency coming to light, but also the alarmingly large number of addition compounds they formed was beginning to be recognized. So the chief exceptions to Frankland's law were being found within a year or so of its formulation.

The second rock on which it tended to founder was the more profound one of language. What was meant by a "latent atomicity"? Its existence could not be detected if it was latent, and the only other way of looking at it was to say that it was the difference between two observed valencies; and this is to say nothing about it at all. In fact, did this theory really differ – except in words – from that of variable valency? Its merit of "explaining" the rule of even number variation was soon to be severely diminished, and it does not seem to have been taken up by the adherents of constant valency, possibly for that reason.

As Mendeléef saw it, it was the thin end of a theoretical wedge, soon to split apart the doctrine of constant valency: "The beginning of this law's collapse has been the assumption of changing value, free affinities and latent valency; by this the basic principles of the law have then lost their support."[82]

2. MOLECULAR COMPOUND THEORY

It is quite obvious that the conception of fixed valency was incompetent to account for all the facts known even when it was put forward. This remains true even in the light of the modes of formulation discussed in the previous section. If an element appeared to have a valency lower than the one regarded as its normal value, this might be accounted for by supposing two (or more) of its valencies became "latent" and satisfied each other. On the other hand, an abnormally high valency might be explained by a dimerization of units or by addenda to the atom concerned forming a chain. But this latter kind of hypothesis necessarily implies that the atom shall be linked to a number of polyvalent atoms or groups; a chain cannot be formed exclusively from monovalent atoms. Therefore molecules like NH_4Cl, PCl_5 and ICl_3, require a different explanation from any advanced so far. For these cases the hypothesis of "molecular compounds" was advanced.

The origin of this idea seems to have been with Kekulé, and his first public expression of it in 1864.[83] As Freund has wisely remarked "it does not seem that Kekulé had been led to it inductively ... but rather that he propounded it with the object of supporting the view held by him concerning the constancy of the binding-capacity of the atoms".[84] Kekulé wrote as follows:

Attraction should be felt even between atoms which are found to belong to different molecules. This attraction provokes the drawing nearer and juxtaposition of the molecules, a phenomenon which always precedes true chemical decomposition. But, it can happen (especially in the cases where double decomposition becomes impossible by the very nature of the atoms) that the reaction stops at this approach; the two molecules adhere together, so to speak, thus forming a group endowed with a certain stability, always less strong than an atomic combination, however. This explains to us why molecular combinations are not formed in vapours, but are decomposed by the action of heat to give back the molecules that formed them.[85]

This was in the paper provoked by Naquet's suggestion that sulphur, selenium and tellurium should be regarded as tetravalent in their chlorides.[86] Kekulé's reply was thus to suggest that instead they might be regarded as molecular compounds and be represented on the lines of the following:

$SeCl^2,Cl^2$ $TeBr^2,Br^2$ PCl^3,Cl^2
NH^3,HCl ICl,Cl^2 $etc.$

Thus Kekulé was suggesting an aggregation by purely physical forces of two chemical entities, these being set free by vaporization. In the following account, the application and results of several experimental tests of this theory will be described.

(i) *"Molecular Compounds" and Vapour Density*

The use of vapour densities in this matter arose with the discovery of anomalous vapour density results. After some early anomalies noted by Dumas,[87] Bineau[88] discovered that when ammonia formed ammonium salts with acids, the vapour showed no contraction. Later, Cahours[89] observed that the vapour density of phosphorus pentachloride decreased with a rise in temperature. Then in 1858, Kopp[90] suggested that both these results might be explicable by dissociation taking place. The general results were confirmed and extended for sal ammoniac[91] and for phosphorus pentachloride.[92]

Meanwhile, the dissociation hypothesis was gaining ground. Actual separation of ammonia and hydrogen chloride (from sal ammoniac) was accomplished by Pebal, using an asbestos plug,[93] and Wanklyn and Robinson used a narrow tube through which they separated by diffusion chlorine and phosphorus trichloride from the pentachloride.[94] Although Deville had first introduced this idea of dissociation,[95] he later had qualms about applying vapour density measurements for vapours condensible above zero,[96] and raised certain other objections.[97] Nevertheless, the dissociation theory received powerful support from Wurtz,[98] who had applied it to amylene "iodohydrate"[99] and "bromohydrate",[100] and from Cannizzaro.[101] Meanwhile the long familiar "nitrogen peroxide" had been submitted to re-examination by Playfair and Wanklyn, and they had concluded that "it exists in two states; that these are two bodies having the same percentage composition as peroxide of nitrogen, but which are polymeric".[102]

Thus by about 1865 a small number of compounds had been recognized as decomposing reversibly on passing into the vapour state. Moreover this was precisely the time in which variation in valency was becoming a matter for contention, and (fortuitously) the compounds examined were just those incapable of formulation in terms of constant valency. In particular, phosphorus pentachloride and salts of ammonium were quite incompatible with trivalent phosphorus or nitrogen. The supporters of Kekulé's hypothesis were jubilant, and in 1868 Wichelhaus could write: "in bodies unquestionably consisting of unitary molecules on the grounds of their vapour density, the valency of the atoms is constant".[103]

It was not long, however, before facts were to emerge to challenge the simple "molecular compound" explanation of anomalous vapour densities. First came the discovery that phosphorus pentachloride did not appear to be completely dissociated at temperatures immediately above those at which it would vaporize. So the work of Wurtz[104] and the researches of Horstmann on this[105] and on ammonium chloride[106] implied partial dissociation quite clearly. The latter wrote concerning phosphorus pentachloride "it is not an undecomposed gas, but a gas in a state of dissociation. However, it does not seem possible completely to hinder the decomposition".[107] This failure to show a clear-cut behaviour was a great difficulty to the Kekulé–Wichelhaus school. Wichelhaus himself was reluctant to concede the force of this argument, appearing to think that it was lessened by a general failure to understand the nature of the processes involved, and seemingly he envisaged some kind of dissemination of solid particles into the vapour.[108]

However, Wurtz strengthened his own case by two simple devices. He measured the vapour density of phosphorus pentachloride at lower pressures, thus giving a lower temperature range in which to work; better results still were obtained when the vapour was diffused into a space saturated with phosphorus trichloride vapour. Under these conditions the vapour density was almost identical with that calculated for a single molecule of PCl_5. "If then in the perchloride, an unsaturated combination, phosphorus shows only three atomicities, as in hydrogen phosphide, it shows here in the protochloride five; in the latter, phosphorus plays the part of a pentatomic element."[109]

This result was followed by the discovery by Thorpe, in 1876, of phosphorus pentafluoride,[110] made by the action of the pentachloride on arsenic trifluoride. Although Thorpe had spoken in favour of trivalent phosphorus in the oxychloride $POCl_3$ only the previous year,[111] he was compelled to admit its pentavalency in the new compound whose vapour density showed no evidence of dissociation at all. He concluded: "the existence of the gaseous pentafluoride, taken in conjunction with the fact that it is perfectly stable even at very high temperatures, is of great interest theoretically inasmuch as this body unequivocally indicates the pentacidity of phosphorus".[112]

Yet even this failed to convince some, we are told.[113] Possibly fluorine was not fully recognized as monovalent, and little was known of its chemistry until Moissan and others began work a few years later. However, Mallet showed in 1881 that the vapour density

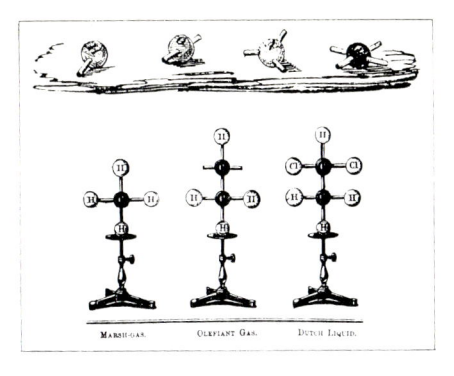

7. Hofmann's croquet-ball models

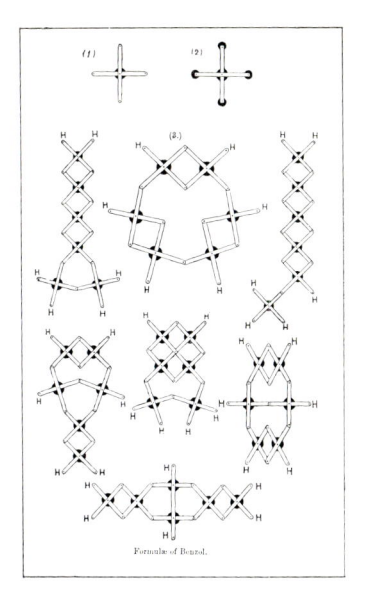

8. Dewar's benzene models

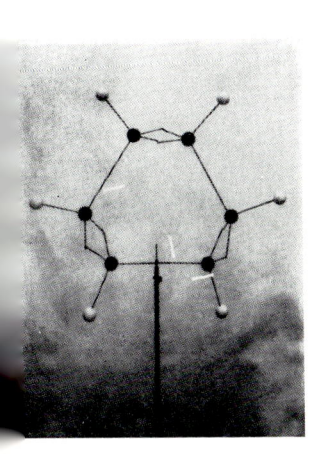

9. Kekulé's benzene model

10. van't Hoff's tetrahedral models, made in 1875 for his fellow-student G. J. W. Bremer

11. *Putney House*, c. 1810. This became, in 1837, the College of Civil Engineering where Frankland conducted his first major experiments on organo-metallic compounds (in 1850), from which emerged his recognition of valency

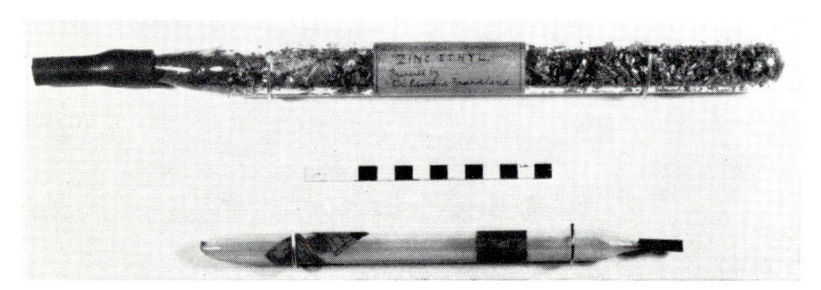

12. *Organo-metallic compounds prepared by Frankland: zinc ethyl (showing a large excess of unreacted zinc) and mercury ethyl*

of hydrofluoric acid was anomalously high,[114] and in 1889 Thorpe and Hambly demonstrated the presence of H_2F_2 in the vapour.[115] Hence Armstrong suggested that "the tendency on the part of fluorine to combine with itself" must now show that "the pentafluoride is by no means necessarily regarded as an atomic compound".[116]

(ii) *Chemical Arguments*

Generally, however, this period saw a marked decline in the emphasis given to "molecular compounds" in chemical discussions, and to understand this we must turn to another line of approach to the problem, that based on chemical reactions. Indeed as early as 1864 Naquet, in his reply to Kekulé, had trenchantly criticized the latter's criterion for a molecular combination:

If one admits with M. Kekulé that the property of not being able to exist in the gaseous state indicates a compound is formed by a simple aggregation of molecules, the largest number of the compounds known at present ought to be regarded as though they did not result from an atomic combination.[117]

Buff also pointed out the uselessness of Kekulé's criterion for non-volatile substances,[118] while Wichelhaus[119] acknowledged the importance of chemical tests, much as Naquet had proposed:

The property which alone characterizes atomic combinations is that they can enter into reactions and undergo double decompositions with other bodies, while the molecular combinations react only by the atomic compounds of which they are made.[120]

Naquet went on to show that most of the substances formulated by Kekulé as "molecular compounds" did in fact undergo double decomposition; these included the perchlorides of phosphorus, selenium and tellurium, and sal ammoniac (PCl_5, $SeCl_4$, $TeCl_4$, NH_4Cl). Iodine trichloride would not behave in this way, however, and Naquet was therefore prepared tentatively to agree that it might be $ICl.Cl_2$.

The example examined by Wichelhaus illustrates the chemical approach to this matter rather well.[121] He was concerned to maintain the trivalency of phosphorus in its "superchloride" (PCl_5), and considered the latter was $PCl_3.Cl_2$. This was borne out by its formation of $POCl_3$ in a number of reactions, as the following:

$$PCl_5 + CO(CH_3)_2 = CCl_2(CH_3)_2 + POCl_3$$

with loss of two chlorine atoms. This "looseness" of the two atoms was compared with the decomposition of PCl_3Br_2 even on melting:

$$PCl_3Br_2 = PCl_3 + Br_2$$

The test was to react this bromochlorocompound with benzoic acid; this reaction he envisaged thus:

$$C_6H_5COOH + PCl_3Br_2 = PCl_3 + C_6H_5COOBr + HBr$$
then, $\quad C_6H_5COOBr + PCl_3 = PCl_2(OBr) + C_6H_5COCl$

the two bromine atoms being removed in the first stage. On the other hand if PCl_3Br_2 is not a "molecular compound" the first stage might involve removal of chlorine instead of bromine:

e.g., $C_6H_5COOH + PCl_3Br_2 = C_6H_5COBr + HBr + POCl_3$
(overall reaction).

In fact, Wichelhaus obtained what he considered was $PCl_2(OBr)$, and therefore regarded his results as confirming the "molecular compound" hypothesis for the chlorbromide, and the trivalency of phosphorus in this series of compounds.

There was also another chemical approach to this problem, which may be generalized thus. Suppose that a compound *ABCD* is regarded as "molecular", and consisting of the two "atomic" units *AB* and *CD*. It will be formed from *AB* and *CD*, but not from *AD* and *CB*:

$$AB + CD = AB.CD$$
$$AD + CB = AD.CB$$

If, therefore, the products from both reactions are identical, the compound is not likely to be "molecular". This was tried in a number of cases. In the controversial field of the quaternary ammonium compounds, Victor Meyer and Lecco found that the same product was obtained *either* from methyl iodide and diethylmethylamine, *or* from ethyl iodide and dimethylethylamine:

$$NMeEt_2 + MeI \rightarrow NMe_2Et_2I \leftarrow NMe_2Et + EtI$$

This clearly demonstrated the improbability of formulae such as $NMe_2Et.EtI$, *etc.*[122] On the other hand, Krüger claimed that different products were obtained from methyl iodide and diethyl sulphide on the one hand, and ethyl iodide and ethyl methyl sulphide on the other,[123] implying that the trialkyl "sulphine iodides" were true "molecular compounds".

It was, of course, common to allot a substance a "molecular" constitution simply if it decomposed very readily, irrespective of arguments drawn from vapour densities. Thus in addition to the examples already quoted, all of which do decompose easily, there was the case of the picrates, a notorious problem for years. Anschütz applied Raoult's cryoscopic method to naphthalene picrate and found the depression of the freezing point of the solution was that to be expected if complete dissociation had taken place, both components being present uncombined. Hence he concluded a "molecular compound" to be present.[124]

Again, Traube applied this designation to oxyhaemoglobin on the basis of its evolution of oxygen under reduced pressure.[125] He also included hydrogen and other peroxides in this category, in the case of the former partly because of the supposed necessity for molecular oxygen to be present when it is formed.[126]

The position, then, was that some substances did seem to exist as "molecular compounds", while for others the evidence was inconclusive or definitely opposed to such a constitution. Not all conceded the force of the chemical arguments, and as with those based on vapour densities, saw loopholes and alternative explanations. Thus, against the molecular formulation of phosphorus pentahalides by Wichelhaus (above), Geuther and Michaelis argued for a simple pentavalency. They discovered that the substance deemed homogeneous by Wichelhaus and written by him as $PCl_2(OBr)$ was in fact a mixture of the oxybromides $POCl_3$ and $POBr_3$.[127]

From the opposite viewpoint Lossen considered that the "identity" of the quaternary ammonium salts of V. Meyer and Lecco (above) might be spurious. Identity of crystalline form would be expected with products as similar as $NMe_2Et.EtI$ and $NMeEt_2.MeI$, he argued, and was no necessary guarantee that they were chemically identical. Moreover he considered exchange of radicals might have occurred in the reaction.[128] The last idea was echoed by Armstrong who also thought isomerization possible.[129]

The situation in the 1870s and 1880s was one of great fluidity, therefore. Most chemists admitted that "molecular compounds" were possible in principle; the problem was where to draw the line. Many of them appeared to think that hydrates and substances of the naphthalene picrate class should be so classified, but even here there was no universal agreement. Few, however, went as far as Kekulé, and some were nearly as uncompromising on the opposite side. Among these, Williamson was quick to point out the difficulties of Kekulé's position at an almost philosophical level:

I cannot see any advantage in representing, as that distinguished chemist proposes to do, terchloride of iodine as $ICl.Cl_2$, pentachloride of phosphorus as $PCl_3.Cl_2$; for by the same right proto-iodide of phosphorus would be PI_3-I_2, and nitrous oxide would be $N_2O_3-O_2$, a change of form which might serve to turn aside the attention of chemists from the fact of the different atomicity of the elements iodine, nitrogen and phosphorus in these compounds, and in ICl, PCl_3, N_2O_3, but which would neither remove the fact nor explain it.[130]

Many years later, when Williamson had to some extent adopted a position of "elder statesman" in chemistry, he reported to the British Association:

The assumption of molecular combination as an unknown something different from chemical combination is open to even more grave objection than those which led us to abandon the dualistic system.[131]

When these words were spoken, it was clear that for some time "molecular compounds" were to many people suggestive of the worst features of the abandoned dualism. It seems incredible that Kekulé, the implacable opponent of "the swindle of dualistic addition",[132] should not have been moved by this. Certainly Wichelhaus was aware of it, and though advocating a "molecular" viewpoint gives this assessment of the situation in 1869:

In general today, chemists have little liking for the assumption of molecular compounds, following the abandonment of the dualistic viewpoint; because a large number of molecules conceived by this as addition compounds have proved to be unitary molecules, all are now decked in unitary formulae, and a uniformity is brought about that does not always seem completely fair. This reaction is going so far that even the double salts and compounds with water of crystallization are classified into the system, with the aid of "change of valency" of the atoms.[133]

Thus it was that, writing in the same year, Blomstrand gave it as his opinion that the solution of the problem of "molecular compounds" was the most pressing problem facing the chemists of the period.[134] The time was ripe for a reappraisal of the whole conception of valency, and highly significant developments were to occur. Of these the first was a growing recognition of a multiplicity of valencies for each element; and from the chaos to which such an oversimplification led there emerged the yet more pregnant doctrine of auxiliary valencies. These two developments will now be considered.

3. THE ADMISSION OF VARIABLE VALENCY

Some confusion exists as to the origins of the theory that an element could have more than one valency. Thus Japp says that, in comparison with Kekulé's idea of fixed valency, "surely Frankland's doctrine of varying valency was earlier in the field!"[135] That Frankland admitted a variation of saturation capacity is beyond doubt (*cf.* p. 188), but it may be questioned whether this was valency in a full sense. Kolbe expressed similar views in 1854.[136] With Couper, however, there is a clear recognition of variation in valency in his distinction between "elective affinity" and "affinity of degree". The latter is a direct acknowledgment of variable valency, and he illustrates it with the two oxides of carbon.[137] It was, however, Naquet who first brought general attention to the idea,[138] Couper's work being largely overlooked. Yet there is reason for supposing that Couper's influence was not altogether absent, but existed as an undercurrent to the thinking of several with whom he had been associated in Paris. Butlerov, for example, admitted variation in valency by even numbers in 1861,[139] and expressed the view that in its oxygen compounds sulphur could have a valency of four or six. Erlenmeyer, who had much in common with Butlerov, expressed the same conviction about sulphur. He wrote:

The atomicity of an element, that is the sum of the equivalents which one atom possesses, is constant, but in relation to different – and probably even to the same – elements it can act now as the total, now as a variable part of the total, with relation to different elements according to their specific affinity.[140]

Writing one sulphur atom as SSSSSS, and with O = 8, he gave these formulae:

```
 SSSSSS      SSSSSS     SSSSSS      SSSSSS
 OOOOOO      OOOO       ClClClClClCl  HH
 ᴗ  ᴗ  ᴗ     ᴗ  ᴗ
```

The influence of Couper may be discerned also in the writings of Wurtz, to which reference will be made later. Meanwhile, Frankland independently distinguished a "stage of maximum stability" from saturation, and worked out some consequences of this distinction.[141]

Apart from the failure of other explanations to account for the wide range of inorganic compounds known, two other factors helped to establish the doctrine of variable valency. The first of these was the publication in 1869 of Blomstrand's *Der Chemie der Jetztzeit*.[142] This was an exposition of the current chemical problems awaiting solution, based openly and avowedly on the electrochemical system of Berzelius. With great clarity and force Blomstrand argued not merely the value of this system, but also its competence to account for this variation in valency which, for Blomstrand, was an established fact. Reaction depends upon the tendency for neutralization of electric polarities, and as these will vary with the reagents, so the saturation capacities will also change and will depend for their magnitude on the nature of the reagents.

This book appears to have had a considerable influence. The other factor which promoted the idea of variable valency was the discovery of the periodic law (*cf.* p. 141). The periodicity of valency involved two main valencies for the periodic Groups V–VII: in the nth Group these were n and $(8-n)$. Also Mendeléef's Periodic Table brought together in different sub-groups elements like sodium and copper, and implied that each should have at least one valency state in common, thus forcing attention on the monovalency of copper, for example. There is no doubt that this discovery assisted those who were advocating variable valency.[143] Thus Wurtz, who in 1864 advocated a chain formula for perchloric anhydride (p. 186), in 1880[144] was writing heptavalent chlorine in perchloric acid as $ClO_3(OH)$.

Variation in the valencies of the non-metals became also admitted for phosphorus, sulphur and others as we have seen; particularly important was the case of oxygen. This first became an issue in 1875, when Friedel obtained an addition compound between dimethyl ether and hydrogen chloride which was partly, but not wholly, dissociated in the vapour phase.[145] He considered that this raised the possibility of tetravalent sulphur in the "sulphine iodides". Molecular formulations, $Me_2S.MeI$ and $Me_2O.HCl$, were ruled out by the evidence available. Friedel thought this hypothesis might also be applied to cuprous and other "lower" oxides, hydrated salts and double salts.[146]

Application of the doctrine of tetravalent oxygen to hydrate formation was made particularly by Wurtz (see p. 186) in 1880.[147] He was, however, following up a suggestion that he had made much earlier, in 1864, when he had supposed that water of crystallization might be held by some kind of "affinity of the second degree".[148] In this, and in his delineation of chemical affinity as "elective force",[149] we may see strong evidence for the influence of Couper in the matter of variable valency.

Tetravalent oxygen also formed the theme of a paper by Heyes[150] in 1888. It was applied by him to hydrogen peroxide,

$$H-O-H$$
$$\|$$
$$O$$

metallic "peroxides", hydrates, carbon monoxide, salts of oxyacids and many other groups of substances. Had the theory of Werner not been so near, there is no doubt that this paper, which has much of value, would have received more attention than was the case.

An example that is still accepted, in a modified form, of tetravalent oxygen, was the hydrochlorides of γ-pyrones, as suggested by Collie and Tickle in 1899.[151]

One of the last suggestions for variable valency before the advent of the electronic theory was the very important case of carbon. In 1904, Nef suggested that this was divalent in carbon monoxide, the carbylamines and the fulminates:[152]

$$O=C \qquad R-N=C \qquad R-O-N=C$$

In the last two cases a quadruple link between nitrogen and carbon was recognized as stereochemically inadmissible, while all three molecules showed the readiness to accept addenda in the manner characteristic of unsaturated substances.

The cases quoted so far have all involved variation of valency among the non-metals. This was more usually recognized than similar variation among the metals, but several chemists did advocate the latter instead or as well. One of these was Wanklyn. In 1869,[153] he proposed recognition of trivalent sodium in an addition compound obtained from sodium and zinc ethyls ($NaZnEt_3$), and also in a compound which he prepared by the action of sodium on ethyl acetate, "sodium triacetyl", $Na^{III}(C^2H^3O)^3$. Wichelhaus pointed out that the former was probably a molecular compound, $NaEt.ZnEt_2$, and consequently few conclusions could be drawn regarding the valency of the sodium in it; the "sodium triacetyl", however, was to be identified with the salt of acetoacetic ester obtained by Frankland and Duppa,[154] which he wrote in the curious form,[155]

$$CH_2Na-CO-CH_2-CO.OC_2H_5$$

Shortly afterwards Wanklyn suggested heptavalent sodium in a compound he had obtained from sodium and alcohol, and which he wrote as,[156]

$$\left.\begin{array}{c} C_2H_5O \\ H \\ C_2H_5O \\ H \\ C_2H_5O \\ H \\ C_2H_5O \end{array}\right\} Na^{VII}$$

In the subsequent discussion at the Chemical Society, Harcourt said this was analogous to the hydrates of sodium hydroxide, while Williamson, in the Chair, urged that distinction should be drawn between atomicity and equivalence, the former being an "unchangeable kind of equivalence", most elements possessing "different replacing values".[157] In other words there was as early as this a recognition of different kinds of binding, whether termed "atomicity" or not. In later years[158] Williamson advocated for many metals a variety of valencies, but this idea can never have been absent. And it is this concept of different species or kinds of valencies that led to the theory of auxiliary valency.

4. THE CONCEPTION OF AUXILIARY VALENCY

There can be little doubt that the slow progress of inorganic chemistry at the end of the nineteenth century was in large measure due to the rigid idea of valency that had started with Kekulé and continued with the advocates of molecular compounds. Eminently suitable for much organic chemistry, this concept acted as a straitjacket to the rest of the science. It is fortunate that there were those bold enough to see this and to question some of the basic assumptions of current thinking. Of these, none was more deeply entrenched than the doctrine of an integral, if not a fixed, combining power for each element. Had non-stoichiometric compounds been recognized earlier, the course of chemical development would have been very different. As things were, a succession of tentative suggestions over a number of years was to be necessary to rouse to unwilling consciousness the idea of auxiliary valency.

In this instance it will be convenient to consider separately the contributions made by individual workers (and speculators) in the field.[159]

(i) *Work between 1875 and 1885*

In the decade following 1875 a movement of thought appears in which the concept of integral values of combination received its first challenge. Particularly in Germany, doubts can be detected on the accepted views on valency, and even on the necessity for the latter at all. The starting point for the new developments was the question of molecular compounds, especially hydrates.

A paper appeared in 1876 by H. Kommrath that seems to have attracted little or no notice at the time (or indeed since). Concerned about "molecular compounds", he wrote:

I distinguish between chemically and physically saturated compounds. Physically saturated compounds (Molecular compounds) do not need at all to be absolutely saturated; they are essentially dependent on the motion relationships of the molecule. Obviously the intensity of the actual force of affinity, as well as the amount of actually changeable, disposable force, is a function of the temperature.[160]

What is interesting about this paper is the suggestion of a varying "disposable force", that is a force of chemical attraction over and above that which results in conventional bond-formation. A conception of this kind was to prove essential for the liberation of inorganic chemistry, especially, from the artificiality of earlier attempts to formulate a large number of compounds. The extent to which Kommrath influenced subsequent thought is debatable. Perhaps this paper is chiefly important in indicating a trend rather than in initiating one.

Four years later Lossen issued a remarkable statement. Taking a stand against the multiplication of theories then occurring, he called for a new empiricism.

The value (WERTH) of an atom is a number expressing how many atoms are found in the binding zone. Since the number of atoms varies in different molecules, *the value of the said polyvalent atoms is also variable.*[161]

This, in a sense, is a retreat to the position taken originally by Frankland, with the difference that correct equivalents would now be used. Valency was just a number. Moreover it was shorn of any necessity to be constant, to vary by even number differences, or to obey any other condition. Multiple bonds were banished, and with them any chance of explaining differences in reactivity in terms of unsaturation. Thus:[162]

The atom-binding in the molecule acetylene, HC–CH, is fully explained by saying that it contains two divalent carbon atoms; similarly, that of benzene by the statement that it contains six trivalent carbon atoms, of which each is joined to one hydrogen atom.

Formulae given to illustrate his ideas include:

$$\begin{array}{c} H \\ \diagdown \\ H \diagup \end{array} C\!-\!C \begin{array}{c} H \\ \diagup \\ \diagdown H \end{array} , \quad O\!-\!O, \quad N\!-\!N, \quad \begin{array}{c} H \\ \diagdown \\ H \diagup \end{array} C\!-\!C \begin{array}{c} O \\ \diagup \\ \diagdown H \end{array}$$

Now there is nothing here on "disposable affinity", but the point is that Lossen was paving the way for such ideas by emphasizing (and doubtless over-emphasizing) the experimental nature of chemistry, and stripping away the needless theoretical accretions it had acquired over the years.

Within a few months of Lossen's paper, there appeared in *Berichte* an account of the views at which Claus had recently arrived.[163] These were in essence a denial of the commonly held idea that in a polyvalent atom a definite number of "separated units of affinity" is always present. Thus while Claus was prepared to concede that in methane the carbon atom exerted a "chemical force of attraction acting as in four single parts", he did not think this was so when carbon was joined to less than four other atoms. Thus he considered carbon disulphide and dioxide:

In the binding of each of these compounds, the force of affinity of the carbon atoms is split only into two, and indeed, because the added atoms are each time of the same kind, into two parts of equal function. In carbonyl sulphide, the latter (division into two parts) is also the case, though these parts are not equal, but of unequal size because of the different nature of the atoms bound (oxygen and sulphur).[164]

Similarly, he could not conceive that:

in the molecule of nitrogen both nitrogen atoms are bound by three divided affinity units that would act from different points of attack, and in different directions.[165]

He finished his paper thus:

To conclude that differences in molecules with absolutely the same arrangement of atoms may be explained by assuming that definite groups of atoms are bound by different valencies to the same elementary atoms is pointless, since the assumption of valencies is as unnatural an hypothesis as their pre-existence in polyvalent atoms, and their degree of action according to definite units of attraction is equally unfounded.[166]

Thus Claus, like Lossen, was attacking the conception of a multiple bond, and above all the "pre-existence" of valencies in an unlinked atom. His paper brought two immediate replies.

First, Lossen returned to the issue, appearing to agree with Claus in most respects, but to wish he had gone farther. Again he urged an empirical treatment rather than one even in terms of a force:

In my opinion it emerges from the law I have given that the value of an atom is a number which expresses how many atoms are directly bound. But the idea of valency is then generally superfluous; the number of acting valencies is of course given by the number and ways of binding the atoms. It is a general law of physics that between two bodies held together there is a force acting that emanates from them both; a special name for the force acting between atoms is needed—in my view – all the less, as our theoretical demonstrations are not at all more easily understood if we speak of the force when we could just as well speak of the bodies themselves.[167]

Secondly, another reply came from Klinger. The chief importance of this was the use of an analogy with magnetism. Just as it is possible for an aggregation of magnets to possess a certain amount of residual magnetic force, so he conceived that in some cases of chemical combination some force may be left over, thereby enabling the substance to give rise to "molecular compounds".[168]

It is now that the focus of interest moved to England for a time. In 1885, Pickering read a short paper to the Chemical Society, which was not published in their *Journal* but was abstracted in the *Proceedings*.[169] The suggestion was put forward that "molecular" or "residual" compounds owed their existence to the non-integral values of the valencies for some atoms. If this were so, there might be a certain amount of residual valency present in a compound with which it might give rise to "molecular compounds".

For example, if hydrogen has a valency of 1, that of oxygen will be a little less than 2, say 1.98. Water would then be written as

$$(H_2^2 O^{1.98})^{+0.02}$$

the upper suffixes representing numerical assessments of the valencies, the final figure being the excess, or "residual valency". Thus water would be unsaturated, but not hydrogen peroxide, which he wrote thus:

$$O \underset{0.98\ |}{\overset{1.0}{\text{———}}} \overset{1.0}{\text{———}} H$$

$$\underset{1.0}{\overset{0.98\ |}{O\text{———}}} \underset{1.0}{\text{———}} H$$

Magnesium sulphate was written as:

$$\left[\begin{matrix} 1.99 & 2.135 \\ Mg & (SO_4) \end{matrix} \right] -0.145$$

The deficit of valency thus arising would be responsible for hydrate formation, for 7 water molecules, each with 0.02 units of excess valency, would bring a total of +0.140 units, thereby very nearly neutralizing the −0.145 units of the anhydrous salt.

Pickering considered, however, that some elements might have nearly integral values, notably among them carbon. In this way the unreactivity of many organic compounds could be explained.

Analysis of these views of Pickering shows that they are simply a quasi-mathematical representation of what was becoming increasingly felt, namely that valency was not to be conceived in terms of a definite number of fixed bonds. But there was one other point emerging in this paper. Pickering went to some trouble to explain that no new kind of force was being envisaged in holding together the "molecular compounds" – only the excess of atomic attraction. This was valuable in undermining the persistent tendency to think in terms of "special cases", a method of approach always likely to hold up the development of a unified science, and particularly undesirable here.

Again, it is not easy to assess the impact made by this paper. No discussion on it was reported in the *Proceedings*, although Pickering did refer to it in a footnote to a paper in the *Journal* the following year.[170] There is, however, one matter of which one may be reasonably sure; the *Proceedings* were "edited by the Secretaries" of the Society, one of whom at that time was H. E. Armstrong, and it is not hard to believe that here Pickering's suggestions would have made some impression. Whether this was considerable or not, is difficult to say, but it was through Armstrong that the next developments were to occur.

(ii) *The Views of H. E. Armstrong*

We are concerned here with three utterances by Armstrong in which he put forward definite opinions on the subject of "residual affinity". The occasion for these was the famous Faraday Lecture of 1881, delivered to the Chemical Society by Helmholtz. This was an impressive plea for a reassessment of chemical ideas in the light of Faraday's Laws of electrolysis. In particular, it contained the suggestion that some chemical compounds were held together by unit charges of electricity, and more will be said of this in connection with electrical theories of valency. But this is not to imply that Armstrong was advocating any such electrical theory; if anything, his contributions were positively harmful to such a viewpoint, first because they were difficult to reconcile with unitary charges, secondly because they were in complete antithesis to the views on electrolysis put forward by Arrhenius. Nevertheless, those ideas that he did propose were of considerable bearing on the theory of "residual affinity".

The first of these main contributions was in 1885. Since this was ostensibly a reply to the lecture by Helmholtz four years previously, one is tempted to enquire why the interval was so long, or alternatively why Armstrong chose this time to reply. One likely reason is that the paper by Pickering had just been given. But also in 1885 another event at the Chemical Society had been the reading of a paper by Baker on combustion of highly desiccated gases.[171] The author had concluded that water was necessary for combustion of carbon monoxide, and this had brought from Armstrong the suggestion that chemical action was "reversed electrolysis".[172] Finally the rapidly accumulating data on conductivity required interpretation, and this Armstrong thought he could give in terms of his theory of "residual affinities".

At the British Association meeting for 1885, Armstrong "took up the gage" thrown down by Helmholtz. In the first place he objected to the latter's dictum that "the atoms cling to these electrical charges, and opposite charges cling to each other".[173] That this was an oversimplification seemed to follow from the differing behaviour in attempted electrolysis of such similarly constituted bodies as silver chloride and hydrogen chloride; also conversion of stannic to stannous chloride should, on Helmholtz's viewing, liberate two free unit charges of positive electricity, while of course it does not.

Next, Armstrong went on to dispute the Faraday Lecturer's classification of electrolytes as "typical" (non-molecular) compounds. His basic point was that water was usually necessary, suggesting that it gave rise to a "molecular compound" with the solute: "electrolysis only takes place when the typical compounds are conjoined and form the molecular aggregate".[174]

Regarding these "molecular aggregates" Armstrong tended to agree with Williamson[175] and many others in denying a fundamental difference between these and "typical compounds". In most of them, the parent molecules survive "in the sense that a hydrocarbon radical, such as ethyl, is preserved intact in an ethyl compound".[176] He was also in agreement with Lossen's definition of valency.[177]

On the formulation of these "molecular compounds" he differed from Williamson in that he considered that the *non-metal*, not the metal, would change its valency. Thus in place of a complex involving nonavalent arsenic in K_2AsF_7, Armstrong preferred:

$$F_3As\begin{smallmatrix}\nearrow F \cdot FK \\ \searrow F \cdot FK\end{smallmatrix}$$

Although most "double salts" were then formulated by guess-work and little more, Armstrong was able to advance a definite chemical reason for his preference; in forming many compounds of this kind the distinctive features of the metal were often retained. Thus the pale green colour due to ferrous iron remains in the double chlorides with potassium, *etc*. Now, the same example was used later to prove the same point, and it is difficult to avoid the impression that Armstrong was trying to provoke discussion rather than to put this forward as a serious argument. For in a great many compounds of apparently the same class sharp changes in colour do occur; the hydration of copper sulphate, the action on this of ammonia, and many reactions of cobalt and chromium compounds suggest that something *is* happening to the metal. However, it does not seem to have aroused any adverse comment, though, as will be seen, this paper and the two that followed it appear as a whole to have made very little impression generally. His most important conclusion was expressed in the following words:

Whatever be the nature of chemical affinity, it is difficult to avoid the conclusion that the "charge" of a negative radical especially is rarely if ever given up at once: that its affinity is at once exhausted. It would also appear that the amount of residual charge – of surplus affinity – possessed by a radical after combination with others depends both on its own nature and that of the radical or radicals with which it becomes associated.[178]

The next year, Armstrong returned to the subject in a paper read to the Royal Society. Again, electrolysis was the opening theme. He launched into a vigorous denunciation of Claus's theory of thermal dissociation, emphasizing again the difference in conductivity shown by water and hydrogen chloride separately and together. He marshalled other evidence, into which it is unnecessary to go here, for his view of "composite electrolytes", *i.e.* those mixtures capable of undergoing electrolysis while their components separately will not. Of these he said:

The influence exercised by the one member of the composite electrolyte upon the other member during electrolysis is at all events mainly exercised by their respective negative radicals, and the extent of the influence thus mutually exerted by these radicals would depend on the extent to which they are still possessed of "residual affinity".[179]

An example of this was hydrogen fluoride which had an anomalously low conductivity at high concentrations, due, he suggested, to the formation of molecular aggregates under these conditions, thus exhausting the residual affinity. Association of HF molecules even in the vapour phase had been noted shortly before.[180]

Armstrong then went on to a fuller discussion of his views on the nature of valency. His starting point was that no truly *saturated* compounds are known. Water, for example, he now had to regard as "an eminently unsaturated compound", in contrast to the views he expressed after hearing Baker's paper. It was impossible otherwise to explain its anomalous physical properties and above all its tendency to give rise to hydrates, particularly with oxygen compounds. Indeed the chief defect of the views of Helmholtz was that his theory:

> does not take into account the fact that the *fundamental* molecules even of so-called atomic compounds *are never saturated*, but more or less readily unite with other molecules to form molecular compounds – molecular aggregates; and unless the application of the theory to explain the existence of such compounds can be made clear, chemists must, I think, decline to accept it.[181]

Again, the case of $Fe_2Cl_4.Cl_2K_2$ was quoted as evidence for the non-participation of the metal in forming new bonds in "molecular aggregates", Armstrong being inclined to draw "an absolute distinction" between hydrogen and the metals on one hand, and the non-metals on the other.

Finally, he suggested that "residual affinity" might be responsible for a large number of chemical reactions, as double decomposition. Initially this was conceived in terms of formation of a "molecular compound" by means of these "residual affinities", and this would then often rearrange to an "atomic" product.

In Armstrong's third paper we have the fullest exposition of the chemical aspects of this theory.[182] Here, too, his views are most heterodox, and it is almost certain that he was overstating his case, in places by a very wide amount. Indeed, he almost says as much in his introduction; his aim is to take up the challenge issued by Helmholtz, and to stimulate discussion of valency on new lines. It is uncertain how far he was successful in this respect; judging by printed papers, such success was very limited.

The behaviour of tetramethylammonium hydroxide and iodide afford good examples. The failure of the latter to undergo facile hydrolysis showed a striking difference from all other iodides *except*

THE CONCEPTION OF AUXILIARY VALENCY 209

those of carbon, e.g., methyl or (better) phenyl iodides. The similarity between the quaternary hydroxide and the strong alkalis, and its difference from compounds like methanol, Armstrong disposed of in the following way: when the latter reacts with acids, it does so reversibly, and with formation of insoluble products. Consequently, any similarity between this and tetramethylammonium hydroxide would tend to be blurred because of these differences. Now this is an astonishing statement. Armstrong completely ignores the possibility that the differences in behaviour could well spring from a fundamentally different kind of structure. It is not to distort historical perspective for us to say that the only *a priori* deduction to be drawn from the very great similarity between quaternary ammonium compounds and the strong alkalis is that of analogous constitutions. Admittedly, the inertness of the iodides raises a difficulty, but one is entitled to ask what Armstrong would have *expected*; applying his own argument that different solubilities make all the difference to the nature of a reaction, we may observe that, given a soluble hydroxide, the action of, say, caustic soda on the quaternary salt would be similar to that on a salt of potassium, and nothing apparently would happen. This is not to assume any conception of ionic halogens, *etc.*, but merely to quote well-known experimental facts. Again, he adduced an additional reason in that "the power to form ammonium compounds ... is not a simple function of the nitrogen atom, but is largely dependent on the nature of the radicals associated with the nitrogen atom".[183] Thus halogenation of the nucleus in aniline reduced its basic strength, acetamide was nearly neutral and phenylhydrazine only a monacid base. Now these facts did receive a ready interpretation on his "residual affinity" theory, but it was too much to assume that they were irreconcilable with any other theory, particularly that of "the pentacidity of nitrogen" in these compounds.

Thus Armstrong concluded that pentavalent nitrogen did not exist in the ammonium compounds, and extended these conclusions to a denial of pentavalent phosphorus in phosphonium compounds, and of tetravalent sulphur in sulphonium compounds. Unwilling to specify exactly what these valencies were, as these could only be found from the hydrides, he decided that they must be regarded as "molecular compounds" where one organic radical was linked to the halogen. Thus, tetramethylammonium iodide was built up of a trimethylamine component linked to a methyl iodide molecule by "residual affinities" on the nitrogen and iodine atoms. In general,

I would define a molecular compound as one formed by the coalescence of two or more molecules, unattended by redistribution of the constituent radicals, and in which the integrant molecules are united by residual affinities.[184]

Using a simple but effective symbolism, he represented the affinity of a monovalent atom thus:

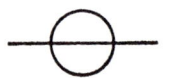

the two projections being the amount of free affinity in that atom. If it was strongly combined with another, there would be very little "residual affinity" left, and so the situation could be represented thus:

If, however, two atoms combined with unequal affinities, that with the greater might have some "residual affinity" left over, as:

The tetramethylammonium iodide he wrote as follows:

"A 'division' of the unit charge somewhat in the manner here suggested" was essential for a chemical explanation of valency.[185]

But opponents of this view would inevitably quote cases where (as has been shown already) evidence for equivalence of valencies seemed almost irrefutable. Undaunted by this, Armstrong dealt with the three most outstanding cases. Phosphorus pentafluoride (see p. 192) he suggested could best be understood in terms of fluorine-fluorine bonds, presumably as $F.P(F-F)_2$. The evidence of Victor Meyer and Lecco (p. 194) for equivalence of all four valencies in quaternary ammonium salts is dismissed with an allusion to the possibility of rearrangement during reaction, as with the sulphur analogues (p. 195). Unable to dispute the logic of Michaelis and la Coste, Armstrong conceded the inter-atomic relations they suggested – but considered the oxygen atom was linked by "residual affinities" to phosphorus, viz.,

$$O \equiv P \begin{array}{l} -C_6H_5 \\ -C_6H_5 \\ -C_6H_5 \end{array}$$

Now this last formula was introduced so unobtrusively that it is possible to miss altogether the fact that it embodied a new conception of "residual affinity". In the previous examples, the "residual affinity" had been something left over from that with which an atom had linked itself on to another; it was thus capable of accounting for some of the cases of anomalously high valencies. But now it was tacitly assumed that a full valency could be satisfied in the same way. Also, Armstrong wrote lead tetraethyl with the same arrangement:

$$\begin{array}{l} H_5C_2 \\ H_5C_2 \end{array} = Pb = \begin{array}{l} C_2H_5 \\ C_2H_5 \end{array}$$

and thought tellurium tetrachloride might be similarly constituted, though he recognized also the possibility of a cyclic structure:

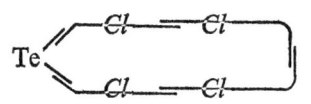

$$Te \begin{array}{c} Cl = Cl \\ Cl = Cl \end{array}$$

This extension to his view, although not specifically defined, was very harmful to it. More questions were raised than answered, e.g., in triphenyl phosphine oxide, why the oxygen did not use its residual affinities to link further with other atoms, why the third "residual affinity" on the phosphorus was not involved, and so on. The most general deduction that could be made from it was that given an n-valent atom it should be possible for it to show $2n$ valencies in some cases; granted that it explains tetravalent lead, it ought also to imply hexavalent aluminium and octavalent silicon, and certainly divalent sodium.

He concluded that if the ideas he had expressed were right, "our views concerning the constitution of the majority of compounds at present rest upon a most uncertain basis".[186] What effects did these publications have on changing that basis?

As has been mentioned already, there is little evidence to suggest any great impact of Armstrong's theories, as here outlined, on fellow-chemists of the time. Indeed a recent biography[187] has very little on this aspect of his work. Fundamentally, two reasons seem to have been responsible. First was the fact that he was deliberately adopting an extreme position in order to provoke further thought, and this was recognized by his contemporaries. Of the weaknesses in his argument that have been mentioned above nobody seems to have taken much notice, probably because they were regarded as fairly obvious. Secondly, Armstrong was taking up a position directly opposed to that of Arrhenius (p. 267) and others with regard to the mechanism of electrolytic conduction. His opposition to the dissociation theory was not without its sound reasons; in particular it seemed to him irreconcilable with the behaviour of composite electrolytes, for which, of course, he supposed the opposite of dissociation, namely the formation of "molecular aggregates". In this view, Armstrong was in a minority although he held it for many years. But it probably meant that views framed in a context of a contra-dissociation theory would meet with a cooler reception than would have been likely before Arrhenius.

However, all this is not to deny them any influence, but only to suggest that this was small. Two papers by Heyes in the *Philosophical Magazine* for 1888 refer to aspects of Armstrong's ideas. Thus:

I would venture to deprecate the use, except with extreme caution, of such phrases as "residual affinities" or "residual valencies", at any rate so long as we fully and strictly accept the fundamental laws of definite, constant and multiple proportions.[188]

In a second paper a few months later, Heyes retained his uncertainty about expressions of this kind, and proposed the term "validity" for cases of disputed valency. He agreed with Armstrong's atomic arrangement for tetramethylammonium iodide, but regarded the iodine atom as either trivalent or "tervalid".[189]

Armstrong's "residual affinities" appeared in a new form in the writings of Thiele (p. 241), and to some extent in his own theory of the benzene ring (p. 252). Their application to inorganic compounds – where Armstrong himself first used them – was continued and extended by Werner on whom, probably, they had their most far-reaching influence.

(iii) *Werner's Theory*

The contributions of Alfred Werner (1866–1919) are at the basis of many present theories of inorganic structure. Like many of the most important hypotheses in science, they represent a discontinuity in the trends of scientific development, although of course they had strong affinities with certain elements of the past.

Werner, who became Professor of Chemistry at Zurich, published his first essay in inorganic structures in 1891,[190] and his most famous paper in 1893.[191] Some of his ideas, however, were not fully formulated until the publication of his book *Neuere Anschauungen auf den Gebiete der anorganischen Chemie* in 1905, translated later into English.[192] He made other contributions to theoretical chemistry, of course, but in these appear his new ideas on valency, based upon the conception of "residual affinity".

It is not the intention to present here a detailed statement of the many applications of Werner's theory, since this has often been done adequately elsewhere and in any case such applications do not present many new ideas on valency itself. Instead, it will be perhaps better to analyse the fundamental axioms that lay behind the immensely successful developments of the theory by Werner and others, and to attempt to relate his work to that of his predecessors in the field.[193]

(a) *Uniform, non-directional valency*

Werner denied that valency was a directed, unit force. As early as the 1891 paper he made it perfectly clear that, for him, it was best conceived as an attractive force acting uniformly from the centre of the atom over all parts of its surface. In his book, two reasons are given for this view. First, the racemization of organic compounds, as mandelic acid at 180°C,[194] was explicable on one of two assumptions; either the radicals changed places, or the valencies did. On the former of these hypotheses it was difficult to see why, if the radicals were liberated, even for a moment, they did not combine to form stable products. He was, of course, using our modern criterion for an inter-molecular rearrangement, and the surprising *chemical* homogeneity of the product justified his conclusion that radicals were not liberated. This could only mean, therefore, that the valencies did change, so refuting the old idea of rigid, fixed bonds. His other argument depended upon the unreactivity of "double bonds" whose existence, as such, he denied.

In these views Werner was breaking no new ground. It has been noted already (p. 202) that Lossen[195] had denied both multiple bonding and constancy of valency, and in this he was joined by Claus.[196] So, despite its merits, Werner felt that "this conception of the unit of valency contains more in it than may be logically deduced from the facts".[197] That he was influenced in this by Claus and Lossen is probable from the approval he gives to them in his book.

(b) *Co-ordination on to metals*

A second distinctive feature of his theory is his recognition that the groups added in the formation of "molecular compounds" were joined to the metal and not to any other part of the molecule. In this he was going back to Williamson's views and rejecting *in toto* those of Armstrong, Blomstrand, *etc*. Only now he had more powerful reasons than Williamson could have brought forward. Thus in the ammonia addition compound of platinic chloride he rejected the usual structure

$$\begin{array}{c} Cl \\ \diagdown \\ \diagup \\ Cl \end{array} Pt \begin{array}{c} Cl-NH_3 \\ \diagup \\ \diagdown \\ Cl-NH_3 \end{array}$$

on the grounds that all chlorine atoms can be removed (in several stages) without removing the ammonia. Two isomeric forms of this compound were known, and to explain this Cleve had proposed two further structures:[198]

$$\begin{array}{c} Cl \\ \diagdown \\ \diagup \\ Cl \end{array} Pt \begin{array}{c} NH_3-Cl \\ \diagup \\ \diagdown \\ NH_3-Cl \end{array} \quad \text{and} \quad \begin{array}{c} Cl \\ \diagdown \\ Cl-Pt-NH_3-NH_3-Cl \\ \diagup \\ Cl \end{array}$$

These however obscured altogether the very similar properties of the two isomers, and also failed to explain the absence of chloride ions (which would be expected with a chlorine "linked" to pentavalent nitrogen).

In short, Werner showed that it was necessary to imagine that the six groups were all attached to the one metal atom. Conventional valency considerations would not permit this, however, and in so doing obscured another factor of which Werner made a great deal. He regarded the new (complex) salts as similarly constituted to the sulphates, formed by adding, *e.g.*, water, to sulphur trioxide. The two reactions

$$SO_3 + H_2O = H_2SO_4$$

and

$$AuCl_3 + KCl = KAuCl_4$$

were regarded as essentially similar, for instance. But while the first was explicable in terms of H–OH adding to an O=S double bond, no analogous explanation was feasible for the latter. This meant that not only the new salts that we term co-ordination compounds, but also well-established substances like sulphates, required fundamentally rethinking. Inorganic chemistry was in fact on the brink of as far-reaching a revolution as its organic counterpart had been experiencing for the past half-century. Werner wrote, speaking of the platinum ammines:

> Hence it is evident that many formulae can be proposed from considerations of valency and the isolated properties of the component elements, but none of these formulae admit of the harmonious correlation of the characteristic properties of the compounds. Moreover, these formulae are unable to give expression, by means of similarly constituted formulae, to the analogy existing between the modes of their formation. This analogy is at least quite as definite as that existing between the addition compounds of SO_3. We therefore arrive at the conclusion that the doctrine of valency is unable to yield satisfactory pictures of the formation and constitution of these compounds directly from their properties, and without reference to the customary ideas on valency.[199]

This abandonment of "the customary ideas on valency", when applied to sulphates and other simple salts, was not to have as serious consequences as might have been expected. The question now arises as to the means whereby Werner proposed to link his groups to the central atom, and this constitutes the third characteristic of his scheme.

(c) *Auxiliary valencies*

Werner was committed to having more groups attached to one metal atom than classical valency would allow. One solution to the difficulty was that advocated by Lossen[200] and Claus;[201] he could have treated the whole matter empirically and merely observed that this number of atoms was so attached. But what he actually did was to combine the best features of the theories of Lossen and Armstrong (pp. 202, 205) and propose that the most likely explanation was that these groups were bound to the metal in defiance of classical laws (Lossen), and that this was accomplished by means of their "residual affinities" (Armstrong). It is to be noticed that Armstrong himself could never have reached this point. He was far too deeply committed to the attachment of the "new" groups to the non-metals. In Werner's own words, the sulphur atom in SO_3, the platinum in $PtCl_4$, the oxygen in H_2O, the chlorine in HCl, and the nitrogen in NH_3:

possess a residue of unsaturated affinity which permits such groups to initially satisfy [*sic*] each other. If the amount of this residual affinity is sufficient to bring about a stable combination between single molecules, then it assumes practically the same rôle as the ordinary valencies, *viz.*, it effects an interdependence between two elementary atoms, and in this way unites two radicals to form a molecular complex.[202]

These auxiliary valencies (*NEBENVALENZEN*) could be quite strong enough to hold together the components of, for example, hydrated platinum (IV) chloride; if they are written as dotted lines, we have:

$$Cl_4Pt\ldots + 2\ldots OH_2 = Cl_4Pt\ldots(OH_2)_2$$

If, however, one of the reactants has a "double bond" (as conventionally written), a rearrangement ensues which produces ultimately a "principal valency compound". *E.g.*,

$$\begin{array}{ccc} O & O & OH \\ \| & \| & | \\ O{=}S\ldots + \ldots OH_2 \rightarrow O{=}S\ldots OH_2 \rightarrow O{=}S{-}OH \\ \| & \| & \| \\ O & O & O \end{array}$$

In this way the long-accepted structure for sulphuric acid is retained. But it will be noted that by this concession, Werner has in fact impaired the value of part of his argument. A key-stone of this is the analogy existing between the two types of reaction, and he is now destroying the analogy between the products. The defect is not serious, however, as similarity of reaction rather than of product is the essence of his reasoning.

These auxiliary valencies were considered to exist in radicals that were capable of existing as independent molecules, as water, ammonia, hydrogen chloride, *etc.* On the other hand principal valencies were regarded as necessary for the linking of hydrogen, chlorine and so on.

(d) *Zones of affinity*

A fourth characteristic of Werner's theory related to the problem of linking by this means to the metal atom. How many such groups could be accommodated? The answer he gave involved the new conceptions of co-ordination number and of spheres of affinity.

He considered the action of ammonia upon trinitritotriamminecobalt (III), in which a fourth ammonia molecule was taken up, but one nitrite group became "ionogenic". The reaction could be written:

$$(H_3N)_3 \, Co\begin{array}{c}NO_2\\-NO_2\\NO_2\end{array} + NH_3 = (H_3N)_3 \, Co\begin{array}{c}NH_3]\,NO_2\\-NO_2\\NO_2\end{array}$$

analogously to

$$\begin{array}{c}H\\H-C-I\\H\end{array} + NH_3 = \begin{array}{c}H\\H-C-N]I\\|||\\H\quad H_3\end{array}$$

Because immense possibilities of isomerism would arise if the four ammonia molecules were not equivalent, Werner rewrote the molecule thus:

$$\begin{bmatrix} O_2N & NH_3 \\ & Co--NH_3 \\ O_2N & NH_3 \\ & NH_3 \end{bmatrix} NO_2 \quad \text{or as} \quad \begin{bmatrix} & (NH_3)_4 \\ Co & \\ & (NO_2)_2 \end{bmatrix} NO_2$$

Arrested by the analogy with the tetramethylammonium compounds he rewrote these as

$$[H_3C...NH_3] I, \text{ or as } \begin{bmatrix} & H_3 & \\ C & & \\ & NH_3 & \end{bmatrix} I$$

The iodine atom, after it ceases to be directly bound to the carbon atom, still remains bound to carbon, and the nitrogen atom, in the addition compound, is bound to carbon by an auxiliary valency. The saturation of the iodine valency, which no longer can take place in the first sphere of attraction of carbon, since all the co-ordination positions are occupied, now takes place outside the first sphere.[203]

In this quotation from his textbook, Werner is employing one of his most characteristic conceptions, that of two spheres of affinity. Introducing this in his 1893 paper he wrote:

If we think of the metal atom as a ball, the six complexes directly bound with the same are found in a sphere described about the latter, and the rest that are found beyond this first sphere lie in a second sphere. For all the compounds being considered, we can propound the general law: The valency of the radical formed by the metal atom and the six complexes bound with it in the first sphere is equal to the difference between the valency of the metal atom and that of the monovalent groups in the first sphere, and wholly independent of the molecules present in the first sphere as water, ammonia, *etc.*[204]

Thus he gave as examples,

$$(Co(NH_3)_6) \text{, valency} = 3-0 = 3$$

$$\begin{pmatrix} & (NH_3)_3 & \\ Co & & \\ & (NO_2)_3 & \end{pmatrix}, \quad \text{,,} \quad = 3-3 = 0$$

$$\begin{pmatrix} & (NH_3)_2 & \\ Co & & \\ & (NO_2)_4 & \end{pmatrix}, \quad \text{,,} \quad = 4-3 = 1$$

Werner has thus suggested two spheres or zones in which combined atoms or radicals can be deemed to exist. The number of addenda capable of existing within the first zone was termed the co-ordination number, and was often, but not always, equal to six. Any further atoms or groups had to go into the second, outer zone. As this was further away from the centre of the atom, groups in

THE CONCEPTION OF AUXILIARY VALENCY 219

this situation were probably "ionogenic" since elements with the greatest atomic volume ionize most easily.

A definite relation existed between the number of auxiliary valencies, the number of principal valencies and the behaviour of the compound:[205]

After saturation of a definite number of auxiliary valencies, each further addition of a molecular component takes place in such a way as to cause the principal valency to transmit ionic properties to the acid residue attached to it.

For illustration he quoted the case of platinum (II) chloride where the co-ordination number is four; successive addition of ammonia gives:

$PtCl_2$ $[PtCl_2(NH_3)_2]$ $[PtCl(NH_3)_3]Cl$ $[Pt(NH_3)_4]Cl_2$

Of the relation between principal and residual valencies, Werner did not say very much. When one is strong in an element, the other is usually strong as well. It does not seem that he envisaged an absolute difference between them, for he allowed intermediate types as in "sodium nitroprusside":

$$\begin{bmatrix} & NO \\ Fe & \\ & Cy_3 \end{bmatrix} Na_2$$

In formulae he differentiated between the two kinds of valency "to contrast sharply defined conceptions", as he considered necessary at a time when "the doctrine of valency is in a transition state".[206] He even went as far as to admit that "the unit of valency is not constant, but varies with the nature of the atoms it joins together, and with the variable amount of affinity on these atoms".[207] Thus in a molecule M–X, if M became saturated by other groups attached to it, X would have a little residual affinity and, in extreme cases, would be able to form compounds. Examples of this were:[208]

$Cl_3S.Cl.AuCl_3$, $Cl_4P.Cl.AuCl_3$ and others.

Although Werner stated these possibilities, in most of his work he was unhampered by considerations of this kind, which would have made for a theoretical system that was so flexible as to be useless as a framework for further ideas. The strength of Kekulé's system was that it was so rigid that, where rigidity was justified (organic chemistry), it was eminently successful. In a similar way Werner's general theory was admirably suited to application in inorganic chemistry in the state in which it then found itself.

(e) *Inorganic stereochemistry*

This last characteristic of Werner's ideas is the most well known. It was in fact their ultimate justification. This was particularly true of his hypothesis of an octahedral arrangement for the valencies when the co-ordination number is six. Assuming that six monovalent groups of atoms are attached to one metal atom, he reasoned as follows:

If we think of the metal atom as the centre of the whole system, then we can most simply place the six molecules bound to it at the corners of an octahedron.[209]

Applying this to a molecule of the type $M(NH_3)_4X_2$, he showed that two isomeric complexes would exist:

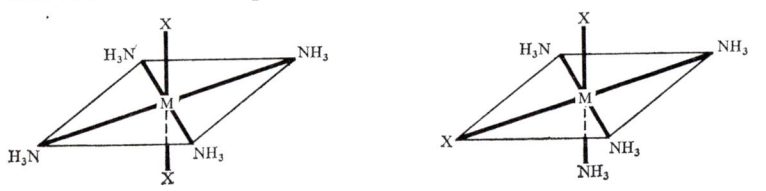

The first experimental evidence for this, however, came in the slightly more complicated substances prepared by Jörgensen[210] from cobalt (III) salts and ethylene diamine. Two isomers were found to exist (one green, one violet), and Werner comments: "This interesting isomerism is a first confirmation of the conclusions drawn from the octahedron."[211] In the same paper he demonstrated how his theory was confirmed by the known isomerism of the platinum ammine chlorides:

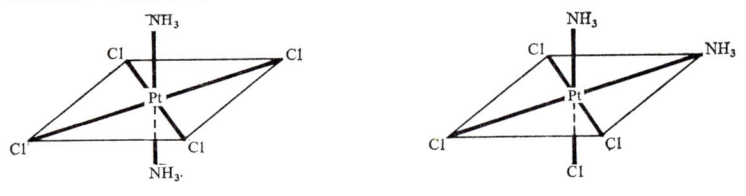

He was able also to show the difference between the *roseo* and *purpureo* salts of cobalt:

$$\left(\text{Co}\begin{array}{c}(NH_3)_5\\H_2O\end{array}\right)X_3 \text{ and } \left(\text{Co}\begin{array}{c}(NH_3)_5\\X\end{array}\right)X_2$$

After this, much other work was done extending the body of evidence for Werner's views. Not only geometrical isomerism, but also optical isomerism was found, exactly as predicted, especially with ethylene diamine and other organic bidentate groups, *e.g.*, in the following example:[212]

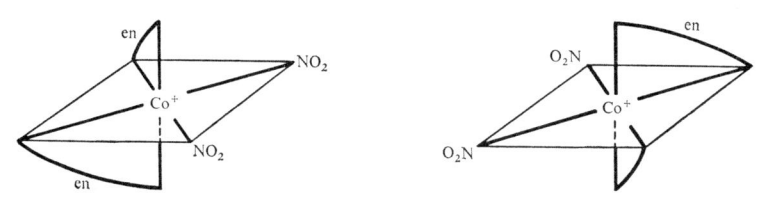

Werner himself was able to prepare the first optically active compound containing no carbon atoms, thereby disproving a suggestion that the presence of carbon had in some way caused the phenomenon:[213]

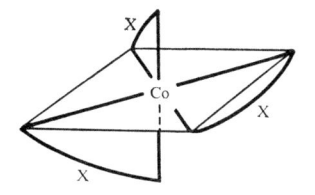

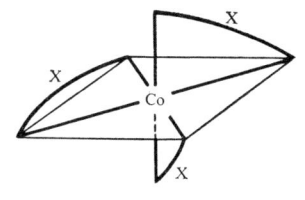

$$x = [Co(OH)_2(NH_3)_4]$$

Now the significance of all this work on isomerism is partly that it confirms Werner's octahedral model. But it does far more than that, for it also justifies his thesis that six groups are attached to one metal atom, and thereby finally settles the issue between chain formulae and what we may now term "co-ordination" formulae. It is interesting to note the reasoning which led him to an octahedral structure; it was just the simplest way of arranging six atoms round one centre. In this respect his argument strikingly anticipates some aspects of present ligand field theory.

Finally, it may be noted that Werner also suggested that in some cases (*e.g.*, platinum) a planar rather than a tetrahedral arrangement was possible.[214] In this way he explained the existence of isomeric ammines of the type:

$$\begin{array}{cc} \text{Cl} & \text{Cl} \\ | & | \\ \text{NH}_3-\text{Pt}-\text{NH}_3 \quad \text{and} \quad \text{NH}_3-\text{Pt}-\text{Cl} \\ | & | \\ \text{Cl} & \text{NH}_3 \end{array}$$

It remains to comment briefly on Werner's influence. This permeated inorganic chemistry in general for many years, but was especially significant in three cases. In the first place Flürscheim utilized his view of each atom having a definite amount of available affinity, most of which was employed in bond formation. A small portion, however, remained as "residual affinity", and he applied this idea to account for an apparent alternation of bond type. If X had a high so-called "affinity demand", relative to hydrogen, the atom to which it was attached would be able to form only weak bonds with its other addenda, and so they in turn would have sufficient affinity to form strong bonds, *etc.* Analogous situations exist when Y has a low affinity demand. This can be represented diagrammatically as follows, thickened lines representing strong links:

$$\text{X}—\text{A}—\text{B}—\text{C}—\text{D} \qquad \text{Y}—\text{A}—\text{B}—\text{C}—\text{D}$$

This alternation he applied to benzene derivatives[215] to account for observed orientations in disubstitutions, *e.g.*,

HSO$_3$

In a number of later publications Flürscheim continued to propagate these views.[216] They paved the way for the electronic theories of conjugated systems, and together with the original views of Werner himself strongly influenced the work of Lapworth. Much of this was in the electronic interpretation of valency (p. 288), but the $\alpha\gamma$ rule of 1901 reflects Werner to some extent.[217] He was able to show that the ready substitution of the γ-hydrogen atoms in ethyl crotonate was parallel with the reactions of the methyl group in *o*- and *p*-nitrotoluene.

Finally, G. N. Lewis has written:

In attempting to clarify the fundamental ideas of valence, there is no work to which I feel so much personal indebtedness as to this of Werner's.[218]

The measure of this debt may perhaps be clear from the facts recorded in Chapter XIV.

Nevertheless, the chief weakness of Werner's ideas was not unnoticed by some workers, amongst them Stewart:

Werner has had to introduce all sorts of subsidiary hypotheses ... thus making the matter extremely complicated. Whether such a steadily increasing complexity tells in favour of any view is for the reader to decide for himself. Progress in science, however, is generally a simplifying process rather than one advancing from complexity to greater complexity.[219]

Thus as with the Theory of Types, so with Werner's theory a series of valuable generalizations was in need of some simplifying factor, and a few far-sighted chemists perceived this. With the rise of the electronic theory Werner's views became incorporated in an even greater generalization.

CHAPTER XI

Valency in Unsaturated Compounds

THE PHENOMENON OF UNSATURATION was no new one. Ethylene and a number of its homologues were familiar substances by the time that the concepts of valency were becoming well known in the early 1860s. Yet compounds of this class were soon to become the centre of some considerable discussion, when the matter of their constitution arose, and were found to raise quite fundamental questions about valency itself. In particular these involved the dispute between advocates of fixed and variable valencies, though matters never reached the intense level of controversy that enlivened inorganic chemistry for half a century.

The nature of unsaturated compounds became of greater interest when it was realized that they could often exist in a combined state as radicals, as well as being free and uncombined. In this connection the work of Wurtz was of primary importance. Having discovered ethylene glycol,[1] he embarked upon a series of experiments involving formation of ethylene oxide and its application in producing the polyethylene glycols[2] and amino-ethanols.[3] In this way the ethylene radical was thoroughly established, and valuable contributions made to the doctrine of polyvalent radicals and its derived concept, valency. Buff also proposed recognition of the "diatomic" radicals C_nH_{2n} at about the same time.[4]

These discoveries helped to focus attention on the *nature* of these hydrocarbons. Frankland also produced the same effect in a discourse on the organo-metallic compounds. Suggesting that the distinction must be made between the "points of maximum stability and saturation", which for many elements were not the same, he showed how the allyl compounds may be based upon the "carbonic oxide" (carbon monoxide) type:[5]

$$C_2 \begin{cases} O \\ O \end{cases} \qquad C_2 \begin{cases} C_4H_5 \\ I \end{cases} \quad (C = 6; O = 8)$$

("carbonic oxide")   (allyl iodide)

This, by strong implication at least, pointed to a connection between the unsaturation of allyl compounds and a carbon atom with an abnormally low valency. When, six years later, Mills protested against different theoretical treatments being given to the two chlorides of iodine and "the precisely analogous case $(C^3H^5)Br$ and $(C^3H^5)Br^3$",[6] he was giving expression to a common belief – almost an intuition – that unsaturated compounds presented fundamentally the same problems as those inorganic examples of anomalously low valency in an element. It is also worth observing that the rise of structure theory raised the same issues in an urgent form, and its progress depended upon their clarification. At the same time it was not until certain structural matters had been settled that a clear understanding of unsaturation could emerge, for its early difficulties were essentially structural.

In the following account the five main views held before the advent of electronic theory will be considered, in approximately their chronological order, though there was a certain overlap between some of them.

1. "GAP-THEORY"

By this name (*Lückentheorie*) Kekulé described the views that he held for most of the 1860s on the nature of unsaturation. They were not his earliest opinions on the matter, but seem to have been the first that he, and others, held with much conviction. He had been dealing with a number of unsaturated acids, and especially the isomerism shown by itaconic, citraconic and mesaconic acids. In the course of discussion of these he said:

Where two atoms of hydrogen are missing, there will be two unsaturated units of affinity of carbon; there will be, so to speak, a gap. One could explain in this way the great facility with which these acids combine with hydrogen and bromine: their free affinities tend to be saturated and the gap to fill.

If one fills these gaps by hydrogen, all the affinities of carbon in the interior of the molecule will become saturated with the same element (hydrogen), and one sees no reason for the existence of an isomer.[7]

In the German version he added a significant footnote, to which we shall return (p. 232), suggesting double bonding could be an alternative explanation.

In the few years following this publication, several other authors used the same idea in their formulae. A celebrated example was Hofmann's model for ethylene which he displayed at the Royal Institution in 1865; it has an unsymmetrical structure, and two "gaps", represented by two spare valencies, on one carbon atom:[8]

$$\begin{array}{c} H \\ | \quad | \\ H-C-C-H \\ | \quad | \\ H \end{array}$$

In 1866 Kekulé's own pupil and assistant, Swarts, sought to clarify the problem of the isomerism of itaconic, citraconic and mesaconic acids. Using the "gap" hypothesis he assumed they differed in the positions at which a carbon atom existed with two unsatisfied affinities, formulating them respectively as:[9]

$$\left\{\begin{array}{l} CO_2H \\ CH_2 \\ C.. \\ CH_2 \\ CO_2H \end{array}\right. \qquad \left\{\begin{array}{l} CO_2H \\ C.. \\ CH_2 \\ CH_2 \\ CO_2H \end{array}\right. \qquad \left\{\begin{array}{l} CO_2H \\ CH_2 \\ CH_2 \\ C.. \\ CO_2H \end{array}\right.$$

The following year, Kekulé presented an account of Swarts's work to the Belgian Academy, and gave a modified description of the views which, he said, were shared by Swarts and himself.[10] Perhaps regarding the last two of the above formulae as identical, he put forward four possible isomers for an unsaturated, dibasic straight-chain acid $C_5H_6O_4$:

$$\begin{array}{cccc}
COOH & COOH & COOH & COOH \\
| & | & | & | \\
CHH & C.. & CH. & CH. \\
| & | & | & | \\
C.. & CHH & CH. & CHH \\
| & | & | & | \\
CHH & CHH & CHH & CH. \\
| & | & | & | \\
COOH & COOH & COOH & COOH \\
(1) & (2) & (3) & (4)
\end{array}$$

In the last two he considered the two isolated affinities would link up to form, respectively, a double bond and a 3-membered ring. Now, he had three substances and four formulae. Assuming no. (4) was the least probable he designated the remaining three to itaconic (1), citraconic (2) and mesaconic (3) acids respectively.

Further, he extended these ideas to fumaric and maleic acids, which he regarded as analogous to (3) and to (1) or (2) respectively; this would make them:

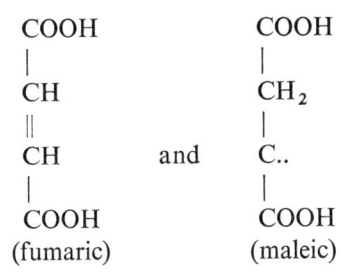

$$\begin{array}{cc} \text{COOH} & \text{COOH} \\ | & | \\ \text{CH} & \text{CH}_2 \\ \| & | \\ \text{CH} \quad \text{and} & \text{C}.. \\ | & | \\ \text{COOH} & \text{COOH} \\ \text{(fumaric)} & \text{(maleic)} \end{array}$$

His choice in this matter was governed by the greater readiness with which maleic acid underwent addition. Some years previously[11] he had suggested that the last-named acids were homologues of the first three. Moreover, Kekulé had encountered the reaction whereby malic acid was converted into a mixture of maleic and fumaric acids.[12] These were noticeably different, and in the absence of clear stereochemical ideas, had to be given different structural formulae; Kekulé's suggestion was reasonable enough:

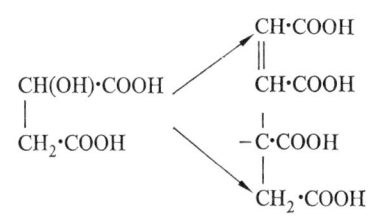

$$\text{CH(OH)·COOH} \atop \text{CH}_2\text{·COOH} \quad \diagup\!\!\!\diagdown \quad {\text{CH·COOH} \atop \|} \atop {\text{CH·COOH} \atop {-\text{C·COOH} \atop \text{CH}_2\text{·COOH}}}$$

The days of van't Hoff's theory were still a short distance away.

A later alternative for maleic acid, $CH_2=C(COOH)_2$, as suggested by Richter,[13] was unacceptable because this would be expected to undergo decarboxylation on heating, which maleic acid does not. Anschütz tells us that about 1880 Kekulé was of the opinion that fumaric was a polymeric form of maleic acid, having by then abandoned his "gap" theory.[14]

This theory was to receive a number of setbacks in the 1860s, but none of these can be regarded as a decisive defeat. Thus in 1861, Butlerov attempted to make methylene from methylene iodide by reduction with sodium amalgam. Unsuccessful in this, he tried more drastic treatment but merely obtained mixtures containing higher alkenes. Therefore he concluded that when methylene is formed, it immediately polymerizes.[15] Attempts were also made to obtain ethylidene ($CH_3.CH:$), *e.g.*, by the removal of HCl from ethyl chloride. Here again there was a uniform lack of success, the inference being drawn that it must be so unstable that it immediately isomerizes to ethylene or is destroyed in some way.[16]

Kekulé must have known of the first of these results when, in 1865, he wrote:

In the group of fatty bodies, the hydrocarbons of the ethylene series can be considered as closed chains. It should be clear that ethylene is the starting member of this series and that the hydrocarbon CH_2 (methylene) does not exist, for it cannot be believed that two affinities which belong to the same carbon atom should be able to link themselves together.[17]

In view of this it is possible to infer that his "gap" theory was propounded with some reluctance.

This quotation from Kekulé shows a curious pattern of reasoning that was often met in this connection. He gives his reason for rejecting the existence of methylene as the incredibility of two affinities existing on the same carbon atom. But how did he know this? Only from the results of experiments like that of Butlerov. The same mental pattern is met in Crum Brown's paper the previous year in which isomerism was discussed in terms of his new graphic formulae:

I do not intend to deny the possibility of this [diatomic carbon], but all we know of such "non-saturated" substances leads to the belief that the atomicity of the carbon radical C_n is reduced, not by one or more of the carbon atoms becoming diatomic, but by the union of the carbon atoms taking place in the way represented by the following graphic formulae.[18]

There then follow some examples of formulae with double bonds between adjacent carbon atoms. He goes on to add that ethylene chloride should be regarded as having a symmetrical structure *because of* the formula for ethylene. In other words, structures are derived from the assumption of double bonds rather than *vice versa*.

Despite this, there is no doubt that the doctrine of multiple bonding was grounded in experiment. Plenty of evidence was available, as will be seen, but here we have an apparently genuine example of a view that arose almost intuitively, later to be justified in practice. Only the exigencies of an unexplained isomerism would have forced Kekulé to fly in the face of this widely felt attitude.

It is hardly surprising that events took this course when it is remembered that structure theory was in its early stages, and time was necessary for the distinction between the two opposing views even to be recognized. It is unhistorical to look at these matters as clear-cut issues, immediately decided by appeal to experiment. And for some, particularly the British chemists Frankland and Crum Brown, the decision was much easier than for others, of whom Kekulé was possibly the chief example. Yet it was only a matter of time before he, and others like him, felt the full pressure of an accumulating weight of structural data. Kekulé's change of opinion may be seen in a joint paper with Zincke in 1870. Arguing "from the standpoint of valency" they concluded that the so-called "chloraceten" of Harnitz-Harnitzky isomeric with vinyl chloride, and therefore of the formula

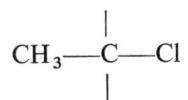

$$CH_3\text{—}\overset{|}{\underset{|}{C}}\text{—}Cl$$

was unlikely to exist. This they showed to be the case.[19] Two years later they were of the same opinion.[20]

It remains to distinguish two hypotheses used under this head; first there is the view of free valencies – or gaps – on adjacent carbon atoms; secondly there is the assumption of isolated valencies on single atoms. In the former case the theory approaches that of a double bond, in the latter that of Frankland's "latent atomicity" (see p. 188). Only when a structure is completely unknown do the views merge. There is little doubt that this was in fact the case at first. The idea of mutual saturation of free affinities on adjacent atoms to form a double bond will be considered in the next section, and raises little difficulty. But the postulation of free affinities on the same atom brings up more complex issues.

The necessary question to ask is why such formulae were ever proposed. With Hofmann's formula for ethylene it was probably a matter of very largely arbitrary choice; in the dehydration of alcohols it was perhaps considered more likely that *both* the hydrogen *and* the hydroxyl came from the same carbon atom. With other cases, however, the authors were driven to their hypothesis by two factors. One, particularly strong with Kekulé and his school, was an abhorrence of variability in valency: all valencies had to be accounted for, even if they were "free"; no other possibility remained for molecules like carbon monoxide, for example. There is good reason for supposing that the existence of that simple anomaly was a powerful influence for the retention of this view.

The other driving force was structural necessity. Thus, faced with the existence of three unsaturated dibasic acids, and assuming straight chains, the proposals put forward by Kekulé and Swarts were highly suitable. They were not then to know that in fact the acids were *branched chain* structures, thus invalidating all their reasoning. As late as 1877, Fittig was impelled by the same cause to adhere to the postulate of divalent carbon for certain unsaturated acids.[21]

To conclude this account of the "gap-theory", it will be as well to summarize the evidence that had begun to accumulate against the possibility of an unsymmetrical formula for ethylene, CH_3CH, and therefore for divalent carbon in it.

(i) Failure to isolate ethylidene[22] and methylene.[23]
(ii) Oxidation of ethylene chlorohydrin to monochloroacetic acid.[24] This implied the presence of a CH_2Cl group in both compounds, and therefore of a CH_2 in ethylene.
(iii) Failure to isolate more than one propylene. If the "gap-theory" were correct, at least two isomers should exist:

$$CH_3.C.CH_3 \quad \text{and} \quad CH_3.CH_2.CH$$

(together with $CH_3CH{=}CH_2$ on the double bond theory). In an attempt to prepare these, Friedel and Ladenburg acted on 2,2-dichlorpropane with sodium, but in fact obtained only the known propylene, $CH_3.CH{=}CH_2$.[25]
(iv) Failure to dehydrobrominate 1-bromo-2-methyl-propene, $(CH_3)_2C{=}CHBr$. Butlerov considered that unless substances like $(CH_3)_2C{=}C$ could exist, this compound should not be dehydrobrominated by alkalis to give a still more unsaturated product. This was found to be so, the product with sodium

ethoxide being an ether (although a cyclopropane had actually been anticipated).[26]

2. MULTIPLE BOND THEORY

The recognition of multiple bonds was a necessary consequence of two things: the acceptance of an unvarying tetravalency for carbon and of a juxtaposition of two carbon atoms in a molecule, each with a deficiency of substituents. But, as has been suggested above, structural certainty was not an indispensable factor, and some early hints at double and triple bonds bear this out.

Probably the earliest of these may be found in Kekulé's most famous paper of 1858 on the nature of carbon; thus he says of certain compounds:

Others contain so many carbon atoms in the molecule that a denser arrangement of carbons must be assumed.... Naphthalene contains still more carbon than benzene. It must be assumed that the carbon in it is arranged in a still denser form, that is, that the individual atoms are still more closely bound to one another.[27]

In this discussion, Kekulé is speaking of alkenes and aromatic compounds together. His supposition that in such cases the carbon atoms are nearer each other has of course been justified by measurements of inter-atomic distances. That is not as fortuitous as it might seem, for present theory explains this shortening of a bond by the assumption of a new kind of interaction between the atoms concerned, and Kekulé, who often spoke of "attraction" between atoms, would have been aware that this increased with decreasing distance. His idea in implying a higher degree of bonding, vague though it was, had a kernel of truth in it.

More precise statements were shortly to come, however. In his *Lehrbuch* (vol. i, 1861), Kekulé repeated his view that carbon atoms were bound more densely in unsaturated compounds,[28] but he also admitted the following possibility:

One atom of a diatomic element can, *e.g.*, combine with another atom of a diatomic element in such a way that the two units of affinity of the one atom are linked with the two units of affinity of the other. The molecule of oxygen, for example, is such a bonding-together of two atoms.[29]

Application of this to carbon did not come yet for Kekulé, however.

In 1861 Loschmidt issued his pamphlet on his own constitutional formulae.[30] These have already been mentioned (p. 97), but it may be added that he recognized two units of affinity joining carbon atoms in, *e.g.*, allyl alcohol:

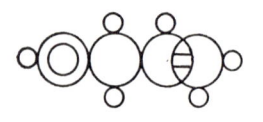

allyl alcohol

Unfortunately, his work was largely unnoticed at the time, and exerted no appreciable influence on the main stream of thought.

The following year Erlenmeyer, with characteristic insight into constitutional matters, saw more clearly how things might be in ethylene and acetylene.

Whereas two oxygen atoms can bind themselves in only one way, which is seen in hydrogen peroxide, carbon is able to link up with hydrogen by means of two atoms in three different ratios, C_2H_6, C_2H_4, C_2H_2. In the first 2×1, in the second 2×2, and in the third 2×3 affinities of carbon are bound together. The corresponding number of hydrogen affinities that can thus unite is $(2 \times 4) - 2$ in the first compound, $(2 \times 4) - 4$ in the second, and $(2 \times 4) - 6$ in the third.[31]

Also in 1862, Kekulé came nearer to an admission of double bonding in unsaturated acids. Having spoken of the possibility of a "gap" in these molecules, marking the unsaturated units of affinity, he added a significant footnote in the German version of the paper:

One can just as well assume that the carbon atoms are to some extent heaped together in such a position that two carbon atoms are bound together with two units of affinity. This is only another form of the same idea.[32]

The last sentence was seized upon by Erlenmeyer, who saw that this double linking necessitated removal of hydrogen from *adjacent* atoms when the acids were formed, but that no such necessity existed with a general assumption of "gaps".[33] Again he was striving for structural clarity a little ahead of Kekulé. At about the same time Butlerov suggested, on the basis of his preparation of ethylene from methylene iodide,[34] that it contained "two CH_2 atoms united with each other".[35]

The first attempt to introduce these ideas other than incidentally was that of Crum Brown. In a paper that is a masterpiece of clarity[36] he introduced to the Royal Society of Edinburgh his new graphic notation, and dealt with a number of cases, real or assumed at that time, of isomerism. Even he, however, fails to give a systematic account of the evidence for double or triple bonds, this not being his main objective. His reluctance to admit "gaps" has already been quoted;[37] it is possible to see why this was so, chiefly by extracting the information from different parts of the paper and piecing it together in an order that has some logical compulsion.

First, there is the constitution of acetaldehyde. Crum Brown considers this had been established by Kolbe,[38] and was universally accepted. In his (and our) atomic weight values he wrote the formula for acetaldehyde thus:

$$\left.\begin{array}{c} CH_3 \\ H \end{array}\right\} CO$$

But phosphorus pentachloride reacts upon this with formation of ethylidene chloride, whose chlorine atoms must therefore replace the oxygen, and thus be attached to the same carbon atom:

$$CH_3CHO + PCl_5 = CH_3CHCl_2 + POCl_3$$

This, however, is not the same product as that obtained from ethylene and chlorine; only one possibility is left for this, ethylene chloride, viz., $CH_2Cl.CH_2Cl$. Consequently the two "spare" affinities to which the chlorine atoms were attached must be on adjacent carbon atoms which makes ethylene

$$CH_2\!-\!CH_2 \text{ or } CH_2\!=\!CH_2$$
$$\;|\quad\;|$$

and the tetravalency of carbon implies the latter.

A point that is rarely made, but which is surely relevant, is that the first formula was ruled right out, not because it left carbon in a low valency state, but because that state was three and not two. Divalent carbon had to be a possibility in view of carbon monoxide, but not until the end of the century was there a suggestion that a valency of three could be permitted[39] (except by a few prepared to admit no limitations whatever on the valencies of an atom).

Once matters had been brought into the open by the employment of a system of notation multiple bonding became generally accepted. Crum Brown himself indicated double and triple bonds as follows:

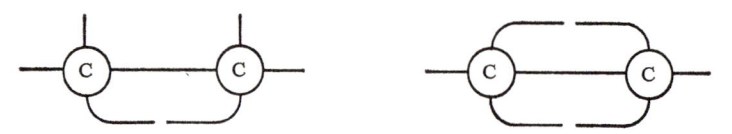

and, with the possible exception of Loschmidt, was the first to grasp the idea clearly enough to be able to give it a simple graphical representation. It has been argued that the use by Butlerov of double bonds in structural formulae (as $CH_2 = CH_2$) led to the widespread adoption of that device in chemistry.[40]

Meanwhile others, however, had produced alternative systems that have failed to meet with general approval. Thus in 1865 Wilbrand represented singly, doubly and triply linked carbon atoms as respectively:[41]

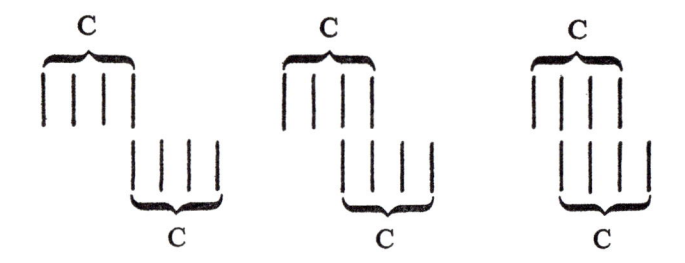

Kekulé himself used an alternative system. In his *Lehrbuch* (vol. i, 1861) the "roll-and-sausage formulae" appeared, and indicated multiple links in a few cases, but not between adjacent carbon atoms; examples include:[42]

carbon dioxide   hydrogen cyanide

In 1865 these formulae were adapted to indicate double bonds between carbon atoms, as in his open and closed chain formulae for benzene.[43]

By 1867 these had been superseded by his new "atomic model", whose advantage over Crum Brown's formulae was that it enabled a better representation of double and triple bonds: "a model of this kind permits the linking of 1, 2 and 3 units of affinity, and, it seems to me, it allows everything that a model is generally in a position to perform".[44] A similar model is depicted in the second volume of the *Lehrbuch* that came out the previous year[45] (see Plate 9).

Also in 1867 came the first formal acknowledgment by Kekulé that free valencies on adjacent carbon atoms would always unite. Rewriting

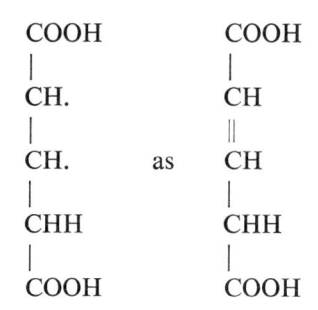

COOH		COOH
|		|
CH.		CH
|		‖
CH.	as	CH
|		|
CHH		CHH
|		|
COOH		COOH

he said that here, "as in all analogous cases, one can with much probability admit that the two atoms of carbon are found in a more intimate combination, so it seems to us".[46]

As a final statement by Kekulé, his joint remarks with Zincke may be quoted for their reference to all three of the theories held by him during the previous dozen years:

The composition and properties of almost all organic compounds, where free carbon affinities or gaps had been previously assumed, can now be better interpreted, one believes, by the assumption of denser carbon binding. Whereas the ease of undergoing addition shown by many compounds had been previously explained by the presence of gaps, it is now more likely that the denser carbon binding is partially loosened in additions of this kind. One now represents nearly all such bodies by formulae which contain doubly or triply bound carbon atoms as ethylene, $H_2C=CH_2$ or acetylene, $HC\equiv CH$.[47]

3. REDUCED VALENCY THEORY

By the year 1870 most chemists accepted the double bond theory for ethylene. How this could be reconciled with an increased reactivity relative to ethane was just not known. Some years had to elapse before the time was ripe for a "strain theory" like that of Baeyer to be suggested. In the 1870s, however, the theory of constant valency became increasingly discredited as the previous chapter has shown. Even the idea of valencies varying by factors of 2 ran into serious difficulties in this decade, and towards the end of it there arose some who advocated a more empirical approach, accepting whatever compounds Nature might offer, and acknowledging no necessity for unvarying valency.

Thus it became possible to regard carbon as trivalent, and its reactivity in ethylene as a manifestation merely of the smaller stability of that state compared with that of its normal tetravalency. The one factor against this was that carbon of all the elements showed such a constancy of valency in all other situations, its monoxide excepted; and, of course, more compounds were known of this than of any other element. Nevertheless, by 1880 some put forward trivalent carbon as the cause of its reactivity in alkenes. Particularly notable were the ideas of Lossen. Rejecting multiple bonds in general, he wrote:

> The atomic binding in acetylene, HC—CH, is fully explained by saying that it contains two divalent carbon atoms; similarly benzene by the statement that it contains six trivalent carbon atoms, of which each is joined to one hydrogen atom.[48]

Consistently with this, he wrote formulae as:[49]

$$\begin{array}{c} H \\ \diagdown \\ C-C \\ \diagup \\ H \end{array} \begin{array}{c} H \\ \diagup \\ \diagdown \\ H \end{array} \qquad O-O \qquad N-N \qquad \begin{array}{c} H \\ \diagdown \\ H-C-C \\ \diagup \\ H \end{array} \begin{array}{c} O \\ \diagup \\ \diagdown \\ H \end{array}$$

So as to leave no room for doubt he added:

> I call oxygen divalent because in its combining zone two atoms of hydrogen, two atoms of carbon, or in short, any two atoms can exist. An atom cannot be twice in the combining zone of a carbon atom; to assume that is nonsense. My way of considering the matter knows no double bonds on polyvalent atoms.[50]

Although others tended to agree with most of this (*e.g.* Claus[51]), most were not at that time prepared to be so definite in their statements on double and triple bonds. But later, when Baeyer's strain theory had waxed and waned, there was a return to the viewpoint of Lossen. Thus in 1902, Hinrichsen joined in the opposition to double bonds, stating that all elementary atoms in a valency state less than the normal were unsaturated.[52] And above all, Werner pointed out the great variation in reactivity in so-called "unsaturated compounds", pointing strongly away from a conception of integral valency, and comparing tetraphenylethylene with styrene, *etc.*; he commented:

> Poly-bound atoms exhibit a degree of saturation the value of which is dependent on the nature of the atoms to which they are joined. This is smaller than that expressed in the structural formula, but greater than the next lower degree of saturation as expressed by the valency formula.[53]

By this time, however, the electronic theory was not far away. Meanwhile, during the interval between Lossen and Werner an alternative viewpoint had come into being – the strain theory of Baeyer.

4. STRAIN THEORY

A curious fact in the evolution of the idea of a double bond was the reluctance to attach such a symbol to those molecules where it would be most useful. For example, Kekulé was prepared to admit it for the oxygen molecule in 1861,[54] but not for unsaturated carbon compounds until 1867.[55] Indeed he applied it to aromatic compounds before alkenes.

No reason seems to have been given for this, but it is likely that the stumbling-block was the great reactivity of ethylene and its homologues. Multiple bonds in many inorganic structures are not centres of any appreciable reactivity, and aromatic compounds are notable for their resistance to addition. It is therefore suggested that Kekulé and others were inclined to regard alkenes almost as a special case, and to explain their intense activity almost in terms of free valencies on the carbon atoms concerned. This would account for the late recognition by some of double bonds in ethylene; and it brings into focus the familiar fallacy that a double bond is a union of greater strength than a single one. At that time, only the alkene (olefin) family seemed to contradict this.

Therefore, when multiple bonding in organic chemistry was recognized a new problem awaited solution: why were alkenes so highly reactive? Several answers were given to this, the most influential being that proposed by Baeyer.

Adolph Baeyer (1835–1917) became a student at Heidelberg in 1853, working at first under the guidance of Bunsen, who occupied the Chair of Chemistry, on the problem of cacodyl. When Kekulé was appointed in 1856 as Privatdocent to take charge of the organic work, Baeyer began to work under his direction; and on moving to the Chair at Ghent, Kekulé took Baeyer with him. They remained in partnership for two years, after which Baeyer moved to a teaching appointment in Berlin where he remained for over twenty years. Contact was not lost, however, and Anschütz refers to letters between the two men from 1860 to 1866, then (1929) in the possession of Baeyer's son. These show a continuation of the close friendship thus maintained by correspondence, particularly on matters of science.[56]

This contact between Kekulé and Baeyer was in large measure responsible for stimulating the latter to his famous "strain theory". During the years at Ghent Kekulé was developing and using his tetrahedral atomic model (p. 234), and Baeyer no less than others was impressed by its simplicity and usefulness. However, twenty years more were to elapse before he was to apply it to his new theory.

In 1885 Baeyer had been Liebig's successor at Munich for some ten years, and had recently turned from his magnificent researches on indigo to a different topic, the derivatives of acetylene. A number of new compounds were prepared, some of them highly unstable, and including polyacetylenes like the following:[57]

$$HOOC-C \equiv C-C \equiv C-C \equiv C-C \equiv C-COOH.$$

Now, it so happened that within the previous three or four years several reactive substances had been prepared that possessed structures involving 3- or 4-membered rings; they included cyclobutane-1,3-dicarboxylic acid[58] and cyclopropane:[59]

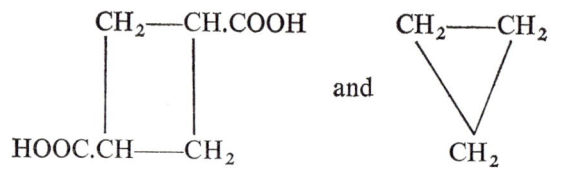

$$\begin{array}{cc} CH_2\text{---}CH.COOH \\ | \quad\quad\quad | \\ HOOC.CH\text{---}CH_2 \end{array} \quad \text{and} \quad \begin{array}{c} CH_2\text{---}CH_2 \\ \diagdown \;\;\diagup \\ CH_2 \end{array}$$

These examples and others were brought within a generalized scheme by Baeyer in 1885. Having outlined the conventionally accepted tenets of stereochemistry at that time, he proceeded thus:

The four valencies of the carbon atom act in directions which unite the mid-point of the sphere with the corners of the tetrahedron and which make an angle of 109°28′ with each other.
The direction of these attractions can undergo a diversion which causes a strain which increases with the size of the diversion.[60]

In the following passage he indicated the source of his ideas:

The meaning of this statement can be easily explained if we start from the Kekulé spherical model and assume that the wires, like elastic springs, are movable in all directions. If now the explanation that the direction of the attraction always coincides with the direction of the wires is also assumed, a true picture is obtained of the hypothesis.[61]

He went on to apply this to ethylene and polymethylenes with arguments now long familiar through frequent repetition. Their confirmation in Perkin's discovery of the stable cyclopentane-1,2-dicarboxylic acid,[62] and modification by the theory of buckled rings put forward by Sachse[63] and Mohr,[64] do not concern us here.[65] Baeyer's importance for the theory of valency lies chiefly in his doctrine that the olefinic double bond is a highly strained, two-membered ring. Of all the theoretical innovations in this paper, however, this was to prove the most ephemeral. The theory of conformational analysis is based on his conception of ring-strain, though it rejects his planar arrangement for saturated rings with six or more ring-atoms. But with this modification, Baeyer's theory exists today in the field of cyclic compounds. It has also been applied to the structure of the diamond in comparatively recent times.[66] Yet the view of ethylene as a 2-membered ring ran into difficulties long before the introduction of π-bonds.

For example, the persistence of unsaturated compounds at a high temperature was one difficulty; labile, strained molecules would be expected to break up under these conditions.[67] Further, on Baeyer's theory a triple bond should be more "strained" than a double, but in fact acetylene underwent addition rather less readily than ethylene.[68] Again, the stability of the carbon-oxygen double bond, for example, was difficult to explain in these terms.[69] But the most puzzling anomaly was the behaviour of conjugated systems. This final point required a new approach, and this was found in the work of Johannes Thiele.

5. PARTIAL VALENCY THEORY

In 1899 Thiele published the first of two papers in which his theory of "partial valencies" was used to account for numerous difficulties in the field of unsaturated and aromatic chemistry. Benzene was still an unsolved mystery, so far as valency was concerned, and an additional puzzle was set by Thiele's own recognition of the phenomenon of 1:4 addition to conjugated systems of double bonds.[70] Both of these difficulties were to yield before Thiele's theory; the former will be discussed in the next chapter, the latter in the following pages.

To explain anomalies in unsaturated systems it was necessary to begin with the simple, normal double bond:

I now assume that in bodies to which a double bond is assigned, actually two affinities of each of the participating atoms are used for binding themselves, but that – owing to the additive power of the double bond – the strength of the affinity is not fully used, and on each atom there is still an affinity residue, or a *partial valency*, an assumption which can also be based on thermal grounds.

This can be expressed in formulae:

$$\overset{\cdots}{C}=\overset{\cdots}{C} \qquad \overset{\cdots}{C}=\overset{\cdots}{O} \qquad \overset{\cdots}{C}=\overset{\cdots}{N} \qquad \overset{\cdots}{N}=\overset{\cdots}{N} \qquad etc.$$

where the sign ... signifies the partial valency. In this partial valency I see the origin of the additive power.[71]

Thiele goes on to show[72] how 1:4 addition may be explained by supposing that in a conjugated chain, the partial valencies not at the ends neutralize each other thus:

$$\overset{\cdots}{C}=\overset{\cdots}{C}-\overset{\cdots}{C}=\overset{\cdots}{C} \qquad becomes \qquad \overset{\cdots}{C}=\overset{\cdots}{C}-\overset{\cdots}{C}=\overset{\cdots}{C}$$

Thus a new double bond results which bears no partial valencies and therefore can be called an *inactive* double bond. . . . If additions enter at the ends of the system, affinity forces become free on the middle carbon atoms, so that the inactive double bonds obtain partial valencies and go over to ordinary active ones:

$$\overset{\cdots}{C}=\overset{\cdots}{C}-\overset{\cdots}{C}=\overset{\cdots}{C} \longrightarrow C-\overset{\cdots}{C}=\overset{\cdots}{C}-C$$
$$\qquad\qquad\qquad\qquad\qquad\quad |\qquad\qquad\quad |$$
$$\qquad\qquad\qquad\qquad\qquad\quad H\qquad\qquad\quad H$$

In view of the impression sometimes created that Thiele was

breaking completely new ground with his concept of "partial valency", it is necessary to point out that his originality lay in its application to *double bonds*. Others before him had postulated valency as a force capable of division to meet the demands of a situation, as Claus[73] and Lossen,[74] and some went so far as to speak in terms of "residual affinities", left over when an atom had formed the bonds required of it by the rules of classical valency theory. The writings of Pickering,[75] Armstrong[76] and Werner[77] have already been discussed in the previous chapter in this connection. The point that does require elucidation is how far, if at all, Thiele knew of, and was influenced by, their views.

It is probable that Thiele was aware of the ideas of Armstrong since the latter had contributed to the theories of benzene structure in which Thiele was interested to a high degree, and which were closely similar to those of Baeyer, in whose laboratory Thiele had studied. The affinity of his views with those of Claus and Lossen suggests some influence here, and this is strengthened by his use of the analogy of magnetism,[78] which was an issue raised by Klinger[79] in the course of the controversy that followed Claus's paper. Pickering's suggestions do not seem to have produced much fruit, except in Armstrong, while Werner's ideas on disposable affinity were still in a partly developed state. However, the fact is that at this time the "quantization" of valency, if one may use the term, was being widely questioned in many quarters, and it is probably best to regard Thiele's conception of it as yet another manifestation of this trend. But it remains true that it was he who applied it first to the problem of the double bond.

That Thiele had grasped a profound truth about unsaturation in general is widely admitted today. His anticipation of quantum-mechanical doctrines is remarkable; thus Dewar, in discussing the molecular orbital picture of butadiene, comments that "Thiele would say that we had reinvented his partial valency theory!"[80] And a more recent author from Thiele's own university has rightly called him "one of the founders of theoretical organic chemistry".[81]

CHAPTER XII

Valency in Aromatic Compounds

ONE OF THE EARLIEST TRIUMPHS – certainly the first major one – of the new structural theory was the solution to the problem of benzene. The hexagon formula of Kekulé, however, created a set of fundamental difficulties for the exponents of the theory of valency that were to remain until the present century. It was ironical that Structure Theory, the child of valency theory, should have rebounded back upon its progenitor quite so soon, and it is a measure of the stature of Kekulé and his colleagues that this setback was allowed to deter them so little.

1. THE ORIGIN OF THE HEXAGON FORMULA

As this subject bears only indirectly on valency theory, being primarily a structural matter, and as the facts are well known in any case, only the briefest summary will be given here.[1]

Apart from a premonition of a cyclic structure by Loschmidt (p. 98), the credit for introducing a six-membered ring formula for benzene must go to Kekulé. This was done in 1865:[2]

When several carbon atoms combine together, they can join in such a way that *one* of the four affinities of each atom is always saturated by *one* affinity of a neighbouring atom. This is how I have explained homology, and in general the constitution of fatty bodies.

Now, one can further admit that several atoms of carbon join together by combining two affinities with two. It can then be admitted that they combine alternately by *one* or *two* affinities.

One could explain these two modes of combination by the periods:

1/1; 1/1; 1/1; 1/1; *etc.*
1/1; 2/2; 1/1; 2/2; *etc.*

If the first mode explains the composition of fatty bodies, the second accounts for the constitutions of aromatic bodies, or at least for the nucleus which is common to all.

In effect, six atoms of carbon in combining according to the law of symmetry, will give a group, which, considered as an *open chain*, will still possess *eight* unsaturated affinities. If one admits, on the contrary, that

THE ORIGIN OF THE HEXAGON FORMULA

the two atoms which terminate this chain combine with each other, one will have a *closed chain,* possessing now *six* unsaturated affinities.

In these papers Kekulé used his own rather strange formulae to which reference has already been made (pp. 98 and 234).

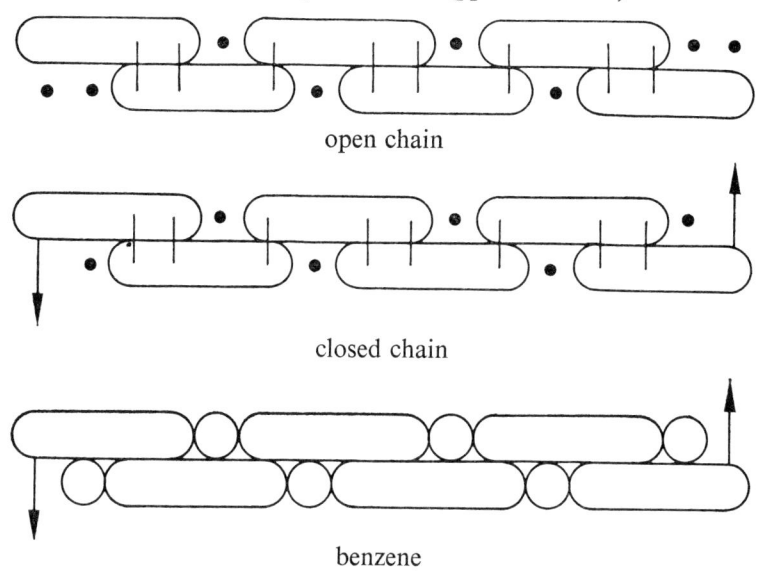

open chain

closed chain

benzene

The famous hexagon formulae (without double bonds) appeared later that year (I),[3] and in 1866 his *Lehrbuch*[4] contained a diagram of his model for benzene, which is equivalent to the modern symbol with alternating single and double bonds (II) (see also Plate 9):

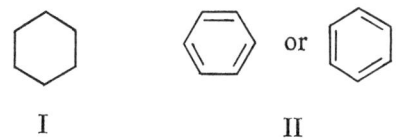

I   II

On the origins of this theory much has been written already. His own account of the "dream" at Ghent[5] is too well known for repetition, although it may be remarked that some new light on a possible cause for the appearance of this symbol in the dream appears in a recent article.[6] Also, to appreciate the historical perspective more accurately, it is worth recalling Armstrong's opinion that the time was ripe for such an announcement, and had Kekulé not made it someone else (probably Fittig) would have done so; and that its immediate impact on chemistry was astonishingly small.[7]

Rigorous proof of this structure came after its announcement, not before. To limit the number of possible isomers of C_6H_6 to a mere handful it is necessary to have evidence for the complete equivalence of all six hydrogen atoms. This was first provided by Ladenburg.[8] The high degree of molecular symmetry involved in the hexagon was further established by the proof by Hübner and others[9] of a symmetrically situated *ortho*-pair and a symmetrically situated *meta*-pair existing for each hydrogen atom. Perhaps the most impressive evidence, however, came from the extensive researches of Körner on orientation of benzene substituents, the results of which were completely in accord with the predictions of Kekulé.[10]

One other structure remained for some time as a rival to that of Kekulé's hexagon: the prism formula of Ladenburg.[11] This was put forward to overcome a difficulty inherent in the hexagon with its alternating bonds, namely, the non-existence of two *ortho*-disubstituted benzene derivatives:

The symmetry of the prism formula prevents this difficulty. Japp[12] has reviewed some of the latter's disadvantages, including (1) its failure to show the analogy between *ortho*-disubstituted benzenes and $\alpha\beta$-disubstituted paraffins, since the *ortho* positions are not connected in the prism; (2) the difficulty of formulating naphthalene, phenanthrene, *etc.* in this way; (3) the non-existence of C_6H_4XY in two enantiomorphous forms, which the prism formula would demand.

The final overthrow of Ladenburg's formula arose with the study of reduction products of *ortho*-disubstituted benzenes. To agree with the number of isomers obtained on introducing a third substituent, the three disubstituted benzenes were written thus:

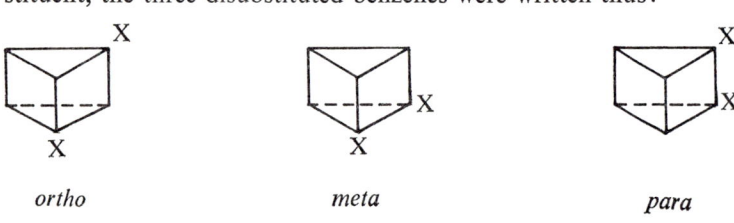

ortho *meta* *para*

Now, Baeyer's reduction of phthalic acid gave a hexahydro-phthalic acid very similar to succinic acid, and clearly having carboxyl groups on *adjacent* carbon atoms. Hence he found that phthalic acid, recognized as a member of the *ortho* series, cannot be represented by the first prismatic formula above.[13] Again, reduction of the *para* isomer, terephthalic acid, would be expected to give two reduction products; only one was obtained.[14]

In the present century, a planar molecule has been decisively indicated by physical methods of analysis.[15]

There has recently been a revival of interest in Ladenburg's ideas in the accumulation of evidence for the existence of a "valence tautomer" of benzene having the structure suggested by Ladenburg and appropriately named *prismane*. It is, of course, not the same as "ordinary" benzene, and several substituted prismanes have been prepared; two of the first cases[16,17] that appear to have been authenticated are indicated below, both being generated by photochemical means:

Finally, the extension of Kekulé's views to other aromatic types must be mentioned. The formulae on p. 246 are modern symbols for the suggestions made soon after Kekulé's hexagon appeared.

In the following discussion, benzene alone will be considered because other aromatic types presented essentially the same aspects, naphthalene — Erlenmeyer, 1886[18]

anthracene — Graebe and Liebermann, 1868[19]

phenanthrene — Graebe and Glaser, 1873[20]
Fittig and Ostermeyer, 1873[21]

pyridine — Körner, 1869[22]
Dewar, 1870[23]

indole — Baeyer and Emmerling, 1869[24]

pyrrole — Baeyer and Emmerling, 1870[25]

and also because these were not regarded as creating so urgent a problem.

It is well known that Kekulé's ring formula soon ran into difficulties. The representation of benzene as cyclohexatriene was open to the objection that it differentiated between the 1:2 and 1:6 bonds. Therefore, two *ortho*-isomers of $C_6H_4X_2$ would be expected, and two *meta*-isomers of C_6H_4XY:

These were not detected, and the existence of only one *ortho*-xylene, for example, was soon pointed out.[26] Also, the representation as a triene was not in accord with the well-known stability of benzene and its reluctance to undergo addition.

On these matters it would appear that a false historical perspective can be rather easily formed, and our modern sensitivity to matters of fine structure may create an entirely false impression on the magnitude of these difficulties in the 1860s and 1870s. Examination of the literature leaves the rather general, but nevertheless definite, impression that most chemists were content to accept the cyclohexatriene structure with few reservations. The problems were small compared with the successes it brought.

The difficulty of the non-equivalence of the 1:2 and 1:6 bonds was the more serious of the two major problems, but even that was not regarded as of great magnitude. Possibly the reason for this may be traced in the suggestion of Victor Meyer that the differences between the two possible *ortho*-isomers would be so small as to be undetectable.[27] Indeed at that time it was not at all clear that *o*-, *m*- and *p*- isomerism necessarily gave rise to as great differences as we recognize today. In the early days of the subject, arguments based upon the non-production of substances necessarily had to be treated with caution.

The previous chapter has shown that ideas on alkenes were by no means crystallized at this time, and that their behaviour often caused surprise. Although aromatic substances were regarded as rather special cases of alkenes, their resemblance to saturated compounds was not so well recognized as to be a cause for concern. Moreover, the novelty of a cyclic structure was at first, one suspects, sufficient to allay doubts engendered by a nonconformity to the behaviour expected from a molecule containing "double bonds".

Armstrong, looking back to that time, was probably right when he said that: "only a few pedants were concerned as to details of structure".[28] Since in this instance details of structure are details of valency, however, the activities of these "pedants" will be our next concern.

2. THE DIAGONAL BOND CONCEPT

(i) *The Dewar Formula*

At a meeting of the Royal Society of Edinburgh shortly after the introduction of Kekulé's hexagon formula, Dewar read a paper on the oxidation of benzyl alcohol, illustrating it with a mechanical model of Crum Brown's system of notation. No less than seven possible structures were illustrated for benzene itself, one of them the Kekulé hexagon and another as follows:

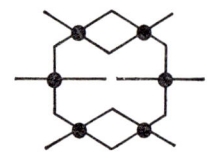

or as more usually written since then

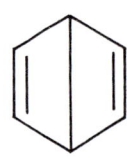

This characteristic formula, with one *para*-bond only, has since become known as "Dewar's structure", though it was not introduced with any more emphasis than any of the other six[29] (see Plate 8).

Now Dewar was at that time a student of Playfair in Edinburgh, but, anxious to become acquainted with Kekulé as a student of his theory, he wrote to him in Ghent, enclosing a model; as a result he was invited by Kekulé to work with him, and he spent a year in his laboratory studying the coal-tar bases.[30] At the same time, Kekulé's assistants included also Wichelhaus and Körner, both of whom were to pursue further the idea of a *para*-bonded structure. Indeed, Dewar himself seems not to have troubled with it any longer. Wichelhaus was in favour of this structure; for it agreed with the polymerization of acetylene:[31]

$$3 \begin{array}{c} CH \\ ||| \\ CH \end{array} \rightarrow \begin{array}{c} CH\text{——}CH\text{——}CH \\ || \quad\quad | \quad\quad || \\ CH\text{——}CH\text{——}CH \end{array}$$

Körner saw in a formula of this kind a means of explaining the transmission of reactivity from position 1 to position 4 – a fact that was difficult to explain on the Kekulé model.[32] The formula was also advocated by Städeler.[33]

The Dewar structure appeared occasionally in certain fused nuclei compounds, where some justification existed in the enhanced reactivity of the positions linked, *e.g.*,

Graebe and Liebermann[34]   Riedel[35]

The bridged ring formula for acridine was used by Riedel to derive a Dewar-type structure for pyridine which, ironically, was opposed to that suggested for that compound by Dewar himself.[36] Following the recent isolation of Dewar benzenes (see below), Dewar pyridines were first identified in 1970.[37]

Structures like this were opposed by Ladenburg,[38] Kekulé[39] and others, and the general objections to "long" bonds of this type were also applicable to other formulae involving diagonal bonds, as those of Claus (p. 250). But Dewar's formula lacked the symmetry[40] so obviously necessary for benzene, and was not generally considered.[41] The advantage of "explaining" transmission of reactivity to the *para*-position – its chief merit – was denied by Armstrong[42] who said that *meta*-substitution was inexplicable by it (not realizing, apparently, that *absence* of certain effects in the *meta* position could conceivably produce substitution there).

In the present century Dewar structures have occasionally appeared as suggested intermediates in chemical reactions. The earliest example of this seems to have been due to Ingold in 1922.[43] But until recently their main rôle has been as hypothetical contributors to the resonance of the benzene molecule (about 20 per cent). However in the 1960s there has been a renewed interest in the possibility of actually isolating substances with constitutions uniquely defined by a Dewar formula. In 1962 van Tamelen and Pappas reported the first genuine Dewar compound (I) by irradiation of its benzenoid isomer, and in 1963 obtained the parent itself, bicyclo[2, 2, 0]-hexa-2,5-diene (II).[44] Both passed easily back to the ordinary benzene. Since then several other Dewar structures have appeared.[45]

The hexamethyl derivative (III) was obtained by a single stage synthesis[46] and is now available in large quantities. The compound (V), though thermally stable to 200°C, could be converted into the prismane mentioned earlier.[47]

$$3\text{Me·C} \equiv \text{C·Me} \xrightarrow{\text{AlCl}_3} \text{hexamethylbenzene} \quad (\text{III})$$

$$(\text{V}) \underset{\text{Acid}}{\overset{\text{Light}}{\rightleftarrows}} (\text{IV})$$

In 1968 "Dewar benzene" itself was prepared by liquid-phase photo-isomerization of "ordinary" benzene.[48]

(ii) *The Claus Formula*

In 1867, Claus put forward his "centric formula"[49] (see next page). This was an excellent symbol representing the high symmetry of the molecule, and incidentally avoiding the inclusion of any olefinic bonds. There is no doubt that the diagonal bonds were suggested independently of Dewar. Claus had been a student of Kolbe, and inherited that chemist's suspicion of excessive dogmatism on structural matters, and possibly also some caution in the reception of Kekulé's ideas. His theories on double bonds (p. 236) convey well his independence of thought, tinged with not a little structural agnosticism. His centric formula is a symbol of his refutation of any implication of double bonds in benzene.

Of this formula Japp has written:

The centrical formula is difficult to criticize; the mode of disposing of the central bonds is entirely without analogy and does not appear to be accessible to the test of experiment. In its application and predictions, the centrical formula is identical with Kekulé's simple hexagon.[50]

However, it was attacked from several quarters. Thus it is impossible to reconcile diagonal bonds of any kind with the van't Hoff theory of tetrahedral carbon assuming a planar ring, as implied by optical inactivity.[51] Again it was objected that on this formulation, bonds 1:4 were the same as 1:6, *i.e.*, that there should be only two isomers of a disubstituted benzene, there being an equivalence between positions 2, 4 and 6 with respect to 1, all being linked directly to the latter.[52] Claus denied this, claiming that the 1:4 bond was longer than the 1:2 or 1:6.[53] Another attack came from a study of the reduction products of benzene derivatives. On addition of two atoms of hydrogen, one diagonal bond should be destroyed, but the other two should remain. Now Baeyer found that when this happened the products were always olefinic.[54] Claus replied that the two remaining *para*-bonds would rearrange themselves into double bonds:[55]

Attempts to decide this by addition to a *para*-disubstituted molecule, as terephthalic acid, were then made; the expectations were:

assuming Kekulé's formula:

assuming Claus's formula:

In fact the product obtained was the *second*.[56] But Baeyer showed that his recent work on reduction of dichloromuconic acid,[57] implying what was later called 1:4 addition, was consistent with a similar behaviour with benzene:

$$HOOC.CH{=}CCl.CCl{=}CH.COOH \xrightarrow[H_2O]{Na/Hg}$$

$$HOOC.CH_2.CH{=}CH.CH_2.COOH$$

The matter was hardly proved by this, however. But one other line of attack remained; for some time compounds had been known for which a bridged ring structure was certainly applicable. The complete dissimilarity between, for example, cyclohexene and camphor implied a significant difference in structure; a diagonal bond had been assumed in camphor, and was therefore missing in cyclohexene:[58]

cyclohexene   assumed formula for camphor

Claus entered with vigour into these controversies, claiming Baeyer was adopting his own views in a different guise (see below), and extending his theory to naphthalene, quinoline and other fused ring systems.[59] But the difficulty of a conclusive refutation, referred to by Japp,[60] meant that most chemists were rather cautious in adopting his formula. It is interesting to see its transient revival this century by Pauling,[61] who, however, felt obliged to abandon it later.[62] Only in modern times was it possible decisively to reject it on the grounds of the impossibly great bond-length, far in excess of the maximum for a stable single bond. This does not apply to the new Dewar valence isomers as the molecules are folded along the 1:4 bond, thereby shortening it. But the Claus structure must be planar.

3. THE CENTRIC FORMULAE

(i) *Lothar Meyer's Formula*

In his *Modernen Theorien* Lothar Meyer put forward the following symbol for benzene:[63]

This represents six free affinities in the centre of the ring. The formula was not generally accepted, possibly because the stability of benzene was hardly consistent with such a picture. But it undoubtedly had an influence on Armstrong who refers specifically to it, emphasizing the difference from his own view.

(ii) *Armstrong's Formula*

It appears that Armstrong proposed his centric formula under the stimulus of Thomsen's *Thermochemische Untersuchungen*,[64] the fourth volume of which he was summarizing in an article in the *Philosophical Magazine*.[65] Thomsen had shown that the heat of combustion of benzene was consistent with the presence in it of 9 single bonds, but not of 3 single and 3 double bonds.[66] This fact, which Ladenburg had taken as a justification for his prism formula[67] and which Thomsen himself used as the basis for his own octahedral model for benzene, Armstrong now applied to a new formula. Doubtless he also saw this as a further extension for his own partial valency theory (p. 205):[68]

I venture to think that a symbol free from all objection may be based on the assumption that of the 24 affinities of the 6 carbon atoms, 12 are engaged in the formation of the carbon-ring and 6 in retaining the 6 hydrogen atoms, in the manner ordinarily supposed; while the remaining 6 react upon each other, – acting towards a centre as it were, so that the "affinity" may be said to be uniformly and symmetrically distributed. I would in fact make use of the following symbol:

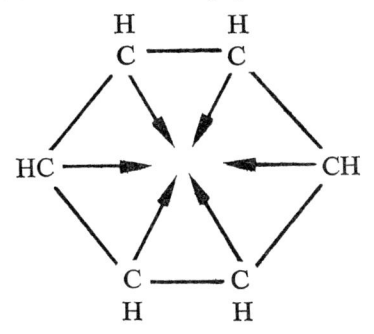

$$\begin{array}{c} H \quad\quad H \\ C \text{———} C \\ HC \longrightarrow \quad \longleftarrow CH \\ C \text{———} C \\ H \quad\quad H \end{array}$$

In this formula, Armstrong is careful to point out the difference from that of Lothar Meyer, as *free* affinities are not visualized. He denies that

any one carbon atom is directly connected with any other atom not contiguous to it in the ring; my opinion being that each carbon atom exercises an influence upon each and every other carbon atom.[69]

(iii) *Baeyer's Formula*

The following year, Baeyer proposed his version of a centric formula:[70]

He was led to this by his own discovery that the first reduction products of terephthalic acid were no longer unreactive to addition, and was thus led to reject the Claus formula.

In keeping with his advocacy of the "strain theory", Baeyer's concept was essentially mechanical:

> If one regards the model of benzene in its *mechanical* relations, one finds two groups of forces expressed by the latter, acting in opposite senses; on the one hand the central valencies, which seek to bring near the atoms to the middle point, and on the other peripheral ones which strive to effect an extension of the ring and of the distance of the atoms from the centre, as a consequence of the strain. These forces are in equilibrium.[71]

There was thus a kind of self-paralysis, and (very significant) an opposition between chemical and physical forces. No question arises of a "smearing-out" of the affinity as Armstrong had visualized, and the two models are in reality quite different, despite their symbolic similarity.

Baeyer, however, did not retain an allegiance to his symbol as did Armstrong. After his recognition of the existence of 1:4 addition in benzene,[72] he returned to the Kekulé symbol for some purposes, and retained his own centric formula only for some others.

These views, however, together with Bamberger's extension of them in his axiom of six central valencies as necessary for aromatic character,[73] were no more than new representations of a familiar problem; they explained nothing; their chief value lay in a correlation of known facts.

4. THE OSCILLATION THEORY

It is now necessary to go back to Kekulé himself. Aware of the deficiencies of his formula in terms of symmetry, at least, and challenged by alternatives even from his own assistants, he brought forth in 1872 a modification that was an attempt to solve the problem of non-equivalence of all the bonds in his doubly-bonded structure.[74]

He started with a concept that he obviously regarded as important, the idea of atomic motion. He had previously written a controversial paper on the relation of this to specific heats,[75] and he was now to apply it to the problem of benzene: "The atoms in the system that we call a molecule must be assumed to be in incessant motion."[76]

THE OSCILLATION THEORY

Asserting that the chemical laws of linking must be observed, and therefore some kind of equilibrium about a fixed point must be postulated, he continued:

> The single atoms of the system rebound on one another with an essentially rectilinear motion, so that they bounce off each other as elastic bodies. That which in chemistry is denoted by valency or atomicity gains now a more mechanical significance; the valency is the relative number of impulses which one atom experiences from another atom in unit time. In the same time in which the monovalent atoms once rebound together in a diatomic molecule, given equal temperature, divalent atoms in a diatomic molecule likewise come twice into collision.[77]

In this statement Kekulé gives one of the rare indications of his views on the nature of phenomena related to chemical structure. Even this, however, is an essentially empirical viewpoint: "valency is the relative number of impulses". It is interesting to compare this with, for example, Frankland's conception of a bond (p. 90).

Kekulé continues:

> Two atoms of tetravalent carbon, if they are bound, as we now say, by one affinity, rebound once from each other in unit time, and therefore in that time during which the monovalent hydrogen once traverses its orbit; in this same unit time they will still encounter three other atoms. The carbon atoms that we now call doubly-linked rebound twice from each other and undergo only two other collisions in the unit time, *etc.*[78]

Applying this to his own benzene formula, he adds:

> Each carbon atom in unit time rebounds three times with other carbon, in fact with two other carbon atoms, once on one and twice on the other. In the same unit time it meets once with hydrogen, which in the same time completes its orbit once.[79]

He then examines the movements of several atoms of the ring in detail, concluding:

> Each carbon atom collides with both the others with which it is joined together equally often, and therefore stands in exactly the same relation to its two neighbours. The usual benzene formula naturally expresses the impulse taking place in unit time, therefore the one phase, and thus one has been misled by the view that 1:2 and 1:6 biderivatives must necessarily be different.[80]

It is an interesting commentary on the state of chemistry at that time that this oscillation hypothesis fell very largely on unreceptive ground. Ladenburg commented: "Kekulé allows two formulae to be valid for the same compound admittedly only in a special case, in which great similarity is present between the formulae", and queried whether Kekulé had really understood the implications of his suggestion.[81] He gave this view no place in his *History of Chemistry*. Again to quote Armstrong, "this oscillation hypothesis never found favour".[82]

It seems that the kinetic theory had not become sufficiently developed for these considerations to be generally popular,[83] and there was doubtless a suspicion of any attempt to undermine the Structure Theory with duplication of formulae for the same compound. Tautomerism was not then a concept, although there can be little doubt that Kekulé's ideas helped to foster it later. At that time, however, it "only served to divert attention from the views based upon sound analogy and to weaken the standing of the double-bond formula".[84]

In 1897, Collie put forward a space-formula for benzene in which the six carbon atoms were imagined at the corners of a regular octahedron, and capable of vibrating about the centre of gravity of the molecule. When they do this, Collie claimed, they pass through an intermediate planar configuration which corresponds to the Kekulé cyclohexatriene formula, the two extremes corresponding to the centric formulae. This was claimed to account for the isomerism of benzene derivatives and also for the space-lattice of the crystalline solid.[85] It was, however, incapable of standing up to physical evidence, as that for the planarity of the benzene ring.

Two comments by workers from the present century will help to place the oscillation theory in some perspective.

Kekulé's idea of the oscillating double bonds in benzene was the closest approximation to the resonance interpretation of benzene and similar molecules achieved during all the time from Kekulé up to 1924.[86]

The revolutionary nature of Kekulé's description of benzene consisted in his recognition that the second half of a double bond, unlike a single bond, is a nomadic entity which cannot always be assigned a unique position in the molecule.... The electronic theory of valency ... shed a flood of light on structural chemistry and revealed the fundamental distinction between localized and non-localized electrons.[87]

5. THIELE'S THEORY

It remains to mention one of the last fruitful pre-electronic theories of the benzene ring. A worker in Baeyer's laboratory, Johannes Thiele, was long familiar with the problems of benzene chemistry, and at the turn of the century he brought to them his own fertile hypothesis of "partial valencies".[88]

As outlined in Chapter XI, this theory afforded an elegant explanation of the phenomenon of 1:4 addition to conjugated systems of double bonds. It was only an extension of this to apply it to a closed system like benzene, and obtain complete saturation of all such partial valencies:

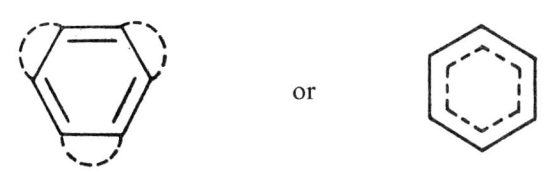

The theory, elegant though it was, ran into difficulty with other cyclopolyenes. Persistent failure to obtain cyclobutadiene led to a realization that it must be intensely unstable,[89] a matter which Thiele could not explain. When in 1911 Willstätter synthesized cyclo-octatetrene[90] he found it an intensely reactive alkene – again a blow to the view of "partial valencies".

cyclo-octatetrene

cyclobutadiene

Nevertheless, Thiele had reached a stage of recognition that it was probably impossible to surpass in the absence of wave mechanics. His views found early electronic interpretations in the work of Kaufmann[91] and of Thomson.[92] In many ways it is possible to agree with a characteristically Russian interpretation of his achievement; this asserts that, from the point of view of bond equivalence, "Thiele's theory of partial valencies represents the actual structure of aromatic substances better than Kekulé's concept of oscillating bonds".[93]

PART FOUR

Valency and Electricity

CHAPTER XIII

The Renaissance of Electrochemistry

THE MODERN INTERPRETATION of valency in electronic terms began just over half a century ago. Yet the foundations had been laid, and many of the essential principles enunciated, 100 years before then. For a period which may be roughly defined as that between 1850 and 1880 the electrochemical theory of Berzelius was in grave disrepute, but by the end of the century its basic axiom, that chemical affinity was fundamentally electrical, was reinstated with, of course, a new precision. The change in its fortunes affected the theory of valency in a profound way, and is the subject of this chapter.

1. THE ELECTROCHEMICAL THEORY OF BERZELIUS

The electrochemical theory of Berzelius originated in the earlier years of the nineteenth century.[1] Like the earlier electrochemical theory of Davy,[2] the views of Berzelius centred around the facts of electrolysis. Elements were classified as electropositive or electronegative,[3] and binary compounds were conceived as arising from a partial neutralization of opposite electricities;[4] excess of either kind of electricity in such a combination would determine its chemical behaviour, combining with other binary compounds or with elements having the opposite kind of electricity in excess.[5] In organic chemistry this same method of analysis of molecules into parts of opposite polarities[6] led to the Radical Theory in the form upheld by Frankland and Kolbe as described in Chapter II.

This simple idea of neutralization of opposite electricities raised one great difficulty. To overcome this Berzelius proposed a supplementary theory of polarity; in this he assumed that atoms were rather like magnets of unequal pole strength, having two poles in which unequal amounts of electricity were concentrated; the predominant pole determined the electrochemical nature of the compound.[7] This early application of Dalton's atomic theory gave to the system proposed by Berzelius a much greater force and comprehensiveness than Davy's less developed theory possessed.

> When heterogeneous atoms combine . . . they appear to adjust or dispose themselves so as to touch with the opposite poles; of which the electricities produce a discharge which causes the phenomenon of elevation of temperature, almost constantly apparent at the time of any chemical combination, and the particles remain combined until their discharged poles are, by some means or other, restored to their former electric state.[8]

To account for combination with more than one atom, Berzelius later added a further idea:

> In the act of joining an atom with several atoms having the opposite electricity predominant the polarity of the latter induces in the atom they combine with a number of polar axes equal to theirs, these axes being either separate or coinciding on the free side with one polar point.[9]

It is hard to grasp the immensity of the influence wielded by Berzelius for several decades, and to understand the hold exerted by his theory. Of its relevance for valency, Palmer has written:

> The idea of a *saturation* (Berzelius used the term "neutralization") rather than a mere electrostatic attraction was demanded even to ensure a constant composition for a given chemical individual; and in this we may see, with some risk of committing an anachronism, the first germ of a theory of valency.[10]

Nevertheless chemistry in general was not ready to receive it, and the whole electrochemical system fell into a steep decline for many years. Before tracing the effects of its ultimate resurgence, we must examine the causes of its temporary, but long, eclipse.

2. THE DECLINE OF DUALISM

Berzelius's theory was essentially dualistic in character: molecules were to be regarded as capable of being broken down into successively 2, 4, 8, *etc.*, constituents, *i.e.*, into pairs of components of opposite polarities. There was thus a fundamental contrast between the atoms in a binary radical, and between radicals in a more complex molecule.

Now such a view of chemistry is eminently suitable for ionic and other inorganic compounds, but not for covalent substances like those predominating in organic chemistry. Yet it was for the latter class that the non-electrical Structure Theory of Kekulé, Couper and Butlerov was so successful and the progress so spectacular. A sturdy independence of the electrochemical theory by those workers concentrating on the study of carbon compounds began to appear, and this was increased by the brilliant investigations that gained in momentum as the century progressed. Of Berzelius's ideas Ketelaar has written:

That his theory of dualism has fallen so completely into oblivion can be attributed to the vigorous development in the nineteenth century of organic chemistry to which his theory did not appear to be applicable.[11]

This general trend was a reflection of a specific difficulty that has been discussed in Chapter III, the problem of substitution of electropositive hydrogen by electro-negative chlorine. If there had been a recognition of a possible diversity of valency types, this would never have been a problem, for it would not follow that what was applicable to one necessarily was valid for the other. More fundamentally still, acknowledgment that an element (hydrogen) could show both positive and negative propensities, though violating the principles of the Berzelian system, would have clarified this situation. However, the refusal of Berzelius to modify his premises led to a general scepticism towards his whole theory.

By about 1860 the electrochemical theory was at its lowest ebb, despite the efforts of a few to keep it alive. This decline in its fortunes was not appreciably altered for another twenty years. Thus in 1873, Clerk Maxwell considered chemical combination to be of a higher order of complexity than electrical phenomena;[12] eleven years later Kolbe, an ardent follower of Berzelius, though admitting that "the chemical force which resides in the atoms, called *chemical affinity* or *attraction*, is evidently related to the force of electricity", rejected the electrochemical theory because it denied the combination of similar atoms.[13] This was a subsidiary factor operating against the theory; with the general acceptance of Avogadro's hypothesis and consequent recognition of diatomic elementary molecules, like H_2, O_2, N_2, Cl_2, *etc.*, came a further blow to dualism.

An interesting indication of the departure from the teachings of Berzelius is the rising tide of scepticism about the atomic ideas that he had so eagerly embraced (p. 145). Still more illuminating were the efforts of some to fill the vacuum left by the departure of the electrochemical theory with alternative explanations of the cause of chemical affinity. The most persistent of these were to be found in the use of the analogy between molecules and the solar system, with gravitation taking the place of electricity as the cause of affinity. The analogy had been in use for a long time, but it gained new emphasis in this period. As the importance of atoms and valency became slowly recognized Frankland could suggest at a British Association meeting that "the atomic theory of Dalton, developed as it has been by the doctrine of atomicity, is rapidly assuming, for chemical phenomena, the position which the theory of gravitation occupies in cosmical science".[14] Two years earlier he had remarked that chemical bonds were probably "as regards their nature, much more like those which connect the members of our solar system",[15] though these words were omitted from later versions of the same passage.[16] A year before this Loschmidt had written favourably of the assumption that "the volume of matter even in a solid or liquid body is vanishingly small compared with the space between, which separates the smallest particles of matter from each other, and that therefore these atoms act on one another only at a distance by forces of attraction and repulsion. This constitution has been intelligently compared with our solar system, in which interplanetary space stands in a similar relationship to the volumes of the sun and planets".[17]

A different substitute for the electrochemical theory existed in a dynamic explanation of valency. In 1869, Buff wrote:

On one view, "the chemical force of affinity is at bottom nothing else than a pure force of attraction, and is, so to speak, of the same quality as the force of gravity". According to this, the affinity of the atoms must be as unchangeable as their weight. On the other view, however, chemical affinity is only a kind of force which is often known as motion, and it is variable like all kinds of motion. ... Perhaps we shall find on recognizing the deep mystery of the nature of chemical affinity that it is a motion of the atoms, more or less hindered in proportion to the number of active affinity units.[18]

A similar idea was expressed by Michaelis, who supposed that atoms oscillated about a mean position under the influence of attractive and repulsive forces.[19] Kekulé's ingenious application of an oscillation theory to benzene has already been mentioned (p. 254).

3. NEW LINKS BETWEEN CHEMISTRY AND ELECTRICITY

The decline of Berzelius's influence was not completely universal, a minority of chemists continuing to think of chemistry in electrical terms. Thus, in 1870, J. P. Cooke, Professor at Harvard, asserted that "the chemists of the new school, in their reaction from dualism, have too much overlooked the electrochemical facts, which are as true today as they ever were".[20] The previous year had seen the publication of Blomstrand's *Chemie der Jetztzeit*, written with the express purpose of reviving the ideas of Berzelius.[21] This exerted a profound influence on Werner and others, though perhaps more in some of the "non-electrochemical" views expressed than in its central theme.

Moreover, by the 1880s the tide was beginning to turn and a new electrochemistry was emerging. For this several factors were responsible, and by the end of the century their influence had contributed strongly to a re-emphasis on the rôle of electricity in chemistry. Chronologically, the first of these was the Faraday Lecture delivered by Helmholtz in 1881 to the Chemical Society of London. Unlike some previous Faraday lecturers, he felt it desirable to return to the man in whose honour he spoke, and to give a new exposition of his teachings.

Now Faraday himself had been powerfully influenced by the electrochemical theory of Davy, and accepted that "chemical and electrical action are merely two exhibitions of one single agent or power".[22] He had also discovered his two laws of electrolysis, and these constituted the starting point of Helmholtz's argument. For some reason the implications of these for valency, like Faraday's view quoted above, seem to have been almost entirely overlooked until this time. Probably the reason may be found in the adherence of most chemists, particularly in Germany,[23] to the views of Berzelius, while Faraday stood more in the Davy tradition. Now, the theory of Berzelius was worked out in far more detail than that of his English contemporary to whom he was indebted for the basic generalization. For many purposes this was an advantage; Davy's ingrained mistrust of theories precluded him from producing one of wide application in matters of detail. But in one respect Berzelius was over-precise, and the generality of Davy's views fitted the facts far better. This was the point taken up by Helmholtz.

The essence of Helmholtz's argument was that Faraday's experiments contradicted the idea of Berzelius that the quantity of electricity in each atom in a compound depends on the mutual electrochemical difference.[24] Taking into account the phenomenon of ionic movement,

Faraday's law tells us that through each section of an electrolytic conductor we have always equivalent electrical and chemical motion. The same definite quantity of either positive or negative electricity moves always with each univalent ion, or with every unit of affinity of a multivalent ion, and accompanies it during all its motions through the interior of the electrolytic fluid. This quantity we may call the electric charge of the atom. . . . Now the most startling result of Faraday's law is perhaps this. If we accept the hypothesis that the elementary substances are composed of atoms, we cannot avoid concluding that electricity, positive as well as negative, is divided into definite elementary portions, which behave like atoms of electricity.[25]

This was extended to non-electrolytes, many of which showed some response, however small, to the action of an electric current. He continued,

I think the facts leave no doubt that the very mightiest among the chemical forces are of electric origin. The atoms cling to their electric charges, and opposite electric charges cling to each other; but I do not suppose that other molecular forces are excluded, working directly from atom to atom.[26]

Of electrolytes, with each unit of affinity charged with one equivalent of positive or negative electricity, he says,

This, as you will see immediately, is the modern chemical theory of quantivalence, comprising all saturated compounds.[27]

Unsaturated compounds are also brought within this scheme, and the basic difference of "molecular compounds" recognized.

Helmholtz's lecture "made quite an unwonted stir among English men of science",[28] and was undoubtedly of great importance in reviving an interest in electrical theories of valency. Thus Lodge, reporting to the British Association in 1885, remarked:

The fact that atoms in electrolytes have a constant charge which is the same for every kind of atom, or at least can only be multiplied by an integer, is so striking, that one is constrained to think whether electricity is something necessarily associated with atoms of matter, whether all electrical actions are simple electrostatics among the atomic charges, and whether no quantity of electricity smaller than an atomic charge can exist.

The notion is repugnant, but it just wants considering; though I should hardly have ventured to suggest it but for the support Helmholtz has given to the view as at least a possible one.[29]

Repercussions of Helmholtz's lecture continued to be felt in England through the 1880s. They were intensified by a second factor that was to underline the importance of electricity for chemists, namely the increasing amount of physical data relating to electrolysis.

The supposition that atoms of electrolytes bore a constant charge would be strengthened by evidence that ionization took place independently of electrolysis, and was not something induced by the electric current. Only then could Helmholtz's ideas receive general application. In fact such evidence was accumulating at this time, and came from two sources. First we may mention the studies on the colligative properties of solutions. Raoult[30] had shown that the depression of the freezing-point of a liquid depended upon the number, rather than the kind, of molecules present, and van't Hoff later gave a generalized picture not only of these results but of other colligative properties such as osmotic pressure, *etc.*[31] However, the difficulty was that electrolytes always gave anomalous results, implying that more molecular species were present than the doctrine of undissociated molecules would allow. A second line of approach came from studies of conductivity, particularly by Kohlrausch,[32] who found that the equivalent conductivity varied with dilution.

These two groups of facts were welded together by the fellow-countryman of Berzelius, Svante Arrhenius. In a memorable paper[33] he showed the necessity of assuming complete ionization of electrolytes at infinite dilution, with partial ionization under other conditions. The degree of ionization could be calculated both from deviations from Raoult's law and from equivalent conductivities, the results being in quite good agreement. These views were advocated by Ostwald[34] and extended by Nernst[35] and others.

Opposition was also encountered, however, most persistently from Armstrong who regarded chemical reaction as "reversed electrolysis"[36] and whose theory of "residual affinity" was in conflict with the ideas of both Arrhenius and Helmholtz. In three papers[37] he expounded his own ideas, thereby giving publicity to both viewpoints, and at the same time initiating the concept of "residual affinity" that was to have far-reaching consequences (p. 205).

A third factor impressing upon chemists the relevance of electricity was the discovery of the electron. For a long time after the invention of the induction coil, the electrical discharge through gases under reduced pressure had attracted the interest of physicists, not least on account of the remarkable behaviour of the "glow" produced. Recently, new possibilities were created by the invention of mercury pumps capable of producing high vacua. By means of these Crookes further investigated the "cathode rays" emanating in the electrical discharge through rarified gases, concluding them to be charged particles of high velocity.[38] In 1897, however, Thomson was able to show that these "cathode rays" were negatively charged particles ("corpuscles"), much lighter than a hydrogen atom.[39] These he believed to be constituents of all matter, and the importance of his work, for chemistry at least, is that it took the postulate of electrical matter beyond the narrow limits of electrolytes.

Even this work, however, did involve the use of an electric current. To remove all suggestions that *this* was the source of the electrons[40] it was necessary to show their production could be independent of a current, and this was demonstrated by the phenomenon of radio-activity. At the turn of the century three kinds of "rays" were noted from radioactive substances, α, β and γ, and the identification of the "β-rays" as fast-moving electrons[41] gave final proof of the existence of electrons in matter quite apart from any conventional electrical phenomena. This fact, coupled with the ever-growing need for a comprehensive theory of valency, was to give the final major impetus to the development of such a theory.

Thus to Blomstrand's revival of Berzelian ideas must be added Helmholtz's advocacy of those of Faraday and thus of Davy – a situation paralleled by the work of Cannizzaro on behalf of Avogadro; when this was followed by the recognition of ionization in solution, the discovery of the electron in gaseous discharges and the identification of it in radioactivity, all the preliminary steps had been taken to an electronic theory of valency. But one other factor must be mentioned, which not only facilitated this development, but explains a curious fact about it. For most of the last half of the century, Germany was unquestionably the leading country for chemical research; in the development of the electronic theory of valency, however, the most significant work was done by English-speaking chemists, some of them fellow-countrymen of Davy and Faraday, and some from the United States.

The reason for this appears to lie in this fact: outside Germany the wave of enthusiasm for organic syntheses was beginning to recede; in Britain, particularly, there had never been as important a centre for this branch as was found, for example, at Heidelberg. Inorganic chemistry, to which many British chemists had made impressive contributions, was labouring under a mass of data too heavy for the slender theoretical supports then available, and in the 1890s it was in a poor state.[42] Physical chemistry, however, was finding a distinctive place in a nation that, since Boyle, had often approached chemistry from physics. The work of Armstrong, Crookes, Thomson and Rutherford manifested the same tendency as that evinced by the reception given to the ideas of Helmholtz. Moreover, the electrochemical views of Berzelius do not seem to have exerted such a hold as on the Continent, and in some respects, as we have seen, these would stand in the way of an *electronic* theory. Any residuum of Davy's ideas still circulating could have been rather more helpful.

One thing is quite certain. In Britain and America the immense volume of synthetic organic work still being turned out by the Germans had no parallel. As Todd has said,

Strong and confident in themselves, the German organic chemists placed little emphasis on theory, and were on the whole unsympathetic to the young and growing science of physical chemistry.... German organic chemistry was by 1900 becoming a top-heavy factual structure which made few moves in theory beyond the original postulates.[43]

Thus for the final stages in the development of valency up to the advent of wave mechanics, it is to British and American chemists that we must turn.

CHAPTER XIV

The Recognition of Electrovalency

ELECTROCHEMISTRY ORIGINATED with the study of electrolytes; it first encountered grave difficulties when attempts were made to apply its teachings to compounds unable to undergo electrolysis. It is not surprising, therefore, that no real progress could be made until a fundamental distinction was recognized between the two classes of compounds. This took place in the early years of the present century, and the first group of substances to yield to the new theory were the electrolytes – the same group that had been the starting point for both Davy and Berzelius.

There were two basic approaches. First emerged a number of suggestions based on a combination of the electronic ideas and the now established fact of the Periodic Law. Later came important refinements springing from the study of atomic spectra.

1. DEVELOPMENTS FROM THE PERIODIC LAW

As Chapter VII has shown, the Periodic Law owed part of its development to the theory of valency. In the first decade of this century the debt was repaid by the unfolding of new aspects of valency from an application of the idea of the electron to the Periodic Law.

The origin of this work was with G. N. Lewis, Professor of Chemistry in the University of California, and may best be stated in his own words:

In the year 1902 (while I was attempting to explain to an elementary class in chemistry some of the ideas involved in the periodic law) becoming interested in the new theory of the electron, and combining this idea with those which are implied in the periodic classification, I formed an idea of the inner structure of the atom which, although it contained certain crudities, I have ever since regarded as representing essentially the arrangement of electrons in the atom. . . .

The main features of this theory of atomic structure are as follows:

(1) The electrons in an atom are arranged in concentric cubes.

(2) A neutral atom of each element contains one more electron than a neutral atom of the element next preceding.
(3) The cube of 8 electrons is reached in the atoms of the rare gases, and this cube becomes in some sense the kernel about which the larger cube of electrons of the next period is built.
(4) The electrons of an outer incomplete cube may be given to another atom, as in Mg^{++}, or enough electrons may be taken from other atoms to complete the cube, as in Cl^-, thus accounting for "positive and negative valence".

In accordance with the idea of Mendeléeff, that hydrogen is the first member of a full period, I erroneously assumed helium to have a shell of eight electrons. Regarding the disposition of the positive charge which balanced the electrons in the neutral atom, my ideas were very vague; I believe I inclined at that time toward the idea that the positive charge was also made up of discrete particles, the localization of which determined the localization of the electrons.

These hypotheses regarding the arrangement of electrons in the atom, while they were discussed freely with my colleagues and in my classes, were given no further publicity. Indeed while this theory of structure seemed to offer a remarkably simple and satisfactory explanation of the process which occurs when sodium combines with chlorine to form sodium chloride, it did not seem to explain chemical combinations of a less polar type, such as occur in the hydrocarbons.[1]

Reference to Lewis's memorandum[2] for 28 March 1902 (see p. 272) shows that he envisaged (though he did not say) that formation of sodium chloride took place by *transfer* of an electron from the sodium to complete the octet of chlorine.

The first *published* statement came in 1904 from Abegg. In 1899, with Bodlander,[3] he had introduced the idea of "electro-affinity", *i.e.* the electromotive force required to strip an electron from an atom, and had pointed out the role of complex-formation in making a strong from a weak electrolyte. He was now concerned with the old problem of "molecular compounds".[4] To explain these he assumed that each element had two kinds of valencies, normal and contra-, whose arithmetical sum was eight. His views may be summarized thus:

TABLE VI

Periodic Group	I.	II.	III.	IV.	V.	VI.	VII.
Normal valencies	+1	+2	+3	+4	+5	+6	+7
Contra-valencies	−7	−6	−5	−4	−3	−2	−1

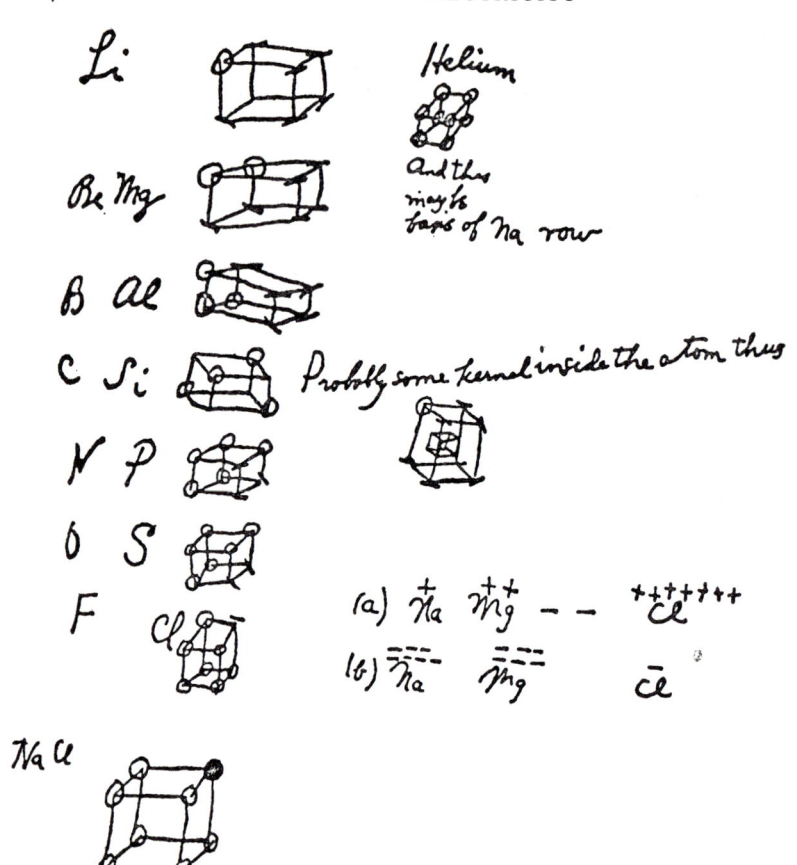

The birth of the octet theory: memorandum of 1902 by G. N. Lewis, from his Valence and the Structure of Atoms and Molecules, *1923, p. 29*

In the formation of "molecular compounds", the contra-valencies might be involved, though in the first three groups they are hardly ever used. His theory was clearly a reflection of Lothar Meyer's observations on the periodicity of valency (p. 141), and also has affinities with the ideas of Berzelius wherein atoms are given opposite kinds of polarity. Its electronic implications are indicated in the words:

The sum 8 of our normal and contra-valencies has therefore the simple significance as the number which represents for all atoms the points of

DEVELOPMENTS FROM THE PERIODIC LAW 273

attack of electrons; and the group number or positive valency indicates how many of the 8 points of attack must hold electrons in order to make the element electrically neutral.[5]

Also in 1904, a paper appeared by J. J. Thomson[6] in which the atom was pictured as a ring of equally spaced electrons oscillating in a sphere of positive electrification. The mathematical consequences of this arrangement were investigated, and the author, limiting himself to the simplest (planar) system, concluded that with a sufficient number of electrons the ring would break into two concentric rings. It was here that the periodic system had an obvious bearing. He saw that a system with one "corpuscle" in the outer ring would behave like the atoms of monovalent electropositive elements, while that with a full ring would have zero valency, the one before it being electronegative and monovalent. Thus a sequence of elements existed closely similar to that of the periodic table. His conclusion was thoroughly dualistic:

When atoms like the electronegative ones, in which the corpuscles are very stable, are mixed with atoms like the electropositive ones, in which the corpuscles are not nearly so firmly held, the forces to which the corpuscles are subject by the action of the atoms upon each other may result in the detachment of corpuscles from the electropositive atoms and their transference to the electronegative. The electronegative atoms will thus get a charge of negative electricity, the electropositive atoms one of positive, the oppositely charged atoms will attract each other, and a chemical compound of the electropositive and electronegative atoms will be formed.[7]

In 1908, an address by Ramsay[8] suggested electrons were atoms of electricity, and that in the formation of salt "the transfer of an electron from the sodium to the chlorine takes place at the moment of combination", thus agreeing with Lewis and Thomson.

In 1916, two papers appeared nearly simultaneously by Kossel[9] (March) and Lewis[10] (April), suggesting that the electrons in an atom are arranged in concentric groups, the outermost always having a maximum of 8, except for the first case when this is 2. Atoms always tended to acquire this maximum number of electrons in the outer groups, and thus to resemble the inert gases.

These ideas, which Lewis had considered fourteen years earlier, were on similar lines in both papers. However, Lewis retained his cubic model, whereas Kossel assumed concentric rings, rather like

Thomson's atom. Their models of the argon atom are given below:

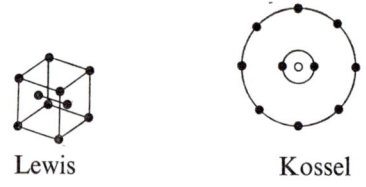

Lewis   Kossel

Lewis observed the similarity between the halogens and hydrogen, in both being able to receive one electron, and between hydrogen and the alkali metals in being able to lose one. His paper extended these considerations far beyond applications only to electrolytes (see next chapter).

Thus by 1916 the doctrine was firmly established of stable electronic arrangements typified by the inert gases, and of the tendency of all other elements to attain to these by gain or loss of the appropriate number of electrons. This would produce ions immediately on combination, and the "molecule" would be held together by electrostatic forces. The difficulty of reconciling the view of Arrhenius with the necessity of total ionization was largely overcome by the concept of ion-pairs, implicit in some early suggestions by Baly and Desch.[11] Thus all the essential features of simple electrovalency were known by this time, and we may briefly note how they had to be modified by the discoveries of spectroscopy.

2. THE IMPACT OF ATOMIC SPECTRA

As a preliminary point it may be noted that the Thomson type of atom had received a serious blow from Rutherford's discovery that α-particles could penetrate relatively large distances into "solid" matter, only being deflected on rare occasions. This clearly suggested that atoms consisted largely of empty space, and led to the formulation of the "Rutherford" atom, where a fairly small positive nucleus is surrounded, some distance away, by the system of electrons.[12]

In 1913/14, Moseley[13] obtained the characteristic X-ray spectra, by bombarding elements with "cathode rays" and determining the frequencies of the emitted X-rays. Moseley's results showed a direct relationship between the frequency of the latter and the atomic number, and he assumed this was the nuclear charge. In this way, the number of electrons (and thus the valency) appeared to depend ultimately upon the nuclear charge. A similar suggestion was made by van den Broek.[14]

Before this time, emission spectra of many elements had been studied, and an increasing mass of data on these was accumulating. Owing above all to the work of Rydberg,[15] certain mathematical properties of these (line) spectra were established. In 1913, Niels Bohr[16] applied to these spectral data the new quantum theory of Planck.[17] This latter concept, arising out of the failure of the law of equipartition to hold, particularly for black body radiation, assumes that macroscopic laws do not necessarily apply on an atomic or subatomic scale, and that bodies emitting or absorbing radiation do not do so continuously but by finite "steps" or *quanta*.[18] By applying this to emission spectra of elements, and assuming that light is emitted only when an electron falls from one energy level to another, Bohr was able to describe a new model of the atom. Basing his idea on the Rutherford atom, Bohr assumed that an electron could occupy only those orbits in which it could have an angular momentum of $h/2\pi$ (where h = Planck's universal constant). In this way he was able to give a quantitative estimate of the spectral lines for a hydrogen atom, in complete agreement with those found experimentally.[19]

The impact of this on electrovalency was not great. In place of "orbits", "shells" with various "sub-shells" were spoken of, and the original models of Lewis and Kossel had to be modified in time, but the main effect of the Bohr atom on valency came in the field of the covalent bond. Until the advent of wave mechanics it gave a most valuable physical picture, and it gave a mathematical basis for the "rule of eight" that was at the root of the early valency theory. With the Bohr theory the atom of the physicists was taken up by the chemists, and the connection between the periodic table and atomic spectra was established.[20]

CHAPTER XV

Extensions of the Electronic Theory to Non-electrolytes

THERE HAS ALWAYS BEEN a deep-seated reluctance amongst chemists to admit a fundamental difference between the various types of valency. From the time of Berzelius to that of Lewis, who confessed "I could not bring myself to believe in two distinct kinds of chemical union",[1] it has been almost an unspoken axiom that all "affinity" is of the same essential nature, however diverse its manifestations may be. Consequently the first attempts to formulate an electronic theory for non-electrolytes, and especially organic compounds, were made along the same lines as those for ionic substances. Later a new approach arose owing to the pressure of facts which could not be accommodated by the dualistic theory, and for a time two schools of thought existed. Their ultimate reconciliation was made possible by the concept of bond polarity as developed particularly from the study of conjugated systems. These three phases, although they overlapped to some extent, will be separately considered in this chapter.

1. ATTEMPTS TO APPLY DUALISTIC IDEAS TO NON-IONIC COMPOUNDS

In 1903, Thomson delivered a lecture "On the Constitution of the Atom"[2] as one of a series of Silliman Lectures given by him at Yale University. They appear to have acted as the first stimulus to two decades of work on this topic in the U.S.A. Some of the views advocated here, although later abandoned by Thomson himself, did influence several American chemists, in particular H. S. Fry, Professor at Cincinnati.

Thomson reminded his audience that:

On the view that the attraction between the atoms in a chemical compound is electrical in its origin, the ability of an element to enter in to chemical combination depends upon its atom having the power of acquiring a charge of electricity.[3]

Thus argon and helium were inert because of the stability of the uncharged atoms. Yet the early ideas of Davy, Berzelius, Faraday and Helmholtz had not then been found as fruitful as the "doctrine of bonds". But Thomson's contention was that the latter was "in one aspect almost identical with the electrical theory". He continued:

The theory of bonds when represented graphically supposes that from each univalent atom a straight line (the symbol of a bond) proceeds; a divalent atom is at the end of two such lines, a trivalent atom at the end of three, and so on; and that when the chemical compound is represented by a graphic formula in this way, each atom must be at the end of the proper number of the lines which represent the bonds. Now, on the electrical view of chemical combination, a univalent atom has one unit charge, if we take as our unit of charge the charge on the corpuscle; the atom is therefore the beginning or end of one unit Faraday tube: the beginning if the charge on the atom is positive, the end if the charge is negative. A divalent atom has two units of charge and therefore it is the origin or termination of two unit Faraday tubes. Thus, if we interpret the "bond" of the chemist as indicating a unit Faraday tube, connecting charged atoms in the molecule, the structural formulae of the chemist can be at once translated into the electrical theory. There is, however, one point of difference which deserves a little consideration: the symbol indicating a bond on the chemical theory is not regarded as having direction; no difference is made on this theory between one end of a bond and the other. On the electrical theory, however, there is a difference between the ends, as one end corresponds to a positive, the other to a negative charge.

These views were proposed before the "octet rule" of Lewis had been made known publicly, and they may be said to represent a physical rather than a chemical approach. It is therefore unjust to press them too far, and infer from them details of what would later be called "octet affiliations". However, they did arouse certain difficulties even then, chiefly those springing from the obvious *differences* between electrolytes and non-electrolytes. To some extent these were minimized by the appreciable conductivities of "non-electrolytes" under abnormal conditions, as of the halogens in non-aqueous media[4] and in the vapour phase at high temperatures,[5] when ions as I^+ (in addition to I^-) were assumed. Further, an electronic explanation was advanced for the variations in ease of ionization; Baly and Desch considered the mobile hydrogen in keto-enol tautomerism as potentially ionic, and this raised acutely the whole question as to the cause of ionization. With regard to salts in general they wrote:

> The bonds of attraction connecting the "ions" together are lengthened by the solvent. When the length of the Faraday tubes is below a certain critical length, the salt is "nonionized". When the average length of the tubes of force is equal to or a little less than the critical length, a few interchanges of ions between adjacent molecules take place, and the salt is partially ionized. When the length of the Faraday tubes is greater than the critical value, then perfectly free interchange takes place between the ions of different molecules, and the salt is completely "ionized".[6]

This was an interesting anticipation of the modern correlation between bond-length and bond-order.

In 1908/9, Fry elaborated Thomson's views, and applied them to benzene.[7] He introduced the idea of *electromerism* or *electronic isomerism* as a deduction from the directed nature of Thomson's valency bonds; a molecule XY could have two forms, depending on the location of positive and negative ends:

$$X^+ \text{------} ^-Y \text{ and } X^- \text{------} ^+Y$$

although these might not be capable of independent existence. Thomson had foreseen this situation, and Fry proceeded to apply the theory to benzene. He considered that an *n*-valent atom could function in $(n+1)$ different ways, according to gain, loss or both of electrons. Thus carbon could have 5 different states:

$$\begin{array}{ccccc} + & + & + & + & - \\ +C+ & -C+ & -C+ & -C- & -C- \\ + & + & - & - & - \end{array}$$

Combination of these in the benzene nucleus leads to six possible electromers:

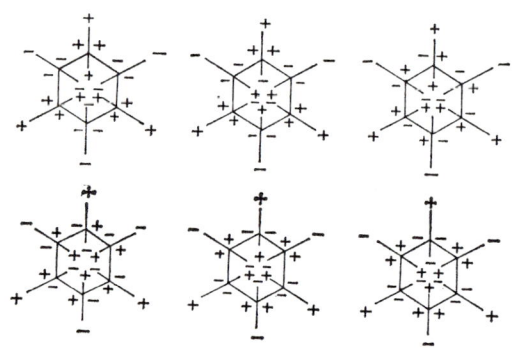

Omitting the centric bonds, it is possible to abbreviate these into one composite formula:

$$\begin{array}{c} H_- \\ H^+ \underset{+}{\overset{-\;+}{\bigcirc}} {}^+_- H \\ H^- \underset{+|}{\overset{-\;+}{}} {}^-_- H \\ H \end{array}$$

thus showing the similarities between *ortho* and *para* positions as opposed to the *meta* positions, and giving an explanation of the Crum Brown–Gibson Rule[8] for orientation in disubstituted derivatives.

A further development of Thomson's idea came from Falk and Nelson in 1909/10. These workers used arrows to represent the direction of transfer of electrons, writing methane and carbon tetrachloride as

$$\begin{array}{ccc} H & & Cl \\ \downarrow & & \uparrow \\ H \rightarrow C \leftarrow H & \text{and} & Cl \leftarrow C \rightarrow Cl \\ \uparrow & & \downarrow \\ H & & Cl \end{array}$$

This led to a difficulty in the case of the alkenes, whose isomerism they were inclined to treat in terms of these unidirectional valencies rather than of stereochemistry. One representation of alkenes appears to imply a neutralization on the two carbon atoms:

$$CRR' \rightleftarrows CRR'$$

Of this difficulty they wrote:

In those cases where one valence proceeds in one direction and one in the other it is assumed that the corpuscles which are transferred are localized on the atoms, as otherwise the carbon atoms would become electrically neutral.[9]

This conception of valency was applied to strengths of organic acids[10] and other data. The most ardent and consistent advocate of this extreme dualistic view to carbon compounds was H. S. Fry, whose ideas have been collected in his book *The Electronic Conception of Valence and the Constitution of Benzene.*[11] However, by the early 1920s there were few supporters left for this theory, which had now run into serious difficulties.

There were, for example, those who thought the theory was *unnecessary*. One of its most useful attributes had been its recognition of a definite oxidation state for each atom ("polar number").[12] Nevertheless Lewis pointed out that:

Oxidation of any element means an increase of its polar number, reduction means a decrease, and this simple system furnishes an adequate method of dealing with all cases of oxidation and reduction. It must be remarked, however, that on account of its very generality this system would apply equally well even if purely fanciful values of the polar number were chosen, provided that the rules required by the fundamental law of the conservation of electricity be observed. Moreover, non-polar compounds may be treated provisionally as polar, and fictitious polar numbers may be assigned without leading to any false conclusions.[13]

Secondly, the dualistic view is *incompatible with physical data*. Thomson, having abandoned his original view, was convinced by the effects of positive rays, the values of dielectric constants and other facts that a very great difference did exist between ionized and non-ionized compounds, and in the latter "intramolecular ionization" was of little significance.[14] Others, as Bates,[15] held the same view, although Fry considered that "the isolated conditions under which the quoted physical phenomena are effected ... are not comparable with the conditions under which the great majority of chemical reactions take place", and therefore are not relevant.[16]

Thirdly, the *unfulfilled predictions* of the theory, especially with reference to electromers, gave rise to scepticism. Thus Bray and Branch[17] pointed out that benzenesulphonic acid could be regarded as a derivative of sulphuric or sulphurous acids, giving sulphur a polar number of 6 or 4 respectively. Hydrolysis under acid conditions favoured the one, under alkaline conditions the other:

$$C_6H_5SO_2OH + H_2O = C_6H_5H + H_2SO_4$$
$$C_6H_5SO_2OH + H_2O = C_6H_5OH + H_2SO_3$$

All attempts to isolate electromers failed, however. (The difficulty is resolved by the mesomerism of the bisulphite ion

$$\overset{-}{O}=S=O \quad , \quad O=S-\overset{-}{O} \quad , \quad \overset{-}{O}-S=\overset{-}{O}$$
$$\quad\;\; |\qquad\qquad\;\; |\qquad\qquad\;\; |$$
$$\quad\;\, OH\qquad\quad\;\, OH\qquad\quad\; OH$$

formed in alkaline hydrolysis, with delocalization of the charge from the sulphur.) Again, Noyes[18] attempted to isolate an electromer of nitrogen trichloride with positive nitrogen, but was unsuccessful:

$$
\begin{array}{cc}
\begin{array}{c}
\quad - \ +Cl \\
N- \ +Cl \\
\quad - \ +Cl \\
\text{(assumed usual electromer)}
\end{array}
&
\begin{array}{c}
+ \ -Cl \\
N+ \ -Cl \\
+ \ -Cl \\
\text{(unknown electromer)}
\end{array}
\end{array}
$$

Thus the extreme form of dualism advocated by Fry and other American chemists gradually fell into disfavour. Before leaving it, however, it is well to note that it stands in a direct generic line with the theory of Berzelius, and is a remarkable anticipation of the concept of polar bonds in use today. Its fundamental merit was its recognition of the unity of valency and its refusal to allow a dichotomy into basically different types. In doing so it gave an electronic interpretation of the characteristic behaviour of aromatic substances, and paved the way for the theories of the English schools in the 1920s and 1930s.

2. INTRODUCTION OF THE CONCEPT OF ELECTRON-SHARING

While Fry and his colleagues were using polarized bonds in their formulae, others were attacking the problem from a different angle. Instead of assuming a transference of electrons, they sought for an explanation of their rôle in some kind of sharing between the atoms that were linked.

In 1907, Thomson (then regarding the atom as a sphere of positive electricity in which the electrons were embedded) proposed that two atoms could be held together by a partial overlapping of their spheres, electrons being situated symmetrically in the region of overlap, thus having "forces electrical in their origin binding the two systems together without a resultant charge on either system".[19] Ramsay, in a lecture the following year,[20] spoke of an electron as "the bond of union between the sodium and chlorine" in sodium chloride, and extended this to molecular hydrogen, chlorine, *etc.*, but did not develop the idea.

These preliminary notions were followed by some contributions from Continental workers, notably Stark[21] and Kaufmann.[22] Both of these produced dynamical models for the atoms and assumed that electrons could function as binding-agents for two atoms between which they were shared, by exerting a force on both positive nuclei. In 1914, Thomson extended his views yet again,[23] and supposed that atoms may be held together by tubes of force from electrons of one to the positive nucleus of the other and by other tubes acting in the opposite direction. His equivalent of a single bond was thus two tubes of force emanating from opposite directions. He wrote benzene as:

[benzene structure diagram]

The next year Parson,[24] a colleague of Lewis, produced a theory of the atom based on magnetic properties, suggesting that atoms could sometimes be held together by sharing a pair of electrons, and emphasizing the importance of the "group of eight".

During these years Lewis had been keeping back from publication his ideas on the "cubic" atom with its tendency to acquire an octet of electrons in each cube. Since 1902 these views had been discussed amongst his colleagues,[25] but their apparent inability to account for hydrocarbons and similar compounds had been a grave difficulty, and had doubtless caused their author to refrain from publication.

Now, however, two factors were obvious that had not been so before. First, it was evident that the efforts of Fry and others to apply a strict dualism to non-electrolytes were running them into serious difficulties, though their position was still being defended with vigour. Secondly, the concepts outlined above had suggested that the electrons could possibly unite atoms by being shared by them. In 1916 Lewis at last published his ideas on the "octet rule", and although Kossel almost simultaneously produced his paper,[26] the American chemist went much further in extending his treatment to non-ionized compounds.[27]

INTRODUCTION OF THE CONCEPT OF ELECTRON-SHARING

To his "rule of eight" (p. 271) Lewis adds a "rule of two"; this asserts the almost universal occurrence of electrons in molecules in even numbers, exceptions being certain metallic vapours at high temperatures, NO, NO_2 and ClO_2. Thus he was led to the importance of the *electron pair* in chemistry. This would explain the tendency for some substances to ionize rather than dissociate into two odd-electron fragments. He quotes the conductivity of molten iodine and of liquid dinitrogen tetroxide as evidence for ionization into, respectively, I^+ and I^-, and NO_2^+ and NO_2^-.

Thus he came to his most far-reaching idea, that of pairs of electrons shared between different atoms and no longer belonging to either exclusively. This sharing will constitute a chemical bond. Thus we have:

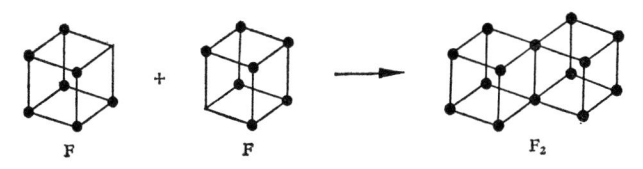

This is in contrast with the electron transfer taking place in polar compounds. But Lewis recognized intermediate states of valency, in which an electron pair was unequally shared between atoms. He represented this (for iodine) as follows:

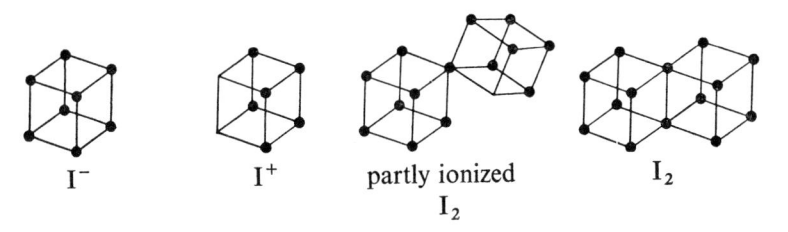

His application of this to polar bonds generally will be considered later (p. 287).

His formulae for ions like those following, where electrons are symbolized by dots, showed for the first time the fundamental similarity between the oxyacids, borofluoric and hydrochloric acids. This had been pointed out by Werner (p. 215), to whom Lewis acknowledges his great "personal indebtedness".[28]

$$\left[\begin{matrix} & \overset{..}{:}\overset{..}{O}: & \\ \overset{..}{:}\overset{..}{O}: & \overset{..}{S} & \overset{..}{:}\overset{..}{O}: \\ & \overset{..}{:}\overset{..}{O}: & \\ & \overset{..}{} & \end{matrix}\right]^{--} \quad \left[\begin{matrix} & \overset{..}{:}\overset{..}{F}: & \\ \overset{..}{:}\overset{..}{F}: & \overset{..}{B} & \overset{..}{:}\overset{..}{F}: \\ & \overset{..}{:}\overset{..}{F}: & \\ & \overset{..}{} & \end{matrix}\right]^{-} \quad :\overset{..}{\underset{..}{Cl}}:^{-}$$

The problem of ammonium is stated thus:

When ammonium ion combines with chloride ion the latter is not attached directly to the nitrogen, but is held simply through electric forces by the ammonium ion.[29]

The formula given is in complete agreement with the characteristics of ammonium compounds emphasized by Werner:

$$\left[\begin{matrix} & H & \\ & \overset{..}{} & \\ H : & N & : H \\ & \overset{..}{} & \\ & H & \end{matrix}\right]^{+}$$

The problem of multiple bonds was at first treated by an extension of the principles laid down for single bonds. A double bond was represented by two cubes with a common face:

Triple bonds, however, cannot be written in this way, and so Lewis modified his cube to a tetrahedron, in which pairs of electrons have been attracted together. This model reflects at once his recognition of the importance of the "rule of two" and of the stereochemical implications of atomic structure. A triple bond is now formed by two tetrahedra with a common face.

Later reflections on the subject led Lewis to abandon both cubic and tetrahedral diagrams, and electron pairs were represented solely by colons.[30] He thus wrote ethylene as

$$\begin{matrix} H & & H \\ \overset{..}{} & & \overset{..}{} \\ H:C & ::& C:H \end{matrix} \quad \text{or as} \quad \begin{matrix} H & & H \\ \overset{..}{} & & \overset{..}{} \\ H:C & : & C:H \\ \underset{.}{} & & \underset{.}{} \end{matrix}$$

INTRODUCTION OF THE CONCEPT OF ELECTRON-SHARING

the latter formula representing an extreme state which may occasionally arise in reactions. Acetylene is similarly written as

$$H : C : : : C : H \quad \text{or as} \quad H : \overset{..}{C} : \overset{..}{C} : H$$

These formulae would represent the great reactivity of ethylene and the rather less reactivity of acetylene. But he added:

> However we choose to visualize this condition of strain, it is evident that, when two atoms attempt to share not one electron pair but two, the molecule does not settle into an inert condition of high stability and low electron mobility, but rather that the system adjusts itself as it best may under the circumstances, and that in this adjustment either one or both of the bonding pairs remains in a state in which the electrons are neither tightly held nor capable of forming with each other a self-contained magnetic system.[31]

Similarly he represents butadiene thus,

$$\overset{H \quad H \quad H \quad H}{\underset{.\ \ \ .\ \ \ .\ \ \ .}{H : \overset{..}{C} : \overset{..}{C} : \overset{..}{C} : \overset{..}{C} : H}}$$

though again he admits that this is "a highly exaggerated representation".[32]

Thus the following ideas are all included in Lewis's theory:
(i) Common cause for both types of valency.
(ii) Fundamental similarity between many kinds of complex acids, with elimination of "molecular compounds" and of a distinction between "primary" and "secondary" affinities.
(iii) Interpretation of reactivity of unsaturated compounds, with a physical explanation of Thiele's "partial valencies".
(iv) Anticipation of polarized bonds without extreme dualistic assumptions.
(v) Explanation of the ammonia/ammonium relationship.
(vi) Elimination of the conception of physically distinct electromers, these differing merely in assumed polarizations of shared-electron bonds.

In his book[33] these ideas are extended and consolidated with great clarity. The reason for their rapid acceptance amongst many chemists may be traced partly to this book, but more to the work of Langmuir. Unable himself to give a detailed account of his new views owing to the exigencies of the 1914–18 war, Lewis comments:

The task was performed, with far greater success than I could have achieved, by Dr Irving Langmuir in a brilliant series of some twelve articles, and in a large number of lectures given in this country and abroad. It is largely through these papers and addresses that the theory has received the wide attention of scientists.... The theory has been designated in some quarters as the Lewis–Langmuir theory, which would imply some sort of collaboration. As a matter of fact Dr Langmuir's work has been entirely independent, and such additions as he has made to what was stated or implied in my paper should be credited to him alone.[34]

In 1919 Langmuir attempted to extend Lewis's theory to all elements, not merely those of the first two rows of the periodic table.[35] He suggested electronic arrangements that have since been shown to be incorrect, and emphasized the importance of the "group of eight" which he renamed "the octet". The two types of valency he tended to distinguish more sharply than Lewis, and to these he gave the names "covalency" and "electrovalency".[36] In a later paper he proposed that:

(i) The electrons in atoms tend to surround the nucleus in successive layers containing 2, 8, 8, 18, 18 and 32 electrons respectively.
(ii) Two atoms may be coupled together by one or more duplets held in common by the completed sheaths of the atoms.
(iii) The residual charge on each atom and on each group of atoms tends to a minimum.[37]

The first of these was shortly denied by Bohr[38] and Bury[39] on spectroscopic and chemical evidence respectively. On this view, accepted today, the arrangement in radon, for instance, is 2, 8, 18, 32, 18, 8. The second statement is of course a repetition of Lewis's view, while the last denies the existence of ions bearing very high charges even though the inert gas structures are preserved. For example it would exclude Cr^{6+}, iso-electronic with argon.

In both Lewis and Langmuir appear formulae with shared pairs that must have originated from one and the same atom, *e.g.*, in the ions of the oxyacids. Attention was drawn to this "borrowing union" by Perkins;[40] it was termed a "semi-polar double bond" by Sugden *et al.*,[41] and a "co-ordinate link" by Sidgwick,[42] from its frequent occurrence in compounds formed when groups co-ordinate on to metal atoms capable of accepting their hitherto unshared electron-pairs.

In this way there arose for the first time a clear physical picture of a covalent bond.

It may be noted that here was a situation closely parallel to that existing in the late 1850s. Then, as 60 years later, the theory of molecular structure was at the cross-roads, having arrived there by two independent routes. The Radical Theory, founded on dualism and so clearly applicable to electrolytes, had given rise to a theory of valency that could obviously be applied to simple inorganic combinations. Its linear descendent, the extreme dualism of Fry, was an extrapolation from the new concept of what became known as electrovalency. On the other hand, the study of organic chemistry had led most chemists of the previous century to the Type Theory, and the achievement of the 1850s was to recognize that polyvalent radicals and atoms could hold together the major components of the types. Now, in the twentieth century, chemists were asking the same fundamental question: by what could the components of an organic molecule be held together? The answer was no longer a polyvalent group, but an electron pair.

Just as the 1850s saw a fusion of two different viewpoints in the concept of valency, so in 1916 the two processes of electron-sharing and electron-transfer were unified by the octet theory. Even here, however, the union was to be troubled with later difficulties, but in a changed form it has survived until today. It is now convenient to turn to a different fusion of the concepts of electrovalency and covalency, the recognition of bond-polarization.

3. THE RECOGNITION OF BOND POLARIZATION

The idea of a polarized bond was no new one even in 1920. The polarity of a molecule was explicit in Berzelius's electrochemical theory[43] and the essence of the dualistic formulae of Fry and others was that a bond also must be regarded as polarized. Lewis, who opposed these views, could not deny some degree of polarity to certain bonds; he wrote:

In the majority of carbon compounds there is very little of that separation of the charges which gives a compound a polar character, although certain groups, such as hydroxyl as well as those containing multiple bonds, not only themselves possess a decidedly polar character, but increase, according to principles already discussed, the polar character of all neighbouring parts of the molecule. However, in such molecules as methane and carbon tetrachloride, instead of assuming, as in some current theory, that four electrons have definitely left hydrogen for carbon in the first case, and carbon for chlorine in the second, we shall consider that in methane there is a slight movement of the charges toward the carbon so that the carbon is slightly charged negatively, and that in carbon tetrachloride they are slightly shifted toward the chlorine, leaving the carbon somewhat positive. We must remember that here also we are dealing with averages, and that in a few out of many molecules of methane the hydrogen may be negatively charged and the carbon positively.[44]

Using these ideas, Lewis showed how chloracetic acid must be a stronger acid than acetic acid. Later, in 1923, the strengths of acids are discussed in these terms in further detail.[45] Lewis was not prepared to speculate how far this polarization effect was propagated down a chain and how far it was transmitted directly through space, but of its reality he was in no doubt. He showed, for example, how it became progressively less marked with distance from the polar atom giving rise to it. Thus α-chloro-acids were stronger than β-acids, and these than the γ-isomers.

Three years later, Ingold designated this permanent effect by the name *inductive effect*,[46] and in 1928 (with Vass) distinguished between its transmission through the carbon chain and through space by the *inductive* and *direct effects* respectively.[47] Thus by 1930 the polarity of unsymmetrical single bonds was a recognized feature of chemical structures.

Meanwhile, however, a series of more complex ideas had been developing around the phenomenon of multiple bonds in general, and conjugated systems[48] in particular. It would be a gross oversimplification to assert even substantial independence between two schools of thought (both in England), but there were two broad approaches which, for the purpose of the present brief survey, it will be convenient to treat separately.

The earlier of these approaches was that associated with the names of Lapworth and Robinson. This had its roots in the former's pre-electronic ideas, together with those of Flürscheim, Hantzsch and Werner, as he himself acknowledged.[49] The first developed form of the theory appeared in a memoir in 1920.[50] Its essence was the existence of a series of alternating polarities down a conjugated chain of carbon atoms, as illustrated by:

$$-\overset{|}{C}=\overset{|}{C}-\overset{|}{C}=O \quad \text{and} \quad -\overset{|}{C}=\overset{|}{C}-C\equiv N \quad etc.$$
$$+\ -\ +\ -+\ -\ +\ -$$

This originated, we are told,[51] with Lapworth's applications of the ionic theory to carbonyl additions, which he had shown[52] involved attack by a negative ion on to carbon, which must thus be positive:

$$\begin{array}{c} +\ - \\ >C=O \\ CN-H \\ -\ + \end{array} \longrightarrow \begin{array}{c} >\overset{|}{C}-\overset{|}{O} \\ CN\ H \end{array}$$

Aromatic substitution was explained by models like the following, representing respectively o/p and m substitution:

Superficially there is a strong resemblance to the theory of Fry, but Lapworth is careful to point out two basic differences. First,

> In attaching the − and + signs to the oxygen and carbon atoms no hypothesis is invoked, nor is it necessary or even desirable to assume that electrical charges are developed on these two atoms (except perhaps at the actual instant of chemical change). The signs are applied, in the first instance, merely as expressing the relative polar characters which the two atoms seem to display at the instant of the chemical change in question.[53]

Secondly, Lapworth assumed that these alternating polarities sprang from the influence of a "key-atom":

> The writer has long held that certain atoms, and especially divalent oxygen and tervalent (negative) nitrogen, tend to produce such an alternation of latent polarities within the molecules in which they occur. . . . The extension of the influence of the directing, or "key-atom", over a long range seems to require for its fullest display the presence of double bonds, and usually in conjugated positions; consequently the principle must find ample scope in the aromatic series where conjugation is the rule.[54]

Fry, it will be recalled, had envisaged an alternation of full charges down even a hydrocarbon chain. Neither Lapworth nor Robinson envisaged more than developments of "anionoid" or "cationoid" character on these atoms.[55]

These views were adopted and extended by Kermack and Robinson,[56] and the curved arrow introduced to indicate electromeric displacements; it became no longer necessary to assume alternating polarities, but changes of this kind were envisaged:

$$-C=C-C=C-C=C- \longrightarrow \overset{+}{-}C-C=C-C=C-\overset{-}{C}-$$

About this time, Lowry put forward a new theory of the double bond: "a double bond in organic chemistry usually reacts as if it contained one covalency and one electrovalency".[57] Thus he wrote:

$$\overset{+}{CH_3.CH}-\overset{-}{O} \quad \text{and} \quad \overset{+}{CH_2}-\overset{-}{CH_2}$$

so giving one atom a complete octet, but denying it to the other. Such an effect was used to account for the behaviour of butadiene and benzene, and many other phenomena. At first it was considered to be a temporary polarization, but Ingold and Ingold,[58] who termed it the *electromeric effect*, also suggested a permanent polarization could occur. This paper, together with one independently produced by Robinson *et al.*,[59] produced a synthesis of the ideas so far current. Later, Ingold classified two polarization and two polarizability effects for what he termed the general inductive and tautomeric mechanisms.[60] Similarly, Robinson gave his generalization of the theory in the form of two lectures to the Institute of Chemistry.[61]

Thus the covalent bond was no longer conceived as a rigid entity with two electrons symmetrically shared between the atoms. A molecule was subject to strains and stresses unimagined twenty years previously, and the valency bonds were more dynamic than static, responding to the demand of a reagent to assist a reaction or remaining as quiescent and still as they had been imagined eighty years before. Problems remained in abundance, not least that of the aromatic ring. The "aromatic sextet"[62] was regarded by its author with some misgivings[63] and by others[64] as quite inadequate.[65] Even the axiomatic tetrahedral disposition of the valencies of the carbon atom was unaccounted for, while, perhaps above all, the reactivity of a "simple" double bond eluded all satisfactory explanation. Chemistry was now ready for the fundamentally new approach embodied in wave mechanics.

CHAPTER XVI

The Coming of Wave Mechanics

1. THE NEW QUANTUM THEORY

FOR THE FIRST twenty years of the present century structural chemistry was relatively undisturbed by the revolution that was changing the face of physics. The quantum theory of Planck (1900), Einstein's Special and General Theories of Relativity (1905 and 1913 respectively), and the discoveries in the early 1900s of radioactivity and of isotopes were each so far-reaching that few men could fail to realize the cataclysmic nature of the change through which science was passing. Yet despite this fact, those chemists who were not actively involved in the new radiochemistry pursued their course with little reference to the upheavals of contemporary physics. As we have seen, the replacement of the Rutherford atom by that of Bohr produced no great modification to the current electrical conception of a chemical bond. Moreover, in Europe at least, much attention was being concentrated upon the chemical production of substances likely to be useful in the 1914 war or its aftermath; to that extent conditions were unfavourable to theoretical speculation. But the chief reason for the small impact of the new physics upon chemistry was simply that, as it stood, it was inapplicable to the chemical problems that were the most urgent. As it now appears, the crying need of theoretical chemistry was emancipation from the corpuscular electron, but the quantum theory and the phenomena of radioactivity tended rather the other way. The truth was that by 1925 physics was at the middle, not the end, of its greatest revolution since Newton. One further convulsive upheaval was necessary before the old trappings could be shed sufficiently to reveal beneath them the ultimate clue to the nature of the chemical bond. The upheaval was caused by the rise of the new quantum theory, or wave mechanics.[1]

Planck's recognition of the quantum of action had been followed by Einstein's view[2] that light consists of small bundles of energy where

$$E = h\nu,$$

h being Planck's constant and v the frequency of the radiation. This corollary to the original quantum theory agreed with the phenomena of black body radiation and the Compton effect,[3] which seemed to demand some kind of particulate explanation; but it lacked the competence of the established wave theory to explain diffraction and reflection. A novel escape from this dilemma was suggested in 1924 by Louis de Broglie, who proposed an acceptance of both undulatory and corpuscular theories of light. Not content with one paradox he proceeded to extend his argument to matter itself and proposed that all elementary particles, including electrons, could be regarded as having a wave character.[4] As he wrote later: "If for a century we have neglected too much the corpuscular aspect of the theory of light in our exclusive attachment to waves, have we not erred in the opposite direction in our theory of matter? Have we not wrongly neglected the point of view of waves and thought only of corpuscles?"[5] The boldness of this suggestion was amply justified by the experimental verification of the wave character of electrons in the diffraction of electron beams through crystals,[6] very thin metal films[7] and line gratings.[8]

This was the age of paradox in physics. First we have the Bohr atom with electrons accelerating according to classical mechanics but radiating in a most non-classical fashion, as laid down by quantum theory; now both wave and corpuscular models, though apparently excluding one another, seemed equally needful for an understanding of the odd behaviour of these same electrons. Then, right on the heels of de Broglie's "wavicles", emerged the new science of wave mechanics, appearing in two guises so different that further problems of reconciliation immediately presented themselves. These latest developments were associated with the names of Heisenberg, Born, Dirac and Jordan on the one hand, and those of de Broglie and Schrödinger on the other. In 1926, Sir Arthur Eddington, writing within twelve months of the first of these advances (a paper by Heisenberg), noted that:

The theory has already gone through three distinct phases associated with the names of Born and Jordan, Dirac, Schrödinger.... You will realize the anarchy of this branch of physics when three successive pretenders seize the throne in twelve months; but you will not realize the steady progress made in that time unless you turn to the mathematics of the subject.[9]

It is indeed only in the realm of mathematics that all these paradoxes could be tolerated, if not fully resolved. Clearly it would divert us still further from the theme of valency to make such a mathematical exploration. Yet some attempt must be made, however qualitatively, to indicate the two streams of thought that arose at this time. Despite the remarks of Eddington just quoted, it is now possible to see that there *were* only two, springing from Heisenberg and de Broglie.

Werner Heisenberg[10] had been a member of Bohr's Institute in Copenhagen. According to W. Wilson, "Heisenberg's work was inspired, directly or indirectly, by the philosophy of Ernst Mach which forbids, even more sternly than did that of Newton, the adoption of hypotheses of any kind or the introduction of notions which lack the quality that may be described as 'observational value' ".[11] So it happened that the paper with which Heisenberg opened this new chapter of physics[12] showed a lofty independence of the models employed in the earlier quantum theory and has all the appearance of an exercise in higher mathematics. From the frequent use of the algebraic device of *matrices*, this treatment of spectral data has become known as *matrix mechanics*. Heisenberg's own paper was followed, as we have seen, by others in rapid succession. The same approach underlay the contributions of Born and Jordan[13] and of Dirac.[14] Of the latter even Eddington could say that it "was highly transcendental, almost mystical",[15] and it is not surprising that this kind of approach evoked only a limited response at first. A more generally appealing presentation of the new quantum theory lay in the wave mechanics of Schrödinger. Although this again was couched in terms that were highly mathematical there was a residuum of classical form (if the old quantum theory can be called classical!) that made it possible to picture at least some of the ideas.

Erwin Schrödinger, who later succeeded Planck in the Chair of Theoretical Physics at Berlin, was in 1926 professor at Zürich. That year saw the appearance of the first[16] of a long series of papers expounding Schrödinger's distinctive ideas on wave mechanics.[17] His starting point was the suggestion by de Broglie of the dual character of the electron and the consequent analogy between optics and mechanics.[18] Since mechanical problems could be translated into optical ones, those on an atomic scale can become restated in terms of undulatory optics. By consideration of a standing wave in three dimensions he obtained an equation involving a finite, continuous function ψ, the co-ordinates x, y and z, and the wavelength. Utilizing de Broglie's relationship:

$$\text{wave-length} = \frac{h}{mv}$$

where m and v are mass and velocity of the particle respectively, he obtained the wave-equation that now bears his name:

$$\frac{\delta^2\psi}{\delta x^2} + \frac{\delta^2\psi}{\delta y^2} + \frac{\delta^2\psi}{\delta z^2} + \frac{8\pi^2 m}{h^2}(E-V)\psi = 0$$

where E and V are respectively the total and potential energies.

At first Schrödinger inclined to the view that the square of the wave-function (ψ^2) was a measure of the electronic density at any one point.[19] Meanwhile, however, Heisenberg had promulgated his Uncertainty Principle[20] which asserts that the position and the momentum of a particle like an electron cannot be simultaneously known. In the light of this, Born[21] suggested that ψ^2 represents not density but probability: the probability of finding the electron at any particular point.

This remains today an acceptable interpretation of the wave-equation. Solution of it for, say, a 90 per cent probability will give a picture of a solid figure within which there is that degree of likelihood of finding the electron. In many respects this differs from the Bohr model. Instead of a tiny particle orbiting in planetary fashion round a nucleus we have a much less tangible electron cloud, blurred at its edges because it is a *probability* function. Also lost is the illusion of motion, for the picture of a moving electron is out of place here. Yet there is still a model, even though it has greatly changed. Indeed it has recently become fashionable (and the process is not entirely frivolous) to construct for didactic purposes atomic models out of toy balloons. The chemist's propensity for visualization is incorrigible.

The Schrödinger wave-equation is now accepted as one of the foundations of modern theoretical chemistry. Yet its immediate impact upon chemistry was small, mainly because of the enormous difficulties in solving it. There was certainly an agreement with Schrödinger's hope that it would be fruitfully applied, but the mathematical hurdles to be surmounted were immense. However, after years of painstaking work Coolidge and James produced an approximate solution for the hydrogen molecule, and found their results in good agreement with experiment.[22] With more complicated systems anything like a complete solution is still impossible, even with the use of high speed computers. Yet much progress has been made in the last thirty years by means of *approximate* methods. Two of these, using different approximations, have been particularly important in the development of the theory of valency, and to these we shall turn later.

Two comments on those early years of the new quantum theory can now be made. In the first place the two approaches of Heisenberg and de Broglie (matrix mechanics and wave mechanics) are not as discordant as was at first imagined. Indeed they lead, ultimately, to the same conclusions, and Schrödinger himself was able to reconcile them in 1926.[23] Fundamentally they differ in the same kind of way as those treatments of electrostatic problems which rely, respectively, on the old action-at-a-distance model and the newer field theory. Secondly, we may see in perspective the achievement of Coolidge and James. Professor D. P. Craig, in an illuminating address on "Theory and Experiment in Chemistry", has pointed out that the importance of this work lay not in "that it enabled the precise realization of experimental results, but that it proved that quantum mechanics has a valid application to molecules".[24]

With this justification for the new quantum theory, much work was performed in the 1930s directed to the approximate solution of the wave-equation for individual isolated atoms. Using the Pauli exclusion principle[25] and the *Aufbau* process, pictures could be obtained of the electronic situations in both ground and excited states of such atoms. The next stage was the extension of these ideas to molecules. To this new movement a sense of urgency was given by the conclusions of a number of workers in organic chemistry, and once again this branch of experimental study yielded ideas of the utmost theoretical importance for chemistry as a whole.

2. STRUCTURAL UNCERTAINTY IN ORGANIC CHEMISTRY

The decade 1924–34 witnessed a rising awareness amongst organic chemists that the "classical" (Lewis-Langmuir) electronic theory of the chemical bond was seriously deficient in a number of important cases. This realization was prompted by new discoveries in organic chemistry and was marked by a firm refusal to save the situation by multiplying *ad hoc* hypotheses. The new ideas which emerged found their eventual justification in the new wave mechanics. But they appeared first as a product of the old chemistry, not the new physics. Once again in the story of valency two independent groups are seen working towards the same goal, the unification of their ideas being ultimately mathematical. This time the two streams of thought arose in Germany and Britain.

F. G. Arndt (b. 1885) was Extraordinary Professor of Chemistry at Breslau in 1924 when, with two co-workers, he published an attempt to clarify the formulation of the γ-pyrones and γ-thiapyrones.[26] The former had been much studied in connection with plant pigments and other matters, and had been long known to have anomalous properties. They lacked most of the reactions of ketones and alkenes to be expected from their classical formula (I, below). Rejecting an earlier suggestion by Collie,[27] Arndt proposed to utilize the idea[28] of a molecule bearing at different points equal but opposite charges, known then in Germany as a *Zwitterion*. Accordingly he postulated the formula II; the γ-thiapyrones were represented by III and IV:

I   II   III   IV

Arndt's decisive step forward was to propose that *both* I *and* II (and also III *and* IV) were needed to describe the molecules adequately. It was evident, for instance, that the solubilities of dimethyl γ-thiapyrone were no more satisfactorily represented by IV than were its chemical properties by III (R = CH$_3$). Hence a new intermediate state was proposed involving *both* classical formulations. He wrote:

With the new views (*Zwitterions*), according to which no new bond across the ring is involved [Collie's theory], but all reduces to a shift in electronic orbitals, one can conceive any intermediate states (*Zwischenstufen*).[29]

These concepts were applied to other heterocyclic systems, but for some years no detailed theoretical exposition appeared. Arndt has recently given the reason:

> During the 1920s every author of organic papers in Germany had to contend with severe rules of utmost brevity; it happened very often that papers were returned for abbreviation, and any organic theory was frowned upon. So I could express theoretical views only in connection with experimental work, and even then only very shortly. In 1927 I tried to publish a general theoretical paper on the idea of "Zwischenstufe" and "Zwischensustand" (resonance), but I was told not to insert such views in experimental papers.[30]

However, the opportunity he sought came in 1930 (after "some bickering with the editors"), and a fairly comprehensive account[31] appeared summing up "the whole theory as far as it could be developed at that time, with no quantum mechanics".[32] *British Chemical Abstracts* gave the following synopsis of the paper:

> In connection with his views on the constitution of 1-alkyl-4-pyridones and similar compounds the author defines a tautomeric equilibrium as characterized by the difference of the point of union of a hydrogen nucleus in the two formulae. An "intermediate stage" is assumed when the two formulae show the same relative position of all atomic nuclei and differ only by "linkings" or "charges".[33]

A general account[34] of this work by Arndt contains also his impressions of its subsequent developments.

> During the years 1924–32 my publications referring to the general chemical theory of Zwischenstufe (resonance hybrid) and Zwischenzustand (resonance) were too scattered and the whole idea too unconventional, to be generally accepted or even universally noticed. It only gained momentum when, in 1933–5, the quantum mechanical work of Pauling, Hückel, and Ingold confirmed it independently and put it on a much sounder physical basis. At that time I was already at Istanbul as a refugee from the Nazi regime...[35]

So, unable to publish in Germany by himself, in 1935 he was joined by his former pupil Eistert in producing a paper[36] that connected his views with current conceptions of resonance. It will now be appropriate to enquire what these other conceptions were.

The decade from 1924, in which these developments were being initiated in Germany by Arndt, was also marked by a parallel movement in England. Ultimately it was the English, rather than the Continental, expression of the idea that was to dominate chemistry.

Partly the reasons for this were political and personal (as in the case of Arndt), but partly the German mistrust of theory, especially physical theory, must be held responsible.[37] For it was from the application of the new science of physical chemistry to the older organic discipline that new light was to break in England and elsewhere. That light was certainly needed. In his memorable obituary notice of N. V. Sidgwick, L. E. Sutton has spoken of the (possibly apocryphal) remark of W. H. Perkin, jr. that "physical chemistry is all very well but it does not apply to organic substances",[38] and of the difficulties in the early 1920s of applying physical methods to organic structural problems; there was little one could do to an organic compound "save to measure its molecular weight and speculate".[39] As the 1920s progressed, however, this situation improved with the gradual adoption of new physical techniques. The parachor of Sugden,[40] the new interest in optical properties as molar refraction, and above all the availability of methods for measuring dipole moments gave powerful tools to all interested in organic structures. Yet it is doubtful if even this advance would have led as quickly to the organic doctrine of "mesomerism" if there had not been another factor in the situation, one that happened to be particularly British. That was the Lapworth–Robinson theory of organic reactivity. We may briefly trace the course of events as follows.

In 1924, the year of Arndt's paper on the γ-pyrones,[41] C. K. Ingold succeeded J. B. Cohen as Professor of Organic Chemistry at Leeds. For the next six years his department produced a steady output of research papers having in common the new interest in reaction mechanism (aromatic nitration, tautomerism and addition reactions were the chief topics). The debt to Robinson is clear enough, and Ingold built extensively on his foundations. He has himself also acknowledged debts to both J. F. Thorpe and H. M. Dawson.[42] Thorpe at that time occupied the Chair of Organic Chemistry at Ingold's former college (Imperial College, London), and had done distinguished work on organic acids and on tautomerism. From him Ingold received a liking for mechanistic problems, and perhaps some of his insight into the importance of the influences that operate in tautomeric phenomena. Ingold's colleague in the Physical Chemistry Chair at Leeds, H. M. Dawson, was doing notable work in reaction kinetics which was also to help in shaping the future: "Dawson taught me a lot of physical chemistry in a quiet way, and I became very interested in his attempts to sort out the kinetic effects of the constituents of electrolytic solutions."[43]

His mechanistic studies at Leeds first led C. K. Ingold, together with his wife, to propose that the conjugative polarizations of Robinson's scheme might have a permanent element in them.[44] One example must suffice. It was known that sulphuryl chloride (SO_2Cl_2) will chlorinate aromatic molecules in such a way as to suggest that the chlorine being inserted is behaving like other *positive* groups, whereas its usual mode is negative (Cl^-). To explain this, the authors supposed that, if not actually positive, the chlorine might be a neutral atom whose electron deficiency would make it simulate to some degree a "positive" reagent. This was not likely to happen if sulphuryl chloride was I (below), for the chlorine atoms would separate as Cl^- ions; therefore II was postulated as a source of the neutral atoms, and the molecule might then exist in an intermediate state between the two:

Shortly after this Ingold proposed a definite test for the permanence of a tautomeric polarization. In dimethylaniline the inductive polarization (arising from the different electronegativities of carbon and nitrogen) would be expected to be as indicated below. But if the electromeric shifts had a time-independent element then the molecule would have a polarization in the opposite direction which would tend to minimize or outweigh the first effect. Since dipole moments are properties of *unreacting* molecules their measured values are functions of permanent electronic effects. Therefore "when the sign of the dipole in dimethylaniline has been determined, it should be possible definitely to answer the question whether tautomeric, like inductive, effects are associated with a permanent displacement of the electrons".[45]

(The arrows point in the direction of electron drift; the +⎯→ symbol indicates the overall direction of the resultant dipole, the head of the arrow signifying the negative, and the + the positive, end.)

This deliberate encouragement to use physical methods to settle issues long regarded as the prerogative of organic chemists was a landmark in what was later to be called physico-organic chemistry. Yet the underlying concept of dipolar molecules goes right back to Berzelius. It was not long before experimental results were seen to confirm the hypothesis of permanent, tautomeric polarizations, and the dipole moments measured decisively settled the matter.[46] Sutton tells us that he was encouraged to begin his work on dipole moments by Sidgwick in 1928. At that time Debye was staying with the latter at Oxford and had suggested this technique (to which he himself had contributed so much) for attacking the problem of thallium ethoxide's structure. In the event, "soon there were so many other things to do that thallium ethoxide was forgotten".[47]

Ingold himself continued to develop and apply the doctrine that was thus confirmed. Indeed even before this, in 1927, he and F. Shaw observed the need to distinguish permanent from time-variable effects in work on kinetics.[48] This distinction was further explained in his paper on tautomerism and aromatic reactivity of 1933.[49] The permanent tautomeric, or conjugative, polarization he now named the *mesomeric effect*, because the actual structures of the molecules concerned were "in between" those of the various classical formulae that could be written. The next year he proposed that the phenomenon be termed *mesomerism*[50] rather than the *resonance* of the quantum chemists who were now reaching similar conclusions from their own side. He wished to imply as clear a distinction as possible between this and *tautomerism*, where isomers did change into each other. This, he maintained, was not the case here as only one species of molecule existed, and the "classical formulae" were merely intellectual scaffolding. So to avoid any suggestion of an oscillation implicit in the word *resonance* he thought it best to coin a new term. In the same year appeared his celebrated paper "Principles of an Electronic Theory of Organic Reactions" in which these and many other important ideas were brilliantly displayed.[51]

There can be no doubt at all that organic chemistry in Britain owed an immense debt to Ingold for these early ideas. We are concerned here only with the concept of mesomerism, but that alone stimulated a vast and valuable amount of research. Its identification in the early 1930s with quantum mechanical resonance enhanced its status still further. But it would be a mistake to suppose that thenceforward it was to be mathematized beyond the reach of the ordinary organic research worker for whom, indeed, it was just beginning to prove so useful. Apart from Ingold and Robinson few men did more to make the new ideas an accepted part of organic chemistry than did N. V. Sidgwick. His encouragement of Sutton has already been noted. In 1931, while in the U.S.A. on a lectureship at Cornell University, he began a warm friendship with L. Pauling whose own characteristic exploration of resonance was just beginning. Thereafter, as Sutton observes, "one of his main tasks was to interpret the concept of 'resonance' between classical structures, and to apply it to various problems".[52] Examples of these applications include the structure of azides and diazo-compounds[53] and his famous "resonance" explanation of the hydrogen bond.[54] His contribution to *Annual Reports* for 1934 asserted that the theory of resonance "must now be taken seriously into account by organic as well as physical chemists".[55] Recalling that by now Robinson was also at Oxford one is not surprised to read: "it is to be presumed that modern organic theories which ascribe the reactivities of molecules to drifts of their linking electrons are to be interpreted with reference to the theory of resonance".[56] And, to quote Sutton once again,[57] "his two Presidential Addresses[58] to the Chemical Society were aimed at converting the chemists of this island to the doctrine of resonance; and they were remarkably effective".

3. ORBITALS AND BONDS

While chemistry, and particularly organic chemistry, was undergoing its own internal revolution, other branches of science were experiencing similar upheavals. Under the stimulus of the new quantum theory advances were being made on several fronts that were to have a direct bearing upon the conception of a chemical bond. Physical studies on the solid state, magnetic behaviour and band spectra yielded valuable grist to the mills of the theoreticians, and gave new quantitative meaning to valency. Anything like a complete solution to the Schrödinger wave equation except in the very simplest cases was not even contemplated. All that could be hoped for was an approximate solution, reached by a combination of mathematical finesse and chemical intuition. The number of arbitrary choices in making approximations was reduced by employing data supplied by the physicists, though inevitably personal preferences could not be excluded altogether. And so this tripartite alliance between chemistry, physics and higher mathematics became yet another demonstration of imaginative co-operation in science.

Out of this complex situation two or three different methods emerged for obtaining a picture of the electronic state of a molecule. Each of the approaches had its own characteristic features and its own valuable "insights". As the pictures became clearer and the mathematics more precise these insights became incorporated in the chemistry of the 1930s, giving quantum mechanical meaning both to the new concepts like mesomerism and more familiar phenomena like unsaturation and directed valency.

(i) *The Valence Bond Approach (V.B.)*

The first of these approaches is based upon a mathematical procedure that constitutes something very like a "thought experiment". One imagines two atoms, complete with all their valency electrons, at first a great distance apart, gradually being brought up close together until they are near enough to form a molecule. The potential energy of this system is now calculated as a function of the interatomic distance and the wave-functions of the valency electrons. There will in general be obtained a curve of the following approximate shape:

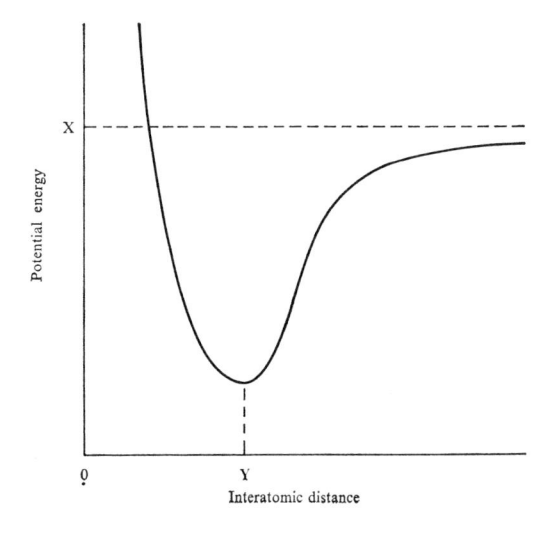

The value of the energy increment X will be the binding energy, while the distance Y will be the length of the valency bond in the stable ("unexcited") molecule. As more energy is added the bond length will vary as the atoms oscillate between the two positions indicated by the two values for interatomic distance at any given energy between A and B. As the energy added reaches X, one of these values approaches infinity, *i.e.* the molecule flies apart. The value of X may be determined experimentally and the agreement between observed and calculated figures is a measure of the validity of this approach. With progressively more sophisticated treatments this agreement has become more and more complete.

In this kind of treatment, a wave-function for the molecule is constructed out of the wave-functions for two electrons, one from each atom in the simplest cases. This process is called electron-pairing, and its success offers part of the quantum-mechanical justification for the Lewis theory of an electron-pair bond. Indeed, the "valence bond" conceived to arise from this process gives the method its name.

It is remarkable that the first appearance of this distinctive approach came within a year of the introduction of Schrödinger's wave equation. In a now classic paper, Heitler and London applied it to the case of the hydrogen molecule, H_2, and obtained excellent agreement with experiment for the bond distance although the binding energy was rather low (3·14 as compared with 4·75 eV).[59] Subsequent authors were able to improve on these results, and a bibliography of calculations on the same molecule has recently been given.[60] The important thing is that in one form or another the valence bond treatment has met with much success in a semi-quantitative and qualitative way, and at a very early stage highly significant deductions were made from it that have affected chemical thinking for 30 years or more.

The most honoured name in the development of valence bond theory from its inception until the present time is that of Linus Pauling (b. 1901). In April 1931 the *Journal of the American Chemical Society* carried the first of a series of papers bearing the general title "The Nature of the Chemical Bond"; this marked the beginning of a rapid flow of publications from the California Institute of Technology at Pasadena under Pauling's name. This particular paper deduced from quantum mechanics the familiar doctrines of tetrahedrally directed valency bonds and free rotation about a single bond, and applied magnetic data as criteria of bond type in complex ions.[61] The modern concept of bond hybridization may be traced to this paper. Within the next few years Pauling was led, by a consideration of many molecules from the V.B. standpoint, to an active devotion to the cause of quantum mechanical resonance which is still commonly regarded as the child of valence bond theory. In a molecule A–B one must take account of the possibility that the bond may be polar to some degree. The extremes could be written (where the dots are the bonding electrons):

$$A : B \quad \text{or } \overset{-}{A} \overset{+}{:} B \quad \text{or } \overset{+}{A} \overset{-}{:} B$$

The wave-function for the molecule must contain terms corresponding to all of these possibilities, and the molecule is then said to show "resonance" between them. Nothing can be inferred from this about the "real" existence of these "structures", for all that is necessary is to acknowledge that only one species of molecule will exist, but partaking of some of the character of each of these contributing, but imaginary, "structures".

These ideas soon appeared in elegant form in Pauling's publications. A paper[62] on one-electron links (as in H_2^+) suggested a resonance between $(H.\ H)^+$ and $(H\ .H)^+$, and one a year later[63] concluded that ionic terms could be ignored for the molecules HCl, HBr and HI but *not* for HF. Within months the concept of resonance was being used to illumine the benzene problem[64] and that relating to the organic azides.[65] Meanwhile the crucial experimental tests were being applied to interatomic distances[66] and heats of formation.[67]

It is impossible in a brief account to do anything like justice to the growth of this highly important viewpoint, still less to expound its multiplicity of applications. It must be sufficient to say that the fusion of quantum-mechanical resonance with the new ideas in organic chemistry gave to science a conceptual tool that has still retained much of its usefulness. As Longuet-Higgins has observed:

In the hands of Pauling and others, the old wine was successfully transferred to the new bottles and re-labelled appropriately. . . . The Robinson–Ingold theory, founded as it was on the maximum of experimental fact and the minimum of hypothesis, emerged from the revolution intact, and more soundly related than ever to physical principles.[68]

Thus as early as 1932 Hinshelwood could write that "the success with which organic chemistry has solved its problems by the aid of its own conceptions has, during the last few years, stimulated theoretical physicists to attempt the translation of these conceptions into the language of quantum physics".[69]

The friendship between Pauling and N. V. Sidgwick, and the latter's contributions to resonance theory from the chemical side, have already been noted (p. 302). Another worker in the same tradition was G. W. Wheland. From 1932–6 he held a Research Fellowship under Pauling at Pasadena, and together they produced an important paper on the resonance of benzene in 1933.[70] Then in 1936/7 Wheland spent a year in England on a Guggenheim Fellowship which enabled him to study with Ingold in London, Sidgwick at Oxford and Lennard-Jones at Cambridge. He was thus very well equipped to become the spearhead of another movement to foster and explore the theory of resonance. This he did with particular effectiveness in his books *The Theory of Resonance and its Applications to Organic Chemistry*[71] and its successor *Resonance in Organic Chemistry*.[72] Meanwhile the successive editions of Pauling's own book,[73] bearing the title of his first major essays in this field, continued to supply generations of undergraduates and others with a synoptic view of chemical structure for which he has earned deep and lasting gratitude.

Yet the theory of resonance, like the V.B. approach with which it is so closely associated, did not reign unchallenged for ever. The alternative molecular orbital approach (which we shall examine shortly) tended to undermine faith in resonance formulations for it appeared to do away with the most objectionable features of the resonance argument, the postulation of hypothetical "structures" for molecules whose actual structure was known to be different from all of them. Indeed the substance of the attack on resonance was essentially epistemological and concerned with problems of "existence" of, for example, the different "structures" of benzene:

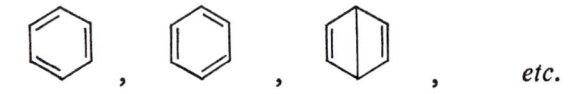

etc.

It must be admitted that the early advocates of resonance did not sufficiently distinguish it from tautomerism, in which different molecular species *are* interconvertible. Thus Hinshelwood observed that it was supposed "that the molecule oscillates rapidly between the two possible structures",[74] while Sidgwick two years later considered a molecule "either as passing from one state to the other with very great frequency (some 10^{15} times per second), or more probably as having a structure intermediate between the two".[75] Despite the disclaimers by later authors, including Pauling and Wheland, the idea of several independently existing resonance "structures" was exceedingly hard to eradicate. Perhaps the word "resonance" was partly to blame. In 1936, Penny and Kynch were deploring its use, with its innuendoes of swinging pendulums, *etc.*, but "the word is now so commonly used that it would be a mistake to attempt to substitute another".[76]

After nearly twenty years of useful life the resonance theory found itself under heavy attack from several quarters. The main reason for this seems to have been the new ascendancy of the molecular orbital treatment, and one of the most trenchant criticisms of the resonance hypothesis, along with its associated V.B. concepts, came in Dewar's *Electronic Theory of Organic Chemistry*, published in 1949.[77] M. J. S. Dewar had worked with Robinson at Oxford and the latter had contributed a Foreword to the book. Dewar's standpoint was clearly indicated in his Preface:

> Previous writers in this field have approached chemistry from the standpoint of the resonance method and their general interpretation of chemical phenomena is now well known. This approach is most unsuitable from the organic chemist's point of view since it involves a new symbolism and a novel and uncongenial outlook. (A cursory examination of the literature shows how difficult it is for chemists to distinguish intuitively between resonance and tautomerism.) For these reasons the molecular orbital method has much to offer.... The author has tried to follow this line of approach.[78]

Other authors since then have adopted a similar attitude to the theory of resonance.

Dewar's attack on resonance was essentially pragmatic. At about the same time a very different assault was launched on grounds that were philosophical and indeed political. In a controversy that had much in common with the Lysenko affair in biology, Soviet science took up arms against the resonance hypothesis (variously associated with the names of Pauling, Wheland and Ingold) and declared it to be ideologically unacceptable. The use of imaginary, ideal "structures" to describe real molecules was incompatible with the dialectical materialism of Marx, Engels and Lenin. In 1949, G. V. Chelintsev published a book on theoretical organic chemistry dispensing with the multiplicity of resonance "structures" and going back to Butlerov's ideal of "one substance, one formula".[79] Thus he wrote benzene as:

In this same year that saw the publication of both Dewar's and Chelintsev's books, a strong attack on resonance came in an article "On a Machist theory in chemistry and its propagandists" by V. M. Tatevskii and M. I. Shakhparanov.[80] Other articles appeared in *Pravda* and elsewhere in celebration of Stalin's 70th birthday demanding reforms in chemical thinking. Matters reached a climax with conferences in 1950 and 1951, in the second of which an all-Union resolution was passed replacing the theory of resonance by a "theory of mutual influences". Denying that the form of a molecule is physically inconceivable, the Soviet chemists concluded that the paradox expressed by resonance was of temporary duration only, arising from the incompleteness of man's knowledge.[81] But for ordinary purposes "the primary practical distinction between the method suggested by the Soviet chemists and the theory of resonance is that the Soviets would be deprived of the use of resonance forms as instructional 'visual-aids' ".[82]

Since that time the heat of the debate has largely been dissipated, but Russian textbooks still avoid the theory of resonance. L. R. Graham considers that official opposition to it may recently have softened, however.[83]

Meanwhile, what of the valence bond theory with which the resonance hypothesis is so closely linked? There can be no doubt that in the years up to about 1945 it dominated inorganic chemistry and exerted a considerable hold on the organic branch. After World War II, however, the new availability of complex, reliable infra-red and ultra-violet spectrophotometers led to an increased interest in the excited (as well as the ground) states of molecules, and in this connection the V.B. method was notably deficient. Hence organic chemistry began to be treated by the alternative molecular orbital method and the trend set by Dewar was followed widely though by no means universally. Technical difficulties were partly responsible for a general reluctance to apply spectrophotometric methods to inorganic materials, although there was more than enough work in organic chemistry to keep all available instruments occupied for a very long time to come. Hence the decade 1945–55 saw the organic branch, rather than the inorganic, turn over to the M.O. approach. At the end of this period, however, even inorganic chemistry was yielding to the probings of the spectrophotometers. In 1950 64 minerals had been examined in this way,[84] and two years later the infra-red spectra of 33 inorganic ions were recorded and used for qualitative analysis.[85] These two papers were followed by others and the "excited state" became a lively topic for the inorganic chemists. Thus A. D. Liehr has asserted that since the early 1950s valence bond theory has been abandoned even by inorganic chemistry,[86] although this suggestion brought a rejoinder from Pauling: "I have not observed a decrease in the value of valence bond theory, especially as applied by the organic or inorganic chemist."[87] With this reminder of the fluidity of chemical concepts, and of the impossibility of assessing events so close to us as this, we shall now go on to deal with the main rival to the valence bond approach, the molecular orbital treatment.

(ii) *The Molecular Orbital Approach (M.O.)*

The foundations for both the V.B. and M.O. methods were laid in the U.S.A., although the influence of European physicists was not small and much valuable work was later done on the Continent. But while Pauling initiated some of the earliest advances for the older theory, a similar service was performed by another American, Robert Mulliken (b. 1896), at Chicago, for the molecular orbital method.

Mulliken's early work was on band spectra. Thanks to the efforts of Pauli and Hund, in particular, the study of these had been simplified and Mulliken's task was thus lightened.[88] He spent much of the summer of 1927 at Göttingen with Hund and it is clear from his own account[89] that he owed much to this encounter. The following year he extended the "isoelectronic principle" of spectroscopy from atoms to molecules.[90] It is not necessary to say more about this principle than to comment that its application to *molecules* marked an imaginative leap forward in the right direction.

The most obvious feature of the M.O. approach is its view of electrons moving in new orbits, called "molecular orbitals", which encircle *several* atomic nuclei. The shape of each of these polycentric orbitals can be calculated by methods closely analogous to those used for isolated atoms. The mathematical treatment involves an approximation termed "Linear Combination of Atomic Orbitals", often known as the L.C.A.O. method. It is sufficient to notice here that, like the V.B. method, it *is* an approximation, although of a different kind. Very briefly, the M.O. method fails to take adequate account of the coulombic repulsion between the two electrons forming a "bond"; it thus underestimates the probability that, if one of the electrons is near one nucleus, then the second is likely to be near the other. In other words it allows too generously for the possibilities

	A: B	*i.e.*,	A^- B^+
and	A :B	*i.e.*,	A^+ B^-

so it may be said to over-emphasize the contribution of ionic "structures" like these. On the other hand, the V.B. method for exactly opposite reasons underestimates ionic contributions.

Robert Mulliken was associated with the development of these ideas from the beginning. Apart from his gift to chemical nomenclature of the term "molecular orbital" his contributions to the theory have extended over the same period as those of Pauling to the V.B. method, and have been deservedly commemorated in the tributes paid in a recent volume.[91]

It is interesting to note that Mulliken drew some of his early ideas from solid state physics as well as spectroscopy. In 1928, Bloch had been led by a study of the theory of metals to propose a delocalization of the electrons responsible for metallic conduction. In arriving at a kind of "infinite orbital" he had introduced a new method, that of L.C.A.O.'s.[92] This was the approach taken up with such success by Mulliken, although Pauling seems to have been the first to apply it to the derivation of a molecular orbital in his calculations on the hydrogen molecule-ion (H_2^+) in 1928,[93] and Lennard-Jones the first to stress the symmetry properties of the constituent wavefunctions.[94] Later, Bloch's method was also embodied in Hückel's molecular orbital treatment of the benzene molecule.[95] (The similarity between the continuous π-electron "cloud" around the ring and the infinitely conducting "cloud" of electrons in a metal has often been noticed.)

Another important antecedent of Mulliken's work on orbitals was the study made by Bethe on atoms embedded in regular solids.[96] From this Mulliken was led directly to his own ideas on the symmetry properties of molecular orbitals. Coulson has pointed out that unfortunately he used a different notation from Bethe and other solid-state physicists, and thus delayed a wider appreciation of his own work. But for this it is likely that modern ligand field theory – so indebted to both Bethe and Mulliken – would have arrived much earlier than it did.[97]

Of the later development of the M.O. theory much must be left unsaid. Hückel gave the first quantum-mechanical treatment of the double bond in terms of M.O.s;[98] his application of it to benzene has already been mentioned. The first attempt to apply the theory to organic reactivity (as opposed to structure) appears to have been made by Pauling and Wheland.[99] In 1939 Mulliken brought together a number of unexplained phenomena (later designated "hyperconjugation") and went on to propose a theoretical treatment in terms of molecular orbital theory.[100] From about 1947 a surge of activity in theoretical chemistry led to the increasing application of M.O. methods to inorganic compounds. In particular, the spectroscopic study of complexes (which involved molecules in excited states) led to an increasing realization of the inadequacy of V.B. methods to deal with them. By now a third contender had arisen – the crystal field theory, based upon the work of Bethe and others in magneto-physics.[101] This considered the electrostatic attraction between, say, the central cobalt atom in a complex, and the six atoms or groups associated with it ("ligands"), these being regarded as ions or as dipoles. This was a striking reincarnation of Berzelius's electrochemical system, although applied to compounds largely unknown to him. But for chemical problems it was too simple, for complexes are not ionic in the way envisaged here. Yet a great step forward was taken when this approach was married to the method of molecular orbitals. The product of this union was found to be devoid of the more objectionable features of both parents. Known as "ligand field theory" it has transformed transition metal chemistry for the last fifteen years.[102] By including in its account orbital overlap it has turned crystal field theory into a viewpoint justified by a wide range of phenomena, including electron spin resonance.

It would be very rash to try to make any historical assessment on developments as recent as this, and still more so on a number of other topics of current interest in the theory of valency. One other point may, however, be legitimately made concerning the relation between the molecular orbital theory and its chief rivals: none of these approaches can claim to be final, or even inherently better than any other. Each approximation has its useful sphere of application, and which is used depends entirely on the nature of the problem to be solved. Even a field of study formerly regarded as amenable to one kind of treatment only may, after some passage of time, turn out to be capable of satisfactory examination from another viewpoint. Nowhere is this at present more the case than in the subject of resonance. Once regarded as the offspring of V.B. theory, and therefore spoken of only in terms appropriate to that theory, it has recently been shown to have its roots well embedded in the M.O. approach also. Some of the familiar concepts of resonance have been shown to be derivable from Hückel's molecular orbital theory[103] (H.M.O.) and a recent writer has concluded: "We deduce that the traditional resonance formulation of electronically excited states of π-electron systems, and the rules that apply, have their roots in H.M.O. theory."[104] As Coulson has observed:

It has frequently been the custom for supporters of one or the other of these theories to claim a greater measure of chemical insight and quantitative reliability for the method of their choice. This is a pity because neither method is complete or fully satisfactory. Fortunately in most of their conclusions the two theories agree, though they reach their conclusions in quite distinct ways. This means that we must not arbitrarily reject one or the other.... There is little doubt but that the molecular-orbital theory is conceptually the simplest. Historically it was developed after the other theory was already established, and for that reason has been a little slower in gaining acceptance. Its present status, however, especially in dealing with excited states, is fully equal to that of the valence–bond theory.[105]

CHAPTER XVII

Some Conclusions

THE HISTORY that has been narrated in the previous chapters has inevitably been predominantly factual, as all history must in the first instance. Yet it seems desirable now to cast a reflective glance over the whole story to see if any general principles may be discerned; to turn (in other words) from the facts to their interpretation. Questions that spring naturally to mind include these: why did the events take place when and as they did? What explanations may be given for the uneven rates of growth of the subject? How far was the history of valency typical of the growth of a scientific idea? Can any methodological generalizations be usefully made?

Complete answers cannot be given to any of these queries, but certain lines of investigation do appear to lead at least in the direction of a partial answer to some of them. The account which follows attempts to examine the history of valency from the points of view of its internal structure and its external relations. While it draws upon the factual material in the earlier part of the book, it makes no claim to be anything more than a personal interpretation. To avoid a multiplicity of notes cross-references to other chapters have been omitted.

1. THE INTERNAL STRUCTURE OF THE HISTORY OF VALENCY

The whole story of valency is the history of an *idea*, altered and developed by empirical facts it is true, but nevertheless an essentially intellectual progress. In this respect it resembles the first century of Daltonian atomism upon which it is based and with whose fortunes it came to be inextricably linked. Like the picturesque atoms of Dalton the chemical "bonds" of Frankland were invented rather than discovered. Concerning the former Caldin has written: "Dalton did not deduce the atomic theory from the laws of chemical combination; he invented it."[1] The same kind of view may be taken of the birth of valency theory which was only in a limited sense a "discovery". This word in science is usually applied to "things" (anaesthetics, fossils, nerves, inert gases) or to events which may first appear as isolated phenomena (like the fogging of a photographic plate) but which subsequently are seen to show regularities expressible as laws (such as those of radioactive disintegration). Frankland's comments on the "general symmetry" of construction of inorganic compounds[2] must surely qualify as a discovery of a regularity. But the concept to which Kekulé and his colleagues were groping, and which Frankland was later to employ so fruitfully, was much more than this. Their goal was a model by means of which the regularities of combining-power might be expressed. Thus in the 1850s the Kekulé–Williamson–Odling school were separated from Frankland not merely by the different traditions they had inherited but more significantly by epistemological barriers. At that time Frankland was not searching for a model but was content with a generalization; not until the next decade were their positions reversed.

The history of valency, therefore, is the development of a philosophical model. So also is nineteenth-century atomism. Now a model in science does not always or even often appear first in its ultimate form. Usually it passes through phases of development in which it is refined and modified as experience dictates. This is particularly true of several major models employed in the last century, and their phases may be labelled as (1) metaphor, (2) mechanism, (3) mathematical abstraction. The first of these stages corresponds approximately to what Mary Hesse has called "conceptual models", which she describes as "imaginative devices to be modified and fitted *ad hoc* to the data" since they are not derived from a causal theory and the relations between their terms are imaginary and arbitrary.[3] Her emphasis is on the logical connection (or lack of it) between model and phenomena. The point now is that *because* of this arbitrariness such models ought to be, and often are, regarded much as we might regard a metaphor in literature. Their appeal will be primarily aesthetic, giving economy, symmetry, harmony, *etc.*, and their validity will not be a matter that can be settled by experiment. Their effectiveness will be governed by psychological rather than logical considerations.

Dalton's atomic theory is a good example. His idea of atoms with characteristic weights was a conceptual model of this kind since all such weights were inaccessible and arbitrary. Hence the clusters of atoms in his formulae could merely be regarded as metaphorical symbols, lacking the approach to objectivity claimed for later formulae. It seems that Dalton himself looked at them like this,[4] and Berzelius certainly did. "It is nothing but a hypothesis, and probably will always remain one",[5] he wrote of the atomic theory.

So long as atomic weights were incapable of reliable measurement this situation remained. Atomism was only a "conceptual model", and its logical deficiencies robbed it of any general recognition as something more than a convenient symbolism. Although such a model may be, in principle, "strongly predictive",[6] in practice this one was not. Writing of Dumas's disenchantment with the atomic theory, Buchdahl has observed:

He objects to Dalton and Swedenborg that their atomic theories merely cover the known facts and predict no new ones. Within the restricted field of the chemical knowledge of the thirties this complaint was bound to remain endemic.[7]

Thus atomism entered into a long "metaphorical" phase which lasted until about 1850. This account of its status may go some way towards explaining why it did not rise to the heights which we (viewing it as much more than a metaphor) might have expected. Only when further reflections on the hypothesis of Avogadro had changed the whole situation of the atomic weights could Dalton's theory even be tested. Moreover, insistence that the theory was only a metaphor may give yet a further clue to Davy's cavalier attitude to it. Scientist-poets will normally keep their metaphors rather strongly away from their science; and they probably recognize them for what they are a little before the majority of their fellow-scientists.

We have looked at the beginnings of atomism not only because the subject impinges directly on its derivative concept, valency, but also because they both developed in the same kind of way. Indeed, several important topics of nineteenth-century science underwent this kind of conceptual change, including valency. The latter began as an extended metaphor in the shape of the theory of types. Again and again Gerhardt and his followers disclaimed any meaning for their types other than pure symbolism; any constitutional significance was vigorously denied. A similar pattern may be discerned in nineteenth-century biology. An article by A. G. N. Flew on "The Structure of Darwinism" has this to say on the emergence of that theory as a specific model:

> The acceptance of Darwin's theory made possible a massive deployment of one model which had been curiously boxed up and impotent for an extraordinarily long time. This was the model of the family.... The various terms appropriate to this model were, apparently, introduced because naturalists noticed analogies which made the idea of family relationship seem apt as a metaphor. But the suggestion that the metaphor might be considerably more than a mere metaphor, that the model could be deployed, seems almost always to have been blocked by the strong resistance of the accepted doctrine of the fixity of species.[8]

Admittedly the chemists and biologists were facing rather different situations. The latter were concerned to replace an existing view by a new one, while their chemical colleagues were evolving an idea where none had existed before. Consequently their "strong resistance" came from a different kind of attitude; not from so-called theological conservatism, but from the fear of scientific men that theories were going to outstrip the facts. Yet in both sciences the metaphors were being transmuted by pressure of the facts into models of greater concreteness and credibility.

The crucial change in the fortunes of chemical atomism came when the doctrine of valency emerged from the accumulation of data in organic chemistry that, by 1850, was crying out for rationalization. Just as Dalton's theory had thrown light upon inorganic chemistry in earlier days, so now the organic branch came under its illumination. But by now the atomic model had been elaborated by the new concept of valency, and had also acquired a more respectable logical status as the atomic weight values emerged as non-arbitrary parameters. For this last reason one could now think much more literally of atoms; while physicists had done this for a long time, their mechanical picture could at last be taken over into chemistry too. To complete the realism of the model only one factor was missing: the links between the atoms themselves. With the arrival of valency not only did the Daltonian atoms gain a sounder logical status; psychologically they were immeasurably more acceptable, for the "mechanical" picture of *combining* atoms was now complete. It did not matter that valency itself was at first only a quantization of combining-power, and its nature completely unknown. It was sufficient that the vague metaphor was now to be replaced by a simple, picturable model of tiny spheres linked by a small number of "bonds". Again, the word "mechanism" seems to be the only one that would adequately fit, particularly in its psychological impact. We are reminded once more of Darwin, for what he "supplied was a series of further working hypotheses on a grand scale as to the *mechanism* by which changes from one species to another could occur".[9]

As the second half of the nineteenth century progressed the spring-and-ball models became widely accepted. But they were never universally popular, and after about 1880 the vividness of the picture began to fade. Even the extension to a 3-dimensional model by van't Hoff, Baeyer and others was not sufficient to save it from the damage inflicted by the variable valency controversy. This in fact came from a deficiency in the model itself. Since valency arose out of a study of carbon compounds, carbon was taken unconsciously as the typical atom. In its constancy of valency and in several other ways it is quite untypical, and models based upon the carbon atom were bound to encounter trouble. By this time also the physicists were in difficulties with their atomic model and a general abandonment of atoms seemed likely.[10] This naturally undermined still further the whole doctrine of valency, including the mechanical picture of it.

Meanwhile from that typically Victorian mechanism, the steam engine, there had arisen the need to study the relations between heat and work, and partly through this the science of thermodynamics was born. Its application to chemistry by Gibbs, Maxwell, Clausius and others had shown the fruitfulness of a purely mathematical approach, unimpeded by models whose validity was always open to doubt. This independence of mechanism appealed widely to those with a distaste for atomism. Again the spirit of *Naturphilosophie* stirred uneasily in Germany, and it found a ready prophet in Ostwald. As late as 1900 he was able to produce a textbook on inorganic and physical chemistry without a single reference to atoms (or, of course, to valency).[11] Emancipation from mechanical models was not complete in chemistry; the organic chemists continued to employ both the atomic concept and its derived theory of valency. As a result the dichotomy between their branch of chemistry and the rest became so great that by the end of the century the subject was beginning to show signs of strain under the influence of such an artificial division. But even so, the mathematization of at least part of chemistry inevitably had its effect on the rest. Ostwald's campaign against atoms had been lost irretrievably by 1903 when the facts of radioactivity began to be understood. But mathematics had come to stay, so that even these newly confirmed atoms had to be submitted to its disciplines. With the discoveries that followed in mathematical physics the pictures of the atom that emerged looked progressively less and less like mechanical models. In its turn valency moved steadily away from the position in which it could be happily imagined in terms of spring-like "bonds". The analogies changed to the unmechanical ones of "clouds". Even electrons, after a short-lived existence as "particles", are now often "delocalized" in a way that no mechanical model could possibly permit. And this is only half the story, because the ionic "bond" has now been stripped of all mechanical accretions, and is now no longer visualized but accepted as another case of action-at-a-distance. It is eminently amenable to mathematical treatment but quite incapable of being represented pictorially. How long it will be before this is true of all valency phenomena remains to be seen. Meanwhile it will be appropriate to conclude with some words towards the end of a recent Tilden Lecture to the Chemical Society on "The Contributions of Wave Mechanics to Chemistry". Following a multiplicity of illustrations couched often in vividly pictorial terms, the Lecturer (C. A. Coulson) observed:

I described a bond, a normal simple chemical bond; and I gave many details of its character (and could have given many more). Sometimes it seems to me that a bond between two atoms has become so real, so tangible, so friendly that I can almost see it. And then I awake with a little shock: for a chemical bond is not a real thing: it does not exist: no one has ever seen it, no one ever can. It is a figment of my own imagination.[12]

Thus the wheel has turned full circle. From a hesitantly produced metaphor the concept emerged as an intricate mechanical model. Thence it has been sublimed into a solution to an equation in mathematics, and with this has come, hesitantly, no doubt, the conviction that all visualizations must be incomplete. What we once suspected we now know: the limitations of our understanding of the external world.

2. THE EXTERNAL RELATIONS OF THE THEORY OF VALENCY

From the internal structure of the history of valency it may be as well to turn, finally, to glance at the relation between events within this field and those outside it. The history of science has been interpreted in general from two main standpoints, designated "intellectualist" and "externalist". The former seeks to explain events in terms of the intellectual structure of science itself, the latter in terms rather of contemporary changes in social, economic, political and other realms external to science. Both these ways of surveying matters may be useful, though the former seems to be more in vogue today. W. F. Cannon appears to share this view, and adds a plea for consideration also to be given to a third factor: "what might be called 'the role of genius' but which I prefer to call 'the uniqueness of the great stylist' ".[13] The story of valency has plenty of "great stylists", and they have featured prominently in earlier chapters. Some account has just been attempted of the internal structure of the development of valency, perhaps approximating to a treatment from the "intellectualist" standpoint. There remains a necessity to comment upon the bearing of "externalist" factors upon events. But since chemistry is only one of many branches of science, and since science is only one part of man's intellectual activity, we must also look for connections between chemistry and other areas of contemporary thought. These two aspects will be looked at in turn.

(i) *Socio-Economic Influences*

The history of valency seems to demand a qualified assent to the view that social and economic pressures can have a significant effect on the growth of a scientific idea. They can, and did, give to the scientist new incentives and new opportunities for raising his intellectual horizons. This was particularly true in the last century, and valency was one of the results of this enrichment of science.

In this development the key factor was that of education. The relation between educational, economic and scientific change was so complex in the nineteenth century that it is often difficult to distinguish cause and effect. But the empirical generalization that progress in chemical theory went hand in hand with adequate chemical instruction, though obvious in the twentieth century, found remarkably little favour in the nineteenth. Where it was perceived the consequences were usually spectacular.

During the important period immediately preceding the formulation of the concept of valency, the growth of synthetic organic chemistry owed an immeasurable debt to Liebig's patient instruction. Long after his teaching was over, the methods of the Giessen laboratory were applied in schools of chemistry appearing all over Europe. The simple fact is that valency arose *directly* out of studies in organic chemistry, made possible by the new methods of practical instruction.

The belated attempts in England to rectify the lack of scientific education there led to the establishment of popular classes in science in the 1850s and 1860s, for which new and popular books had to be written. This was the situation in which, largely thanks to Frankland, the utility of valency first became generally appreciated. This usefulness was, of course, a pedagogic one at first; it became predictive later. As we have seen both our notation and our nomenclature of valency may be traced to this time. One might therefore have expected great things of English chemistry, but, alas, the vision did not last and the impetus was gradually spent. Science and technology generally shared this decline. At the Great Exhibition in 1851 Britain was awarded most of the prizes, but at the Paris Exhibition in 1867 the figure was only 10 out of 90; "public opinion at last began to be impressed by the inadequacy of the provision for technical education"[14] and in 1881 the Royal Commission on Technical Instruction was appointed. But by then, for chemistry at least, it was too late and no further major contributions to the theory of valency came from England for the rest of the century.

Once again it is to Germany that we must look for the next developments. Of the emergence of German technical chemistry Charles Wilson has written: "The material basis of the chemical industry which grew from relatively nothing in the 1860s to the largest contemporary industry of its kind lay in the rich salt and potash deposit of Prussian Saxony. Its scientific basis was the system of state education.... The German fine chemicals and synthetic dyestuff industry was the shining example of an industry based on British and French invention converted to a virtual monopoly for Germany by education and enterprise."[15] So it is not surprising that many significant developments in chemistry (particularly organic) from about 1870 to 1900 came from Germany. Only in the twentieth century with its insistence on universal scientific education could chemistry be expected to advance on a broad international front.

There are, however, social factors even more fundamental than education, not least of these being the presence or absence of war. One must surely accord some significance to the fact that the great upsurge of chemical activity that took place in the years before 1848 occurred at a time of international peace in Europe almost without precedent in its duration. Perhaps this is a major reason for the impression that 1850 marks the "massive watershed between 'Newtonian' and 'modern' " science.[16]

Even the 1848 war could not entirely obliterate the pattern of chemical education in which a young aspirant to chemical honours would expect to travel and to study under masters of his subject in several different foreign universities. "Cross-fertilization" remains an apt, if overworked, description of the process to which ideas became subjected. One need only speculate on the course of events if Frankland had not been to Marburg or if Kekulé had not spent those years in London. Obviously the prolonged peace was not the *only* factor favouring the wandering chemical scholar; he was clearly indebted to the improvements in communications also. But the absence of international conflict does seem to have been particularly important. Moreover, it did not only affect the mobility of individuals; there is clear evidence that at least sometimes it affected their psychology as well. When war eventually came their outlook betrayed its effect.

Perhaps the biologist Virchow is the best example of this. For him the 1848 revolution symbolized the medical reforms he believed inevitable. It is in connection with Virchow that E. Mendelsohn has written: "seldom can one identify a period in history when the ideas at the very basis of a science are so clearly interacting with those of the wider social scene".[17] Chemistry lacks the drama of a Virchow, yet there are examples. Davy, recoiling from the horrors of the French Revolution, was nevertheless deeply committed to "freedom" and rejected a slavish adherence to tradition in science. And then Kolbe, fighting a desperate rearguard action against the new structural chemistry, displayed in his later years (according to Armstrong[18]) a bitterness and monomania that owed much to the war of 1870. These are but two examples of how the issues of war and peace affected, although indirectly, the progress of science.

(ii) *Intellectual Influences*

When the history of valency is considered in the general context of nineteenth-century science, certain connections become apparent that a close preoccupation with chemistry alone may fail to reveal. Why this should be so is debatable, but it is true that valency emerged under the stimulus of ideas that were certainly not confined to chemistry. At least five characteristics of this kind may be discerned.

The first of these was an idea implicit in much of early nineteenth-century science, the belief in the worth of collecting enormous masses of data. It has been suggested that this sprang from the convictions of Baconianism, and that the aim was essentially inductive. Perhaps astronomy gave the lead here with the star catalogues of Herschel and others. If so it was rapidly followed by geology and biology, while the young science of organic chemistry gave evidence of a similar tendency. It was the accumulation of facts in this last field that lent urgency to the task of finding a successor to the failing electrochemical theory of Berzelius and thus led to valency. Yet at a deeper level it may be wondered just how typical was the approach of the organic chemists. If colleagues in other fields were genuinely inspired by the Baconian motive, *their* inspiration would seem to have been otherwise. In so far as a clear philosophical purpose can be detected from their writings it was usually the modest task of testing a hypothesis already formulated; this might concern the generality of a synthetic method, the correctness of a particular formulation, the possibility of new reactions, and so on. These papers do not even pay lip-service to the Baconian ideal, and may certainly be absolved from the charge[19] of hypocritical attachment to it. One suspects that much of the real driving-force behind the early extensions of organic synthetic methods came from just the sheer joy of discovery. The relevance of this motive has not been lost even today, at least among those who practice the same art. However, whatever may have been the basic psychological or philosophical cause, the chemistry of carbon compounds partook fully in the contemporary characteristic of copious fact-collecting.

The accumulation of data constituted a challenge to chemistry, and it responded in terms of an idea that had already proved its usefulness elsewhere. The Theory of Types, both an extended analogy and a classificatory system, had its counterpart in geology and in biology. As P. Cook has neatly remarked, "it was not enough that compounds were; as with flowers and fossils of the nineteenth century they had to be like".[20] Perhaps this may explain the first characteristic more convincingly than the Baconian myth does. Facts were collected for the aesthetic pleasure of fitting them into their typical families, and of seeing the various sets approach completion. In chemistry the Theory of Types certainly illustrated this general attitude. It is interesting that Gerhardt's term *homologous series*[21] is itself an echo of the biologists' *homology*.

A third characteristic of early nineteenth-century chemistry, vital in the concept of valency, was of course its atomism. Again external connections may be traced. A recent paper by F. G. Kilgour has shown "that the idea of atomism was a productive concept in several nineteenth-century sciences and, thereby, lent some unification to the period".[22] Examples are given of the theories of *cells*, of *germs* and of Mendel's "exactly similar factors" (or *genes*); even physics with its anti-atomistic field theories capitulated (in the last month of the century) with Max Planck's *quantum of action*.

Nor was this all. Nineteenth-century science witnessed a steep decline in theories of vitalism, as the names of Schwann, Vogt, Pasteur and others bear witness. As the century progressed the discredited *Naturphilosophie* was giving way to a thoroughgoing materialism. At the same time that organic chemistry was getting rid of a *Lebenskraft*, physiology was being reduced to physics and chemistry. From the "religious vitalism" of Berzelius, to the "uncertain vitalism" of Liebig and the "ordered vitalism" of Gerhardt,[23] carbon chemistry passed ultimately and gradually to the materialism of structure theory. If, however, vitalism had persisted as an acceptable "explanation" for organic phenomena, the need for physical models of atoms and bonds would hardly have arisen.[24] Thus it is possible to see the theory of valency in the same tradition as the atoms, not of Dalton, but of Democritus.

A final aspect of nineteenth-century science may be noted, for this also had a profound effect on the theory of valency. It was the tendency already noted for mathematics to lend increasing abstraction to a model (like a chemical bond). The advent of field theory in physics was followed shortly after by the acceptance of thermodynamics into chemistry. The former discredited the "material" ether, and the latter laid bare new general laws that were independent of all mechanism. In the 1850s papers were appearing in which careful distinctions were made between equations that were tied to a particular theory of matter and those that were not. It soon became possible for the thermodynamic viewpoint to assume such an importance that the atomic controversy fell into the background. The new insights diverted attention away from considerations of valency and structure. Only the organic chemists (and not all of these) were generally committed to the latter, and the division between organic and physical chemistry became wider as the century progressed. Not until the vindications of atomism of the early twentieth century did the two branches come together again, and then it was to form more powerful an alliance than chemistry had yet seen. And now the same thing that happened to the "bond" has happened to the atom and the electron. They too have been transformed from their first crude conceptions, and they also are most fruitfully imagined in mathematical terms.

Periodical Publications Consulted

Abhl. Akad. Berlin: Abhandlungen der Königlich Preussischen Akademie der Wissenschaften, Berlin
Abhandl. Königl. Sächs. Ges. Leipzig: Abhandlungen der Königlich Sächsischen Akademie der Wissenschaften
Akad. Handl.: Akademisk Afhandlung, Uppsala
Ambix
Amer. Chem. J.: American Chemical Journal
Amer. J. Phys.: American Journal of Physics
Amer. Scientist: American Scientist
Anal. Chem.: Analytical Chemistry
Angew. Chem.: Angewandte Chemie
Ann. Chim.: Annales de Chimie
Ann. Chim. Phys.: Annales de Chimie et de Physique
Ann. Phil.: Annals of Philosophy
Ann. Physik: Annalen der Physik
Ann. Rep. Chem. Soc.: Annual Reports of the Progress of Chemistry
Ann. Sci.: Annals of Science
Annalen: Annalen der Chemie [und Pharmacie]
Arch. Hist. Ex. Sci.: Archive for History of Exact Sciences
Arch. int. d'Hist. Sci.: Archives internationales d'Histoire des Sciences
Ber.: Berichte der deutschen chemischen Gesellschaft
Brit. Chem. Abs.: British Chemical Abstracts
Brit. J. Hist. Sci.: British Journal for the History of Science
Brit. J. Phil. Sci.: British Journal for the Philosophy of Science
Bull. Acad. roy. Belg.: Bulletin de la Classe des Sciences, Académie royale de Belgique
Bull. Soc. chim. France: Bulletin de la Société chimique de France
Chemistry
Chem. Brit.: Chemistry in Britain
Chem. Centr.: Chemische Centrastelle für öffentliche Gesundheitspflege
Chem. Comm.: Chemical Communications
Chem. News: Chemical News
Chem. Rev.: Chemical Reviews
Chem. Weekblad: Chemisch Weekblad
Chymia
Compt. rend.: Comptes rendus hebdomadaires des Séances de l'Académie des Sciences
Edin. New Phil. J.: New Philosophical Journal of Edinburgh
Educ. Chem.: Education in Chemistry
Engineer: The Engineer
Engl. Mech.: The English Mechanic and World of Science
Gazzetta: Gazzetta chimica Italiana
Hist. Sci.: History of Science
Hist. Stud. Phys. Sci.: Historical Studies in the Physical Sciences
Isis
J. Amer. Chem. Soc.: Journal of the American Chemical Society
J. Chem. Educ.: Journal of Chemical Education
J. Chem. Phys.: Journal of Chemical Physics
J. Chem. Soc.: Journal of the Chemical Society
J. de Phys.: Journal de Physique, de Chimie, d'Histoire naturelle et des Arts, etc.
J. für Chem. und Phys.: Journal für die Chemie und Physik
J. Hist. Ideas: Journal for the History of Ideas
J. nat. econ. Sci. Palermo: Giornale di Science naturali ed economiche di Palermo
J. Org. Chem.: Journal of Organic Chemistry
J. prakt. Chem.: Journal für praktische Chemie

J. Roy. Inst.: *Journal of the Royal Institution*

J. Roy. Inst. Chem.: *Journal of the Royal Institute of Chemistry*

J. Russ. Phys. Chem. Soc.: *Journal of the Russian Physical and Chemical Society*

J. Soc. Arts: *Journal of the Society of Arts*

J. Soc. Chem. Ind.: *Journal of the Society of Chemical Industry*

Jahresbericht: *Jahresbericht über die Fortschritte der physischen Wissenschaften*

Jap. Stud. Hist. Sci.: *Japanese Studies in the History of Science*

Laboratory: *The Laboratory*

Lancaster Guardian: *The Lancaster Guardian*

M. & B. Lab. Bull.: *May and Baker Laboratory Bulletin*

Mem. Chem. Soc.: *Memoirs of the Chemical Society*

Mém. de Math. et de Phys.: *Mémoires de Mathématique et de Physique des Savants étrangers*

Mem. Proc. Manchester Lit. Phil. Soc.: *Memoirs and Proceedings of the Manchester Literary and Philosophical Society*

Nature

New Biol.: *New Biology*

Nic. J.: *Nicholson's Journal of Natural Philosophy, Chemistry and the Arts*

Notes & Rec.: *Notes and Records of the Royal Society*

Nuovo Cim.: *Nuovo Cimento*

Phil. Mag.: *Philosophical Magazine*

Phil. Trans.: *Philosophical Transactions of the Royal Society*

Philippine J. Sci.: *Philippine Journal of Science*

Phys. Rev.: *Physical Review*

Phys. Today: *Physics Today*

Phys. Zeitsch.: *Physikalische Zeitschrift*

Physis

Proc. Acad. Nat. Sci.: *Proceedings of the Academy of Natural Sciences,* Philadelphia

Proc. Chem. Soc.: *Proceedings of the Chemical Society*

Proc. Roy. Inst.: *Proceedings of the Royal Institution*

Proc. Roy. Soc.: *Proceedings of the Royal Society*

Proc. Roy. Soc. Edin.: *Proceedings of the Royal Society of Edinburgh*

Qu. Rev.: *Quarterly Reviews of the Chemical Society*

Quart. J. Sci.: *Quarterly Journal of Science*

Reader: *The Reader*

Rép. de Chim. pure: *Répertoire de Chimie pure et appliquée*

Rep. Brit. Assoc. Adv. Sci.: *Report of the British Association for the Advancement of Science*

Rev. d'Hist. Sci.: *Revue d'Histoire des Sciences*

Rev. Mod. Phys.: *Reviews of Modern Physics*

Russ. Chem. Rev.: *Russian Chemical Reviews*

Schweigger's Journal

Science

Science News

Sci. Prog.: *Science Progress*

Sitzungsber. Akad. Wiss. Wien: *Sitzungsberichte der Akademie der Wissenschaften zu Wien*

Smithsonian Inst. Pub.: *Smithsonian Institution Publications*

Stud. Romanticism: *Studies in Romanticism*

Tetrahedron Letters

Trans. Farad. Soc.: *Transactions of the Faraday Society*

Trans. Roy. Soc. Dublin: *Transactions of the Royal Society of Dublin*

Trans. Roy. Soc. Edinburgh: *Transactions of the Royal Society of Edinburgh*

Uspekhi Khim.: *Uspekhi Khimii*

Vocational Aspect: *Vocational Aspect*

Voprosui Filosofii: *Voprosui Filosofii*

Zeitsch. anorg. Chem.: *Zeitschrift für anorganische und allgemeine Chemie*

Zeitsch. f. Chem.: *Zeitschrift für Chemie*

Zeitsch. phys. Chem.: *Zeitschrift für physikalische Chemie*

Zeitsch. Phys.: *Zeitschrift für Physik*

Zhur. obshchei Khim.: *Zhurnal obshchei Khimii*

Notes and References

The numbers in square brackets at the top of each page of Notes refer to the pages in the text on which references to these Notes appear.

CHAPTER I

Page
3
1. H. Butterfield, *The Origins of Modern Science*, London, 1950, p. 189.
2. J. J. Berzelius, *Ann. Phil.*, 1816, *8*, 263.
3. From an account of his travel in 1861 presented to Kazan University, quoted in *Centenary of the Theory of Chemical Structure*, ed. B. A. Kazansky and G. V. Bykov, Moscow, 1961, p. 136.

4
4. It has, however, been suggested that in the chemical industry "significant differences between the countries ... became more pronounced as the years passed and finally produced distinct characteristics" (L. Haber, *Proc. Chem. Soc.*, 1958, 246).

5
5. K. D. C. Vernon, *J. Roy. Inst.*, 1963, *39*, 396.
6. A. Volta, *Phil. Trans.*, 1800, *90*, 403.
7. W. Nicholson and A. Carlisle, *Nic. J.*, 1800, *4*, 179.
8. Davy's "Personal Notebook, no. 13C", now in the library of the Royal Institution.
9. The effects of these lectures are described in *Phil. Mag.*, 1801, *9*, 281.
10. H. Davy, *Elements of Chemical Philosophy*, London, 1812.
11. Idem, *Phil. Trans.*, 1807, [97], 1.

6
12. Anne Treneer, *The Mercurial Chemist*, London, 1963, p. 93.
13. The nature of this stimulus, as suggested by Davy's Notebooks at the Royal Institution, has been discussed by the present author in *Ann. Sci.*, 1963, *19*, 255.
14. H. Davy, *Phil. Trans.*, 1808, [98], 1.
15. Edinburgh, 1st ed. 1802; 2nd ed. 1804; 3rd ed. 1807.
16. J. Thomson, *Ann. Phil.*, 1813, *2*, 445n; 1814, *3*, 337.

7
17. See note 13, and R. Siegfried, *Isis*, 1967, *58*, 236.
18. J. Thomson, *The History of Chemistry*, London, 1831, vol. ii, p. 293.
19. J. Davy (ed.), *The Collected Works of Sir Humphry Davy*, London, 1840, vol. v, p. 328n.
20. See, *e.g.*, J. Z. Fullmer, *Chymia*, 1962, *8*, 163; and note 13 above.
21. *E.g. Jac. Berzelius Bref*, ed. H. G. Söderbaum, Uppsala, 1912, vol. i, part 2, p. 23 (letter to Berzelius, 1811).
22. J. Kendall, *Humphry Davy*, *"Pilot" of Penzance*, London, 1954, p. 155.
23. J. C. Gregory, *A Short History of Atomism*, London, 1931, p. 56. See also D. M. Knight, *Stud. Romanticism*, 1966, *5*, 185; *idem, Atoms and Elements, A Study of Theories of Matter in England in the Nineteenth Century*, London, 1967.

8
24. See note 9.
25. Sir L. Woodward, *The Age of Reform, 1815–1870*, Oxford, 2nd ed., 1962, p. 564.

Page
8   26 H. M. Jones and I. B. Cohen, *Science before Darwin*, London, 1963, p. 7.
    27 *Berzelius und Liebig, ihre Briefe von 1831–1845*, ed. J. Carrière, Munich and Leipzig, 1893, p. 134.
9   28 A. Avogadro, *J. de Phys.*, 1811, *73*, 58.
    29 A. Crum Brown, *Phil. Mag.*, 1869, [3], *37*, 395.
    30 *Engineer*, 1869, *27*, 251.
    31 *Ibid.*, 1867, *24*, 483. In 1869 the Chemical Society found it necessary to debate the truth of atomism. On this see W. H. Brock and D. M. Knight, "The Atomic Debates: 'Memorable and interesting evenings in the Life of the Chemical Society'", *Isis*, 1965, *56*, 5; and W. H. Brock (ed.), *The Atomic Debates: Brodie and the Rejection of the Atomic Theory*, Leicester, 1967.
10 32 W. V. Harcourt, *Rep. Brit. Assoc. Adv. Sci.*, 1831, *1*, 28. The importance of the *popular* education movement in the development of valency is discussed in a later chapter (p. 103).
11 33 Sir P. Linstead, *Notes & Rec.*, 1962, *17*, 22.
    34 *Ibid.*, p. 29. On Hofmann's work see J. Bentley, *Ambix*, 1970, *17*, 153.
    35 The decline of English science is one of the issues discussed by E. Mendelsohn in "The Emergence of Science as a Profession in Nineteenth-Century Europe", ch. I in *The Management of Scientists*, ed. K. Hill, Boston (U.S.A.), 1964.
    36 The relevant page of Davy's laboratory notebook is reproduced in facsimile by Sir Harold Hartley, *Proc. Roy. Soc.*, A, 1960, *255*, facing p. 163, and also in his *Humphry Davy*, London, 1966, facing p. 41.
    37 See note 13.
    38 See note 14.
    39 *Phil. Trans.*, 1810, [*100*], 231; 1811, [*101*], 1.
    40 For later ideas see W. V. Farrar, "Nineteenth Century Speculations on the Complexity of the Chemical Elements", *Brit. J. Hist. Sci.*, 1965, *2*, 297; and D. M. Knight (see note 23).
12 41 See C. A. Russell, *Ann. Sci.*, 1963, *19*, 117, 127.
    42 *Jac. Berzelius Bref* (see note 21) shows a continuous stream of correspondence between the two men from 1808.
    43 From a brief account of it in a paper by Wollaston, *Phil. Trans.*, 1808, [*98*], 96. On Berzelius's atomism see also C. A. Russell, "Berzelius and the Development of the Atomic Theory", in *John Dalton, 1766–1844*, ed. D. S. L. Cardwell, Manchester, 1968, p. 259.
    44 A. J. Grant and H. Temperley, *Europe in the Nineteenth and Twentieth Centuries (1789–1938)*, 5th ed., London, 1939, p. 186.
    45 One beneficent legacy of phlogiston was a renewed preoccupation with the phenomena of combustion. Berzelius saw this (*Traité de Chimie*, Paris, 1831, vol. iv, p. 554), and identified "chemical" and "electrical" fire.
13 46 See, *e.g.*, G. Hennemann, *Naturphilosophie im 19. Jahrhundert*, Freiburg/München, 1959.
    47 Beethoven admired Goethe, setting to music his "Egmont", and used Schiller's "Ode to Freedom (or Joy)" in his last symphony.
    48 H. C. Oersted, *Phil. Mag.*, 1820, *56*, 395.
    49 T. J. Seebeck, *Abhl. Akad. Berlin*, 1822–3, 264.

Page
13   50 L. P. Williams, *Hist. Sci.*, 1962, *1*, 7.
- 51 See R. Sharrock, "The chemist and the poet: Sir Humphry Davy and the preface to Lyrical Ballads", *Notes & Rec.*, 1962, *17*, 57.
- 52 The assertion (see note 50) that Coleridge was a disciple of Kant has been corrected by W. F. Cannon (*Hist. Sci.*, 1964, *3*, 32), who points out that the influence was rather that of Schelling, another founder of the *Naturphilosophie*. For an opposed view, and a valuable discussion of Davy's "Kantianism", see T. Levere, *Affinity and Matter*, Oxford, in the press.
- 53 Sharrock, *op. cit.*, p. 59.
- 54 Thus in 1807 Coleridge wrote to Dorothy Wordsworth "Davy supposes that there is only one power in the world of the senses; which in particles acts as chemical attractions, in specific masses as electricity, and on matter in general, as planetary gravitation" (Treneer, *op. cit.* [see note 12], p. 104). This belief in an underlying unity of nature moved Davy as it was later to move Faraday (see note 50). The materiality of light and heat, and the concept of matter as a resultant of attractive and repulsive forces may be discerned in some of Davy's private notebooks from his early days. Even Davy's distrust of atomism may owe something to the *Naturphilosophen*, for it shared with them an overt hostility to all mechanical models for natural phenomena. This hostility to atomism lasted into the present century in the case of Ostwald who founded his *Journal der Naturphilosophie* in 1904. On the Idealism of Davy (and Oersted), see D. M. Knight, "The Scientist as Sage", *Stud. Romanticism*, 1967, *6*, 65. On Ostwald see *J. Chem. Soc.*, 1904, *85*, 517.
- 55 See J. R. Partington, *A History of Chemistry*, London, 1962, vol. iii, p. 674.

14   56 Berlin, 1802.
- 57 Paris, 1803.
- 58 Breslau and Hirschberg, 1792–4.
- 59 J. J. Berzelius, *Larbök i Kemien*, Stockholm, 1808–30.
- 60 Idem, *Traité de Chimie*, Paris, 1831, vol. iv, p. 532.
- 61 As expressed in Beethoven's "Pastoral" symphony, for instance (1808).
- 62 W. Prandtl, *Deutsche Chemiker*, Weinheim, 1956, p. 7.

15   63 *Annalen der Pharmacie* (1832–9), becoming *Annalen der Chemie und Pharmacie* (1840–73), and then *Justus Liebig's Annalen der Chemie* (from 1873 on).
- 64 Wöhler's phrase (*Briefwechsel zwischen J. Berzelius und F. Wöhler*, ed. J. Braun and O. Wallach, Leipzig, 1901, vol. i, p. 604), of 1835.
- 65 *J. Soc. Arts*, 1866, *14*, 626.

16   66 A. von Humboldt, *Versuche über die gereizte Musket und Neroenfasser, etc.*, Posen and Berlin, 1797, vol. i.
- 67 J. W. Ritter, *Beweis dass ein beständiger Galvanismus den Lebensprocess im Thierreich begleite*, Weimar, 1798.
- 68 Hennemann, *op. cit.* (see note 46), p. 36.
- 69 W. Whewell, *History of the Inductive Sciences*, London, 1847, vol. iii, p. 195.
- 70 On the chemical importance of *Naturphilosophie*, see D. M. Knight, "Steps towards a Dynamical Chemistry", *Ambix*, 1967, *14*, 179.
- 71 A. Wurtz, *Dictionnaire de Chimie*, Paris, 1869, vol. i, p. 1.

Page
17 72 Nevertheless the German chemists were in many cases able to accept Lavoisier's theory, not lagging behind those of other countries (Partington, *A History of Chemistry*, London, 1962, vol. iii, p. 493).
    73 See M. P. Crosland, *The Society of Arceuil. A view of French Science at the time of Napoleon I*, London, 1967.
18 74 C. L. Berthollet, *Researches into the Laws of Affinity*, trans. by M. Farrell, London, 1804, p. 45.
    75 C. M. Guldberg and P. Waage, *Études sur les Affinités Chimiques*, Christiania, 1867.
    76 L. G. Proust, *Ann. Chim.*, 1797, *23*, 85.
    77 See C. A. Russell, *Ann. Sci.*, 1963, *19*, 117.
19 78 W. A. Smeaton, *J. Roy. Inst. Chem.*, 1958, *82*, 656.
    79 See N. J. Graves, "Technical Education in France in the Nineteenth Century", *Vocational Aspect*, 1964, *16*, 148.
    80 The Type Theory was introduced there by H. F. P. Limpricht (*Grundriss der organischen Chemie*, Brunswick, 1855–6).
    81 In Chapters II and III respectively.
20 82 The influence of "positivist" thinking on French chemists continued well beyond the heyday of Laurent and Gerhardt. The last great exponent of this trend was probably M. Berthelot (1827–1907), who, for much of his life at least, was an agnostic in religion and in matters of molecular composition. When he was appointed to the first Chair of Organic Chemistry at the Collège de France the Journal of the Society of Arts (*J. Soc. Arts*, 1865, *13*, 687) jubilantly announced "his appointment is regarded as an advantage by the friends of positive as opposed to speculative science". A brief discussion of the influence of Comte on Williamson is to be found in an Appendix to *The Atomic Debates*, ed. W. H. Brock (see note 31).

CHAPTER II

22 1 G. de Morveau, *Méthode de Nomenclature Chimique*, Paris, 1787, p. 38.
    2 A. L. Lavoisier, *Traité de Chimie*, Paris, 1789, vol. i, p. 209.
    3 H. Davy, *Phil. Trans.*, 1810, [*100*], 231–57; 1815, [*105*], 203.
    4 P. L. Dulong, *J. für Chem. und Phys.*, 1815, *17*, 229.
    5 J. L. Gay-Lussac, *Ann. Chim.*, 1815, *95*, 161.
    6 C. A. Russell, *Ann. Sci.*, 1963, *19*, 117, 127.
23 7 W. G. Palmer, *Valency, Classical and Modern*, Cambridge, 1929, p. 6.
    8 N. V. Sidgwick, *Electronic Theory of Valency*, Oxford, 1927, p. 51.
    9 J. B. Dumas and P. Boullay, *Ann. Chim. Phys.*, 1828, [2], *37*, 15–23.
    10 J. J. Berzelius, *Annalen*, 1832, *3*, 282 (letter to Wöhler and Liebig, dated 2 Sept. 1832).
    11 *Idem*, *Jahresberichte*, 1833, *13*, 190–7.
    12 J. Liebig, *Annalen*, 1834, *9*, 1.
24 13 A view originally put forward by Ampère as far back as 1816: *Ann. Chim. Phys.*, 1816, [2], *2*, 16n. "Ammonium" also played an important part in the conceptual schemes of Davy and Berzelius at one time. For recent views, including evidence for its existence, see J. S. Wan, *J. Chem. Educ.*, 1968, *45*, 40.
    14 F. A. Kekulé, *Lehrbuch der organischen Chemie*, Erlangen, 1861, vol. i, p. 64.

Page
24 15 J. Liebig and F. Wöhler, *Annalen*, 1832, *3*, 249–82.
    16 J. J. Berzelius, *ibid.*, p. 282.
    17 J. B. A. Dumas and E. M. Péligot, *Ann. Chim. Phys.*, 1834, [2], *58*, 5.
    18 A. A. T. Cahours, *Annalen*, 1838, *30*, 228.
25 19 J. Liebig, *ibid.*, 1838, *25*, 3.
    20 See note 11.
    21 R. W. Bunsen, *Ann. Physik*, 1837, *40*, 219; *42*, 145; *Annalen*, 1841, *37*, 1; 1842, *42*, 14; 1843, *46*, 1.
    22 L. C. Cadet, *Mém. de Math. et de Phys.*, 1760, *3*, 633. See J. S. Thayer, "Cadet's fuming liquid", *J. Chem. Educ.*, 1966, *43*, 594.
26 23 J. J. Berzelius, *Jahresberichte*, (a) 1839, *18*, 487; (b) 1841, *20*, 526; (c) 1842, *21*, 503; (d) 1845, *24*, 640.
27 24 H. E. Roscoe, *J. Chem. Soc.*, 1900, *77*, 518.
    25 E. Frankland, *Phil. Trans.*, 1852, [*142*], 439.
    26 Kekulé, *op. cit.*, vol. i, p. 64.
28 27 C. Gerhardt, *Ann. Chim. Phys.*, 1839, [2], *72*, 184.
    28 (a) *Mem. Chem. Soc.*, 1847, *3*, 386; (b) *Annalen*, 1848, *55*, 288; J. R. Partington (*A History of Chemistry*, London, 1964, vol. iv, p. 505n) observes that a reflux condenser had been used previously by Mohr (*Annalen*, 1836, *18*, 232).
29 29 H. Fehling, *Annalen*, 1844, *49*, 91.
    30 See note 28 (a).
    31 J. B. A. Dumas, *Compt. rend.*, 1847, *25*, 383.
    32 H. Kolbe, *J. Chem. Soc.*, 1850, *2*, 157; *Annalen*, 1849, *69*, 258.
31 33 Kekulé, *op. cit.*, vol. i, p. 75.
    34 H. Kolbe, *Handwörterbuch der Chemie*, Braunschweig, 1848, vol. iii, pp. 177, 185, 442.
32 35 J. Liebig, *Annalen*, 1839, *30*, 139.
    36 H. Kolbe, (a) *Annalen*, 1850, *75*, 216; (b) *J. Chem. Soc.*, 1851, *3*, 372.
34 37 C. A. Russell, "The Influence of Frankland on the Rise of the Theory of Valency", *Actes of Xth International Congress of the History of Science*, Ithaca, 1962, p. 883; C. A. and S. P. Russell, "Frankland in Lancashire: the making of a chemist", in preparation.
    38 See note 28a. On Playfair see A. Scott, *J. Chem. Soc.*, 1905, *87*, 600.
    39 Autobiographical *Sketches from the Life of Sir Edward Frankland*, London, 1902, p. 166.
    40 See note 28b.
35 41 See note 12.
    42 Frankland, *op. cit.* (see note 39), p. 173.
    43 R. Bunsen, *Annalen*, 1837, *42*, 45.
    44 E. Frankland and H. Kolbe, *J. Chem. Soc.*, 1848, *1*, 60. The reaction is still occasionally used; *e.g.*, G. W. Miller and F. L. Rose, *J. Chem. Soc.*, 1965, 3367.
    45 E. Frankland, *Experimental Researches in Pure, Applied and Physical Chemistry*, London, 1877, p. 47.
    46 *Idem*, *J. Chem. Soc.*, 1851, *3*, 50n.
    47 *Idem*, *op. cit.* (see note 45), p. 47.
36 48 The date is given as 1847, obviously in error, in Autobiographical *Sketches* (see note 39), p. 99, but is correctly quoted as 1848 in *Experimental Researches* (see note 45), p. 63. On Queenwood see D. Thompson, *Ann. Sci.*, 1955, *11*, 246.

Page
36   49 Autobiographical *Sketches* (see note 39), pp. 78–9, 174.
37   50 E. Frankland, *J. Chem. Soc.*, 1850, *2*, 263.
     51 *Ibid.*, 1850, *2*, 263.
38   52 *Ibid.*, 1850, *2*, 297.
     53 *Ibid.*, 1851, *3*, 30.
     54 *Ibid.*, 1851, *3*, 50.
     55 A. W. Hofmann, *ibid.*, 1851, *3*, 121.
     56 E. Frankland, *ibid.*, 1851, *3*, 322.
39   57 Idem, *op. cit.* (see note 45), p. 65.
     58 B. C. Brodie, *J. Chem. Soc.*, 1851, *3*, 405.
     59 See note 36.
     60 E. Frankland, *J. Chem. Soc.*, 1850, *2*, 298.
40   61 Idem, *op. cit.* (see note 39), p. 186.
     62 Idem, *Phil. Trans.*, 1852, [*142*], 417.
     63 Idem, *op. cit.* (see note 39), p. 187.
     64 Idem, *J. Chem. Soc.*, 1854, *6*, 57.
     65 Idem, *Phil. Trans.*, 1852, [*142*], 423.
41   66 *Ibid.*, pp. 439–40.
42   67 *Ibid.*, p. 440.
     68 *Ibid.*, p. 440.
     69 *Ibid.*, pp. 441–2.
43   70 For this rather complicated matter see pp. 35, 153. It is interesting to recall that when eventually free radicals were recognized (M. Gomberg, *Ber.*, 1900, *33*, 3153) the concept of valency no longer needed them, and, indeed, received them with some embarrassment. The discovery of triarylmethyl radicals had a negligible effect on the general understanding of valency. On the other hand, the recognition of free methyl and ethyl (F. A. Paneth *et al.*, *ibid.*, 1929, *62*, 1335; 1931, *64B*, 2702) came at a more opportune moment, for the monolithic classical theory of structure was being undermined by the encroachments of theoretical physics. With the work of Kharasch and others in the 1930s free radicals became an accepted part of chemical theory. See also A. J. Ihde, "The history of free radicals and Moses Gomberg's contributions", *Free Radicals in Solution*, London, 1967, p. 1. Gomberg's identification of the dimer as hexaphenylethane has recently been corrected (H. Lankamp, W. J. Nauta and C. Maclean, *Tetrahedron Letters*, 1968, 249).

CHAPTER III

45   1 *E.g.* J. B. A. Dumas, *Ann. Chim. Phys.*, 1833, [2], *52*, 400, *etc.*
     2 J. J. Berzelius, *Jahresberichte*, 1832, *11*, 210. These formulae, unlike those on p. 28, are based on correct atomic weights.
     3 *Ibid.*, 1840, *19*, 370.
     4 L. H. F. Melsens, *Compt. rend.*, 1842, *14*, 114.
     5 J. J. Berzelius, *Traité de Chimie*, Paris, 2nd ed., 1849, vol. v, p. 43.
     6 *E.g.* A. Laurent, *Chemical Method*, trans. W. Odling, London, 1855, p. 204.
     7 F. R. Japp, *J. Chem. Soc.*, 1898, *73*, 109.
46   8 C. L. Berthollet, *Ann. Chim.*, 1789, *1*, 30; 1798, *25*, 233.
     9 H. Davy, *Phil. Trans.*, 1815, [*105*], 203.

[46–57] NOTES AND REFERENCES

Page
- 46
  - 10 P. L. Dulong, *J. für Chem. und Phys.*, 1815, *17*, 229.
  - 11 T. Graham, *Phil. Trans.*, 1833, [*123*], 253.
  - 12 J. B. A. Dumas and J. Liebig, *Annalen*, 1838, *76*, 113.
  - 13 C. Gerhardt, *Ann. Chim. Phys.*, 1843, [3], *7*, 129.
  - 14 J. B. A. Dumas, *ibid.*, 1834, [2], *56*, 113, 140.
  - 15 *Annalen*, 1840, *33*, 179, 259.
  - 16 H. V. Regnault, *ibid.*, 1840, *34*, 45.
  - 17 J. B. A. Dumas, *Compt. rend.*, 1838, *6*, 647, 695; *ibid.*, 1842, *14*, 199n. On Dumas and Laurent see T. H. Levere, *Ambix*, 1970, *17*, 111.
  - 18 A. Laurent, *Ann. Chim. Phys.*, 1836, [2], *61*, 125.
- 47
  - 19 L. H. F. Melsens, *Compt. rend.*, 1842, *14*, 114.
  - 20 A. W. Hofmann, *Annalen*, 1850, *74*, 174.
  - 21 A. W. Williamson, *Rep. Brit. Assoc. Adv. Sci. (Trans. of Sections)*, 1850, *20*, 65.
  - 22 C. Gerhardt, *Ann. Chim. Phys.*, 1839, [2], *72*, 184.
  - 23 *Idem, ibid.*, 1853, [3], *37*, 331.
  - 24 H. Kolbe, *Handwörterbuch der Chemie*, Braunschweig, 1855, vol. vi, p. 802.
- 48
  - 25 See also Chapter I, note 82.
  - 26 P. Cook, *Science News*, 1959, *53*, 35.
  - 27 E. Frankland, *Phil. Trans.*, 1852, [*142*], 440.
- 49
  - 28 Misquoted as 1854 by P. E. Verkade, *Proc. Chem. Soc.*, 1958, 205.
  - 29 On Kekulé see R. Anschütz, *August Kekulé*, 2 vols., Berlin, 1929.
- 50
  - 30 *Ibid.*, vol. i, p. 956.
  - 31 K. R. Webb states that Kekulé's *first* paper was in 1854 (*J. Roy. Inst. Chem.*, 1957, *81*, 729), but this ignores his work on amyl hydrogen sulphate (*Annalen*, 1850, *75*, 275) and two joint publications with von Planta (*ibid.*, 1853, *87*, 1 and 364).
  - 32 Anschütz, *op. cit.*, vol. i, pp. 69–70.
  - 33 See note 21.
- 51
  - 34 A. Laurent, *Ann. Chim. Phys.*, 1846, [3], *18*, 266.
  - 35 A. W. Williamson, *J. Chem. Soc.*, 1852, *4*, 352.
  - 36 C. Gerhardt, *ibid.*, 1853, *5*, 127.
  - 37 A. W. Williamson, *ibid.*, 1852, *4*, 350.
  - 38 *Ibid.*, p. 351.
- 52
  - 39 *Ibid.*, p. 353.
  - 40 E. Frankland, *Phil. Trans.*, 1852, [*142*], 417.
- 53
  - 41 A. W. Williamson, *Proc. Roy. Soc.*, 1854, *7*, 13.
  - 42 *Ibid.*, p. 11.
  - 43 *Ibid.*, p. 14.
- 54
  - 44 *Ibid.*, p. 135.
  - 45 Anschütz, *op. cit.*, vol. i, pp. 50–1.
  - 46 A. von Baeyer, *Gesammelte Werke*, Braunschweig, 1905, vol. i, p. 15.
  - 47 W. Odling, *J. Chem. Soc.*, 1855, *7*, 1.
- 55
  - 48 *Idem, Proc. Roy. Soc.*, 1855, *2*, 66.
  - 49 *Idem*, Preface to *Outlines of Chemistry*, London, 1870.
  - 50 F. A. Kekulé, (a) *Proc. Roy. Soc.*, 1854, *7*, 37; (b) *Annalen*, 1854, *90*, 309.
- 56
  - 51 *Idem, Proc. Roy. Soc.*, 1854, *7*, 37–8.
  - 52 *Idem, Annalen*, 1854, *90*, 314–15.
- 57
  - 53 *Idem, Proc. Roy. Soc.*, 1854, *7*, 40.
  - 54 Anschütz, *op. cit.*, vol. i, pp. 55–6.

Page

57
55 F. R. Japp, *J. Chem. Soc.*, 1898, *73*, 121.
56 E. W. Hiebert, *J. Chem. Educ.*, 1959, *36*, 324.
57 A. Wurtz, *Ann. Chim. Phys.*, 1855, [3], *43*, 492.
58 Idem, *Compt rend.*, 1856, *43*, 199.
59 F. R. Japp, *J. Chem. Soc.*, 1898, *73*, 112–13.

58
60 F. A. Kekulé, *Annalen*, 1857, *104*, 129. For clarity Kekulé's barred symbols are here replaced by modern ones.
61 H. Limpricht and L. W. J. von Uslar, *ibid.*, 1857, *102*, 239. Limpricht, in Wöhler's laboratory at Göttingen, was the first to introduce the Type Theory into Germany (*ibid.*, 1858, *106*, 134).
62 O. Mendius, *ibid.*, 1857, *103*, 39.

59
63 F. A. Kekulé, *ibid.*, 1857, *104*, 139.

61
64 *Ibid.*, *105*, 177.
65 *Ibid.*, p. 181.
66 *Ibid.*, p. 129.

62
67 *Ibid.*, 1858, *106*, 129.
68 *Ibid.*, 1857, *101*, 200.

63
69 Hiebert, *op. cit.* (see note 56), p. 326.
70 L. Schischkoff, *Annalen*, 1857, *101*, 213.
71 *Ibid.*, p. 204.

64
72 *Ibid.*, pp. 204–5.
73 E. von Meyer, *A History of Chemistry*, trans. G. McGowan, London, 2nd ed., 1898, pp. 309–10.
74 *J. Chem. Soc.*, 1898, *73*, 122.
75 *J. Chem. Educ.*, 1959, *36*, 326.

65
76 *Rep. Brit. Assoc. Adv. Sci. (Trans. of Sections)*, 1855, *25*, 62.
77 *Annalen*, 1858, *105*, 279.
78 *Ibid.*, 1858, *105*, 283.
79 *Ibid.*, 1857, *104*, 133n.
80 *Ibid.*, p. 145.

66
81 This particular gratitude to Wurtz probably springs from the latter's emphasis of the concept of polyatomicity, which until his paper of 1856 (see note 58) had hardly ever been used. *Cf.* A. Wurtz, *A History of Chemical Theory*, trans. H. Watts, London, 1869, pp. 146 and 206–7.
82 *Annalen*, 1857, *104*, 136. Italics mine.

68
83 *Ibid.*, p. 152.
84 *Ibid.*, p. 153.

69
85 *Ibid.*, p. 154.

70
86 W. V. and K. R. Farrar, *Proc. Chem. Soc.*, 1959, 287.
87 F. Rochleder, *Annalen*, 1852, *7*, 64.
88 F. A. Kekulé, *Ber.*, 1890, *23*, 1305.

71
89 *Ibid.*, p. 1306.
90 See note 88.
91 See note 58.
92 A. S. Couper, *Compt. rend.*, 1857, *45*, 230.

72
93 *Idem*, (a) *ibid.*, 1858, *46*, 1107–10; (b) *Edin. New Phil. J.*, 1858, *8*, 213–17.
94 *Edin. New Phil. J.*, 1858, *8*, 214.
95 See note 105.
96 F. A. Kekulé, *Annalen*, 1861, *117*, 145.
97 H. Kolbe, *ibid.*, 1860, *115*, 157.

Page
72   98 *E.g.*, H. Schreib, *Ber.*, 1880, *13*, 465.
     99 R. Anschütz, *Annalen*, 1885, *228*, 308; 1906, *346*, 286.
     100 See note 29.
     101 R. Anschütz, *Proc. Roy. Soc. Edin.*, 1909, *29*, 193.
     102 F. R. Atherton, *Phosphoric Esters and Related Compounds*, London, 1957, p. 75.
     103 A. G. Pinkus, P. G. Waldrop and W. J. Collier, *J. Org. Chem.*, 1961, *26*, 682.
73   104 A. G. Pinkus and P. G. Waldrop, *ibid.*, 1966, *31*, 575.
     105 A. S. Couper (a) *Compt. rend.*, 1958, *46*, 1157; (b) *Phil. Mag.*, 1858, [4], *16*, 104; (c) *Ann. Chim. Phys.*, 1858, [3], *53*, 469. (Reprinted in *Alembic Club Reprint*, no. 21, Edinburgh, 1933.)
74   106 *Compt. rend.*, 1858, *46*, 1157.
     107 Anschütz, *op. cit.* (see note 101), p. 202.
77   108 See note 105b.
     109 See note 105c.
79   110 F. A. Kekulé, *Compt. rend.*, 1858, *47*, 378.
80   111 See note 101.
     112 A. Wurtz, *Rép. de Chim. pure*, 1858, *1*, 49.
     113 A. M. Butlerov, *Annalen*, 1859, *110*, 51.
     114 D. F. Larder denies that there is evidence that Crum Brown even knew Couper or had read his paper (*Ambix*, 1967, *14*, 115n, 118n).

CHAPTER IV

81   1 A. W. Williamson, *J. Chem. Soc.*, 1852, *4*, 353.
     2 A. W. Williamson and G. Kay, *Proc. Roy. Soc.*, 1854, *7*, 135.
     3 F. A. Kekulé, *Annalen*, 1857, *104*, 129.
82   4 Article "Aequivalent und Aequivalenz" in the *Neues Handwörterbuch der Chemie*, by H. von Fehling, Braunschweig, 1871, vol. i, p. 77. The German terms he refers to were used by R. A. C. E. Erlenmeyer (*Zeitsch. f. Chem.*, 1860, *3*, 539; 1864, *7*, 628).
     5 C. Schorlemmer, *Rise and Development of Organic Chemistry*, London, 1879, p. 45.
     6 A. W. Hofmann, *Introduction to Modern Chemistry, Experimental and Theoretical*, London, 1865, p. 169. So also L. Meyer, *Die modernen Theorien der Chemie*, Breslau, 1864, p. 76.
     7 W. Odling, *Tables of Chemical Formulae*, London, 1864.
     8 A. Laurent, *Chemical Method, Notation, Classification and Nomenclature*, London, trans. W. Odling, 1855.
83   9 E. Frankland, *Lecture Notes for Chemical Students, embracing Mineral and Organic Chemistry*, London, 1866.
     10 *Idem*, *e.g.*, *J. Chem. Soc.*, 1866, *19*, 372.
     11 C. T. Kingzett, *Chemical Encyclopedia*, London, 5th ed., 1932, p. 944.
     12 E. Frankland, *Phil. Trans.*, 1852, [*142*], 440.
     13 A. S. Couper, *Phil. Mag.*, 1858, [4], *16*, 104.
     14 A. W. Hofmann, *Proc. Roy. Inst.*, 1865, *4*, 414.
     15 Frankland, *op. cit.* (see note 9), p. 19.
     16 A. W. Williamson, *J. Chem. Soc.*, 1869, *22*, 328.
84   17 See note 2.

Page
84
18 F. A. Kekulé, *Proc. Roy. Soc.*, 1854, *7*, 37–8.
19 See note 3.
20 F. A. Kekulé, *Bull. Acad. roy. Belg.*, 1860, [2], *10*, 71.
21 A. Wurtz, *Bull. Soc. chim. France*, 1864, [2], *1*, 33.
22 F. A. Kekulé, *Lehrbuch der organischen Chemie*, Erlangen, 1861 and subsequent years.
23 See note 20, *etc.*
24 A. W. Williamson, *J. Chem. Soc.*, 1864, *17*, 211.
25 E. Frankland, *e.g.*, *Rep. Brit. Assoc. Adv. Sci.* (*Trans. of Sections*), 1868, *38*, 34.
26 B. C. Brodie, *J. Chem. Soc.*, 1868, *21*, 367.
27 *Reader*, 1865, 372.
28 A. Wurtz, *Bull. Soc. chim. France*, 1864, [2], *2*, 247.
29 R. A. C. E. Erlenmeyer, *Zeitsch. f. Chem.*, 1864, *7*, 628.
30 S. R. Bottone, *Engl. Mech.*, 1872, *14*, 448.
31 *Nature*, 1874, *9*, 420.
32 J. F. Heyes, *Phil. Mag.*, 1888, [5], *25*, 223.
33 W. Odling, *ibid.*, 1858, [4], *16*, 40.

85
34 See note 27.
35 Hofmann, *op. cit.* (see note 6), pp. 168–9. Hofmann's collaborator, F. O. Ward, appears to have been responsible for the terminology (*Chem. News*, 1866, *13*, 262).
36 H. E. Roscoe, *Nature*, 1869, *1*, 165.
37 *Idem*, *Elementary Chemistry*, London, 1871, 3rd ed., p. 172.
38 J. P. Cooke, *The New Chemistry*, London, 1874, p. 235 *etc.*
39 J. B. Stallo, *Concepts of Modern Physics*, London, 1882, p. 36 *etc.*
40 A. Wurtz, *Elements of Modern Chemistry*, trans. W. H. Greene' London and Philadelphia, 1881, p. 222 *etc.*
41 Frankland, *op. cit.* (see note 9), p. 19.
42 *Idem*, *Lecture Notes for Chemical Students*, London, 2nd ed., 1870, vol i, (Inorganic), p. 19.
43 *Idem*, *Experimental Researches in Pure, Applied and Physical Chemistry*, London, 1877, p. 145.

86
44 C. H. Wichelhaus, *Annalen*, 1868, supp. vol. vi, p. 257.
45 (a) R. Anschütz, *August Kekulé*, Berlin, 1929, vol. i, p. 249;
(b) J. C. Gregory, *A Short History of Atomism*, London, 1931, p. 119.
46 C. H. Wichelhaus, *Ber.*, 1868, *1*, 77.
47 F. A. Kekulé, *Zeitsch. f. Chem.*, 1867, [2], *3*, 217.

87
48 D. and K. C. Bailey, *An Etymological Dictionary of Chemistry and Mineralogy*, London, 1929, p. 293.
49 *Oxford English Dictionary*, 1929 edition.
50 A. Findlay, *Spirit of Chemistry*, London, 1930, p. 66n.
51 *Ann. Winton*, 1276, p. 120: quoted in W. Stubbs, *Select Charters*, Oxford, 9th ed., 1913, p. 422.
52 *E.g.*, "*valencie*, puissance, might", in H. C(ockeram), *The English Dictionarie: or an Interpreter of hard English Words*, London, 1626, Part I.
53 Article "Aciditat" in the *Neues Handwörterbuch der Chemie* (see note 4), p. 52. The older form WERTHIGKEIT is often met in the nineteenth century.

Page
54 J. A. Geuther, *Lehrbuch der Chemie gegründet auf die Werthigkeit der Elemente*, Jena, 1869.
55 *Engl. Mech.*, 1869, *8*, 561.
56 See note 49.
57 *Engl. Mech.*, 1869, *10*, 272n. Sprague was interested in scientific nomenclature and introduced several new terms for electrical units. "Equivolt", the only one to survive, exists today solely as the telegraphic address of *The Electrical Times* (J. Mellanby, *The History of Electric Wiring*, London, 1957, p. 19).
58 *Engl. Mech.*, 1869, *10*, 222.
59 *Ibid.*, p. 198.
60 A. G. V. Harcourt, *Rep. Brit. Assoc. Adv. Sci.* (*Trans. of Sections*), 1875, *45*, 36.
61 *Encyclopaedia Britannica*, 9th ed., 1876, p. 473.
62 E. Frankland and F. R. Japp, *Inorganic Chemistry*, London, 1884, p. 58.
63 *Sketches from the Life of Sir Edward Frankland*, London, 1902, pp. 186–7.
64 See note 43.
65 It is interesting that in England the title of C. A. Coulson's book *Valence*, Oxford, 1952, has not persuaded later writers to conform to that usage. A distinction between "valence" and "valency" suggested by J. N. Friend, *The Theory of Valency*, London, 1909, p. ix, has never gained support: "valency" was "the power possessed by atoms generally to combine with one another", while "valence" was "the actual force or bond effecting that union".
66 H. C. Longuet-Higgins, *Educ. Chem.*, 1965, *1*, 66.
67 R. B. Heslop, *J. Roy. Inst. Chem.*, 1964, *88*, 282.
68 F. A. Kekulé, *e.g.*, *Lehrbuch der organischen Chemien* (see note 22), 1866, vol. ii, p. 257.
69 E. Frankland, *J. Chem. Soc.*, 1866, *19*, 377–8.
70 *Idem, op. cit.* (see note 9), p. 25.
71 *Idem, op. cit.* (see note 42), p. 25.
72 *Idem, op. cit.* (see note 43), p. 10.
73 *Ibid.*, p. 209; *cf. Phil. Trans.*, 1855, *145*, 274.
74 *Ibid.*, p. 233; *cf. Phil. Trans.*, 1857, *147*, 78.
75 *Ibid.*, p. 287; *cf. Phil. Trans.*, 1862, *152*, 178.
76 Frankland, *op. cit.* (see note 9), p. 19.
77 *Idem, op. cit.* (see note 42), p. 18.
78 *Ibid.*
79 "Urban", *Engl. Mech.*, 1869, *9*, 222.
80 For further speculations relating to Frankland's use of "bonds", see C. A. and S. P. Russell, "Edward Frankland in Lancashire: the making of a chemist", in preparation.
81 A. W. Williamson, *J. Chem. Soc.*, 1869, *22*, 299.
82 *Cf.* G. Porter, "The Chemical Bond since Frankland", *Proc. Roy. Inst.*, 1965, *40*, 384.
83 *Chemistry*, 1960, *33*, no. 6, 1.

CHAPTER V

1 W. Odling, *J. Chem. Soc.*, 1855, *7*, 1.

Page
93
2 F. A. Kekulé, *Annalen*, 1857, *104*, 129.
3 *Idem*, *Lehrbuch der organischen Chemie*, Erlangen, 1861 vol. i, pp. 93–4.
4 A. Wurtz, *Ann. Chim. Phys.*, 1855, [3], *44*, 306.
5 E. Frankland, *J. Chem. Soc.*, 1866, *19*, 372.

94
6 A. W. Hofmann, *Introduction to Modern Chemistry*, London, 1865, p. 168.
7 A. W. Williamson, *J. Chem. Soc.*, 1864, *17*, 421.
8 *Idem*, *Phil. Mag.*, 1865, [5], *29*, 464.
9 See note 3.

95
10 F. A. Kekulé, *Annalen*, 1872, *162*, 309.
11 C. Gerhardt, *Traité de Chimie*, Paris, 1856.
12 F. A. Kekulé, *Annalen*, 1857, *104*, 129.
13 E. Frankland, *J. Chem. Soc.*, 1866, *19*, 372.

96
14 F. A. Kekulé, speech delivered at Bonn, 1 June 1892; printed in vol. ii of Anschütz's biography (see note 20), p. 950.
15 *E.g.*, W. Odling, *J. Chem. Soc.*, 1855, 7, 1.

97
16 J. Loschmidt, *Chemische Studien*, Vienna, 1861.
17 *Idem*, *Sitzungsber. Akad. Wiss. Wien*, 1865, *52*, Section 395.

98
18 F. A. Kekulé, *Lehrbuch der organischen Chemie*, Erlangen, 1861, vol. i.
19 *Ibid.*, p. 165.
20 R. Anschütz, *August Kekulé*, Berlin, 1929, 2 vols.

99
21 F. A. Kekulé, (a) *Bull. Soc. chim. France*, 1865, [2], *3*, 98; (b) *Annalen*, 1866, *137*, 129.
22 *Idem*, *Bull. Soc. chim. France*, 1865, [2], *3*, 100; *Annalen*, 1866, *137*, 134.
23 Anschütz, *op. cit.*, vol. i, p. 305. Ironically, Kekulé's own symbols received their share of abuse, *e.g.* by Kolbe who derided them as "bread-rolls" (*J. prakt. Chem.*, 1881, *24*, 375).

100
24 A. Wurtz, *Leçons de Philosophie Chimique*, Paris, 1864.
25 A. Naquet, *Principes de Chimie fondés sur les théories modernes*, Paris, 1865.
26 C. W. Blomstrand, *Die Chemie der Jetztzeit*, Heidelberg, 1869.
27 Anschütz, *op. cit.*, vol. i, p. 298.
28 *Idem*, *Proc. Roy. Soc. Edin.*, 1909, *29*, 193.
29 T. S. Wheeler, *Proc. Chem. Soc.*, 1959, 221.
30 P. E. Verkade, *Proc. Chem. Soc.*, 1958, 208; J. H. S. Green, *J. Roy. Inst. Chem.*, 1958, *83*, 522, 524; H. C. Brown, *J. Chem. Educ.*, 1959, *36*, 109.
31 See, *e.g.*, T. S. Wheeler and J. R. Partington, *The Life and Work of William Higgins, Chemist*, Oxford, 1960; O. T. Benfey, *J. Chem. Educ.*, 1959, *36*, 319n.
32 *A Comparative View of the Phlogistic and Anti-phlogistic Theories*, London, 1789.

101
33 See Wheeler and Partington, *op. cit.*, p. 140.
34 Thesis for M.D., Edinburgh, 1861: "The Theory of Chemical Combination." On this, see D. F. Larder, *Ambix*, 1967, *14*, 112.
35 A. Crum Brown, *Trans. Roy. Soc. Edin.*, 1864, *23*, 707.
36 W. A. Tilden, *The Progress of Scientific Chemistry*, London, 1913, p. 167.

[101–8] NOTES AND REFERENCES

Page
101 37 A. Crum Brown, *Proc. Roy. Soc. Edin.*, 1865, *5*, 429; *ibid.*, 1869, *6*, 518.
102 38 Anschütz, *op. cit.*, vol. i, p. 292. Nevertheless, the French paper (see note 21a) has a footnote reference to Brown's formulae. Kekulé did not consider them important.
     39 See note 21b.
     40 E. von Meyer, *A History of Chemistry*, trans. G. McGowan from 2nd German ed., London, 1898.
     41 A. Ladenburg, *Lectures on the History of the Development of Chemistry*, trans. L. Dobbin from 2nd German ed., Edinburgh, 1900.
103 42 J. L. and B. Hammond, *The Bleak Age*, London, 2nd ed., 1947, p. 162.
     43 By the 1860s there was considerable middle-class support also for popular education. See further C. A. Russell, "Popular chemical education in England, 1860–1880", *Actes of XIth International Congress of the History of Science*, Warsaw and Cracow, 1965, *4*, 54.
     44 Meyer, *op. cit.* (see note 40), p. 586 *et seq.*; H. E. Roscoe, "Scientific Education in Germany", *Nature*, 1869, *1*, 157; 1870, *1*, 475.
104 45 E. Frankland, *Lecture Notes for Chemical Students, embracing Mineral and Organic Chemistry*, London, 1866, p. v.
     46 *Idem*, letter dated 4 June 1866. Extract reproduced in *J. Chem. Soc.*, 1923, *123*, 3425.
     47 *Idem*, *Lecture Notes for Chemical Students* (see note 45), p. 18.
     48 A. W. Hofmann, *Proc. Roy. Inst.*, 1865, *4*, 414.
105 49 *Rep. Brit. Assoc. Adv. Sci. (Trans. of Sections)*, 1867, *37*, 42.
     50 J. P. Cooke, *First Principles of Chemical Philosophy*, London, 1870.
     51 *Engl. Mech.*, 1872, *14*, 589.
     52 W. G. Valentin, *Introduction to Inorganic Chemistry*, London, 1872.
106 53 W. A. Miller, *Elements of Chemistry, Theoretical and Practical*, vol. ii, *Inorganic Chemistry*, London, 1874.
     54 *Nature*, 1873, *7*, 160.
     55 *Rep. Brit. Assoc. Adv. Sci. (Trans. of Sections)*, 1873, *43*, 63 and 65.
     56 J. C. Brough, *The B-Hive*, 1868. Quoted in J. Kendall, *At Home among the Atoms*, London, 1929, p. 123, and by A. Scott, *J. Chem. Soc. (Trans.)*, 1916, *109*, 346.
     57 The connection between "popular" education and the spread of the theory of valency has not been sufficiently appreciated. The use of valency and of graphic formulae in the *English Mechanic* from 1869 is evidence for the acceptability of these concepts to working-class readers. By 1871, graphic formulae were required for the Department of Science and Art's examinations (*Nature*, 1871, *5*, 27). Frankland was the examiner in chemistry.
     58 J. Wilbrand, *Zeitsch. f. Chem.*, 1865, *8*, 685.
107 59 F. A. Kekulé, *ibid.*, 1867, [2], *3*, 214. On Kekulé's atomic model see also pp. 107, 162, 234–5, 238–9 and 346 n. 29; and Plate 9.
     60 *Idem*, *Ber.*, 1869, *2*, 365. This objection (that graphic formulae implied planar molecules) had already been met by Brown's distinction between "chemical" and "physical" positions (p. 102).
     61 J. Dewar, *Proc. Chem. Soc.*, 1897, *13*, 239.

CHAPTER VI

108 1 R. Anschütz, *August Kekulé*, Berlin, 1929.

Page
108 2 H. E. Armstrong, *Nature*, 1930, *125*, 807.
     3 Unpublished letter from R. Anschütz to H. E. Armstrong, dated 22/7/1930, now at University College, London.
     4 Anschütz, *op. cit.*, vol. i, pp. 554–69; originally written 1883. See also note 23.
109 5 R. Anschütz, *Proc. Roy. Soc. Edin.*, 1909, *29*, 193.
     6 *Sketches from the Life of Sir Edward Frankland*, London, 1902, p. 187.
     7 E. Frankland and H. Kolbe, *Annalen*, 1857, *101*, 257.
     8 E. Frankland, *Experimental Researches in Pure, Applied and Physical Chemistry*, London, 1877, p. 148.
     9 F. A. Kekulé, *Ber.*, 1890, *23*, 1305.
    10 See note 4.
    11 Anschütz, *op. cit.* (see note 1) vol. i, p. 540.
    12 F. R. Japp (Kekulé Memorial Lecture), *J. Chem. Soc.*, 1898, *73*, 97.
110 13 A. von Baeyer, *Gesammelte Werke*, Braunschweig, 1905, vol. i, p. 15.
111 14 F. A. Kekulé, *Compt. rend.*, 1864, *58*, 510.
    15 Reproduced in Anschütz, *op. cit.*, vol. i, pp. 145–9.
    16 F. A. Kekulé, *Compt. rend.*, 1858, *47*, 378.
    17 *Idem, Annalen*, 1857, *104*, 129.
    18 *Idem*, Letter to Wurtz, 15 Feb., 1859, reproduced in Anschütz, *op. cit.*, vol. i, p. 147.
    19 *Idem, Lehrbuch der organischen Chemie*, Erlangen, 1861.
    20 A. von Baeyer, *Annalen*, 1858, *107*, 257.
    21 Anschütz, *op. cit.*, vol. i, p. 156.
    22 H. Kolbe, *Zur Entwicklungsgeschichte der theoretischen Chemie*, Leipzig, 1881; reprinting of *J. prakt. Chem.*, 1881, [2], *23*, 305, 353, 489; *24*, 374.
    23 F. A. Kekulé, *Annalen*, 1883, *221*, 230.
    24 Anschütz, *op. cit.*, vol. i, pp. 540 (benzene theory) and 554 (valency theory). See also note 4. Recently republished as *Cassirte Kapital aus der Abhandlung über die Carboxytartronsaüre und die Constitution des Benzols*, Weinheim, 1965.
    25 A. Ladenburg, *Vortrage über die Entwicklungsgeschichte der Chemie der letzteren hundert Jahren*, Braunschweig, 1869.
    26 H. Kopp, *Die Entwicklung der Chemie in der neuen Zeit*, Munich, 1873. See also *J. Chem. Soc.*, 1898, *73*, 117.
112 27 See note 12.
    28 See note 2.
    29 F. A. Kekulé, *Ber.*, 1890, *23*, 1304–5.
113 30 H. E. Armstrong, *Nature*, 1930, *125*, 809.
    31 E. Frankland, *Phil. Trans.*, 1852, [*142*], 417.
114 32 *Idem, op. cit.* (see note 8), p. 145. Kekulé (Anschütz, *op. cit.*, vol. i, p. 561) translated "equivalence" as "Valenz".
    33 Frankland, *op. cit.* (see note 6), pp. 186–7.
    34 *Ibid.*, p. 153.
    35 *Ibid.*, p. 154.
    36 *Ibid.*
    37 Anschütz, *op. cit.*, vol. i, p. 565.
    38 E. Frankland, *Phil. Trans.*, 1852, [*142*], 417.
    39 H. Kolbe, *Ausfuhrliches Lehrbuch der organischen Chemie*, Braunschweig, 1854, vol. i, p. 23.

Page		
114	40	Anschütz, *op. cit.*, vol. i, p. 566.
115	41	See A. J. Ihde, "The Karlsruhe Congress: a Centennial Retrospect", *J. Chem. Educ.*, 1961, *38*, 83.
	42	Anschütz, *op. cit.*, vol. i, p. 565.
116	43	A. von Baeyer, *Annalen*, 1858, *105*, 274.
	44	E. Frankland, *Phil. Trans.*, 1852, [*142*], 441.
	45	*Idem, op. cit.* (see note 8), p. 153.
	46	Anschütz, *op. cit.*, vol. i, p. 568.
117	47	E. Frankland, *Phil. Trans.*, 1852, [*142*], 441.
	48	Anschütz, *op. cit.*, vol. i, p. 561.
	49	Japp, *op. cit.* (see note 12), pp. 116–17.
118	50	F. A. Kekulé, *Annalen*, 1857, *104*, 129.
	51	Anschütz, *op. cit.*, vol. i, pp. 147, 554.
	52	*Ibid.*, p. 554.
	53	*Ibid.*, p. 51.
	54	In 1869, however, he said: "oxygen binds the two atoms together" in water or potassium hydroxide (*J. Chem. Soc.*, 1869, *22*, 355).
119	55	A. Wurtz, *Rép. de Chim. pure*, 1858, *1*, 24n.
	56	See note 18.
	57	F. A. Kekulé, *Annalen*, 1854, *60*, 308.
	58	Anschütz, *op. cit.*, vol. i, p. 148.
	59	A. Wurtz, *Ann. Chim. Phys.*, 1859, [3], *55*, 479.
120	60	W. Odling, *J. Roy. Inst.*, 1855, *2*, 63.
	61	Anschütz, *op. cit.*, vol. i, p. 109.
	62	Frankland, *op. cit.* (see note 8), p. 154.
	63	*Idem, J. Chem. Soc.*, 1866, *19*, 37.
	64	*Idem, Proc. Roy. Soc.*, 1865, *14*, 198.
122	65	H. Kolbe, *J. prakt. Chem.*, 1881, [2], *23*, 374.
	66	*Ibid.*, p. 374n.
	67	*Ibid.*, 1881, [2], *24*, 398.
	68	Anschütz, *op. cit.*, vol. i, p. 558.
123	69	A. Ladenburg, *History of the Development of Chemistry*, trans. from 2nd German ed. by L. Dobbin, Edinburgh, 1900, p. 236.
	70	Anschütz, *op. cit.*, vol. i, p. 556.
	71	H. Kolbe, *Ausfuhrliches Lehrbuch der organischen Chemie*, Braunschweig, 1854, vol. i, p. 742.
	72	Anschütz, *op. cit.*, vol. i, p. 558.
	73	*Ibid.*, pp. 559–60.
124	74	E. Frankland, *Proc. Roy. Inst.*, 1858, *3*, 540.
	75	*Idem, op. cit.* (see note 6), p. 189.
	76	See note 16.
	77	A. S. Couper, *Compt. rend.*, 1858, *46*, 1157.
	78	F. A. Kekulé, *Annalen*, 1858, *106*, 209.
125	79	*Idem, Compt. rend.*, 1858, *47*, 378. *Cf.* also H. L. Buff, *Annalen*, 1856, *100*, 223.
126	80	C. Gerhardt, *J. Chem. Soc.*, 1853, *5*, 127.
	81	A. W. Williamson, *Proc. Roy. Soc.*, 1856, *7*, 13.
127	82	In a letter to J. E. Marsh, incorporated by him in his Obituary Notice on Odling: *J. Chem. Soc.*, 1921, *119*, 559.
	83	Kolbe, *op. cit.* (see note 71), vol. i, p. 23.
	84	J. Walker, *J. Chem. Soc.*, 1923, *123*, 3425.

M

Page
128
85 H. E. Armstrong, *Nature*, 1930, *127*, 807.
86 See note 4.
87 E. von Meyer, *A History of Chemistry*, 2nd English ed., trans. from 2nd German ed. by G. McGowan, London, 1898, p. 598.
88 Ladenburg, *op. cit.* (see note 69), p. 240.

129
89 A. Wurtz, *A History of Chemical Theory*, trans. H. Watts, London, 1869.
90 See note 19.
91 W. A. Miller, *Rep. Brit. Assoc. Adv. Sci.* (*Trans. of Sections*), 1865, *35*, 24.
92 E. Frankland, *Lecture Notes for Chemical Students*, London, 1866, p. 32.
93 Oration at the Lancastrian Frankland Society, *Lancaster Guardian*, 26 Jan. 1934.
94 Anschütz, *op. cit.*, vol. i, p. 39.
95 Letter by Frankland to his wife, 1 May 1867; reproduced in Frankland, *op. cit.* (see note 6), p. 400.

130
96 See note 31.
97 See note 4.
98 Frankland, *op. cit.* (see note 92), p. 20. (*Cf.* ch. X.)

131
99 von Meyer, *op. cit.* (see note 87), p. 337.
100 E. Frankland, *Sketches* (see note 6), pp. 287 *et seq.*
101 See note 43.
102 F. A. Kekulé, *Annalen*, 1857, *104*, 134n.
103 *Idem*, *Annalen*, 1854. *60*, 308.
104 Frankland, *Phil. Trans.*, 1855, *145*, 274.
105 Japp, *op. cit.* (see note 12), p. 131.
106 A. Wurtz, *Ber.*, 1880, *13*, 7.
107 Ladenburg, *op. cit.* (see note 69), p. 233.

132
108 See note 39.
109 Frankland, *Sketches* (see note 6), p. 193.
110 E. Frankland and H. Kolbe, *Annalen*, 1857, *101*, 257.
111 Ladenburg, *op. cit.* (see note 69), p. 234.
112 H. Kolbe, *J. prakt. Chem.*, 1881, [2], *23*, 365.
113 See note 92.
114 E. Frankland, *Lecture Notes for Chemical Students*, London vol. i (Inorganic), 1870; vol. ii (Organic), 1872.

133
115 G. C. Foster, *Rep. Brit. Assoc. Adv. Sci.*, 1859, *29*, 21–2.
116 See note 16.
117 *E.g.*, A. M. Butlerov, *Annalen*, 1868, *146*, 260.
118 Wurtz, *op. cit.* (see note 89), p. 159 (note).
119 Ladenburg, *op. cit.* (see note 69), p. 255.
120 Published as a book, London, 1959.
121 *E.g.*, H. C. Brown, *J. Chem. Educ.*, 1959, *36*, 104.
122 See note 19.
123 Japp, *op. cit.* (see note 12), p. 138.
124 H. E. Armstrong, *J. Soc. Chem. Ind.*, 1929, *48*, 916.

CHAPTER VII

137
1 See C. A. Russell, *Ann. Sci.*, 1963, *19*, 127.
2 J. J. Berzelius, *Traité de Chimie*, Paris, 1831, vol. iv, p. 569.

Page
137   3 J. W. Döbereiner, *Ann. Physik*, 1817, *56*, 331; *ibid.*, 1829, [2] *15*, 301.
      4 J. B. A. Dumas, *Ann. Chim. Phys.*, 1859, [3], *55*, 129.
      5 A. E. B. de Chancourtois, *Le Vis Tellurique, classement naturel des corps simples ou radicaux obtenu au moyen d'un Système de Classification helicoidal et numérique*, Paris, 1863.
      6 W. Odling, *Quart. J. Sci.*, 1864, *1*, 642; *Dictionary of Chemistry*, ed. H. Watts, London, 1865, vol. iii, p. 975.
      7 J. A. R. Newlands, *Chem. News*, 1863, *7*, 70; 1864, *10*, 59, 94, 240; 1865, *12*, 83, 94; 1866, *13*, 113, 130.
      8 See W. A. Smeaton, "Centenary of the Law of Octaves", *J. Roy. Inst. Chem.*, 1964, *88*, 271, and J. W. van Spronsen, *Chymia*, 1966, *11*, 125, and *The Periodic System of Chemical Elements*, Amsterdam, London and New York, 1969.
138   9 W. Odling, *Phil. Mag.*, 1864, [4], *27*, 115.
     10 A. W. Williamson, *J. Chem. Soc.*, 1864, *17*, 211.
     11 W. A. Miller, *Rep. Brit. Assoc. Adv. Sci.*, (*Trans. of Sections*), 1865, *35*, 24.
     12 *Idem, Elements of Chemistry, Theoretical and Practical*, London, 4th ed., 1867, vol. i, p. 33.
139 13 E. Frankland, (a) *Lecture Notes for Chemical Students, embracing Mineral and Organic Chemistry*, London, 1866, p. 32; (b) *J. Chem. Soc.*, 1866, *19*, 372.
     14 *Idem*, (a) *Lecture Notes for Chemical Students*, vol. i, *Inorganic Chemistry*, London, 1870, p. 32; (b) *Experimental Researches in Pure, Applied and Physical Chemistry*, London, 1877, p. 25.
     15 H. E. Roscoe, *e.g.*, *Phil. Trans.*, 1868, *158*, 1; 1869, *159*, 679.
     16 See note 11.
     17 Miller, *op. cit.*, (see note 12) London, 5th ed., ed. by H. McLeod, 1874, vol. ii, *Inorganic Chemistry*.
     18 See, *e.g.*, Paul Kolodkin, *Dmitri Mendeléièv, et la Loi Périodique*, Paris, 1963; W. P. D. Wightman, "Mendeleeff and the Periodic Law of the Chemical Elements", *J. Roy. Inst. Chem.*, 1958, *82*, 688.
     19 J. L. Meyer, *Die modernen Theorien der Chemie*, Breslau, 1st ed., 1864.
140 20 *Idem, Annalen*, 1870, supp. vol. vii, 354.
     21 D. I. Mendeléef, *J. Russ. Phys. Chem. Soc.*, 1869, *1*, 60; (a) abstract in *Zeitsch. f. Chem.*, 1869, [2], *5*, 405; (b) German translation in Ostwald's *Klassiker* no. 68, p. 20.
     22 *Idem, Zeitsch. f. Chem.* (see note 21a), p. 406.
     23 *Idem, Klassiker* no. 68 (see note 21b), p. 25.
     24 *Idem, Annalen*, 1871, supp. vol. viii, 227–8.
     25 J. L. Meyer, *ibid.*, 1870, supp. vol. vii, 358.

CHAPTER VIII

143   1 F. A. Kekulé, *Ber.*, 1869, *2*, 329.
      2 *Idem, Lehrbuch der organischen Chemie*, Erlangen, 1861, vol. i, p. 157.
144   3 *Ibid.*, pp. 157–8.
      4 *Idem, Annalen*, 1858, *106*, 14n.
      5 A. M. Butlerov, *Zeitsch. f. Chem.*, 1864, *7*, 513, etc.
      6 L. Playfair, *J. Chem. Soc.*, 1863, *16*, 274.
145   7 See W. H. Brock (ed.), *The Atomic Debates*, Leicester, 1967.

Page
145
8 B. C. Brodie, *Phil. Trans.*, 1866, *156*, 781.
9 S. D. Tillman, *Nature*, 1872, *6*, 172.
10 F. A. Kekulé, *Laboratory*, 1867, *1*, 303.
11 A. W. Williamson, *J. Chem. Soc.*, 1869, *22*, 328.
12 *Idem, Nature*, 1873, *8*, 406.
13 A. Wurtz, *Compt. rend.*, 1877, *84*, 1349.
14 M. Berthelot, *ibid.*, p. 1189.

146
15 W. V. and K. R. Farrar, *Proc. Chem. Soc.*, 1959, 289; formulae by Dalton and Kolbe are interchanged in the text.
16 D. N. Kursanov, M. G. Gonikberg, B. M. Dubinin, M. I. Kabachnik, E. D. Kaverzneva, E. N. Prilozhaeva, N. D. Sokolov, and R. K. Friedlina: "The present state of the Chemical Structure Theory", *Uspekhi Khim.*, 1950, *19*, 529; English trans. by I. S. Bengelsdorf in *J. Chem. Educ.*, 1952, *29*, 2.
17 E.g. *Centenary of the Theory of Chemical Structure*, ed. B. A. Kazansky and G. V. Bykov, Moscow, 1961 (Papers by Butlerov, Couper, Kekulé and Markovnikov with Introduction and Notes). The present author has commented on this in a review in *Arch. int. d'Hist. Sci.*, 1962, *15*, 195. See also G. V. Bykov in *J. Chem. Educ.*, 1962, *39*, 220.
18 W. V. and K. R. Farrar, *Proc. Chem. Soc.*, 1959, 287.

147
19 A. M. Butlerov, *Zeitsch. f. Chem.*, 1861, *4*, 549.
20 R. A. C. E. Erlenmeyer, *e.g. Zeitsch. f. Chem.*, 1862, *5*, 221.

148
21 G. V. Bykov, *History of the Classical Theory of Chemical Structure* (in Russian), Moscow, 1960.
22 *Idem, Proc. Chem. Soc.*, 1960, 211: Farrar and Farrar expressed themselves "in complete agreement" with this statement.
23 See note 18.
24 *Ibid.*

149
25 A. Ladenburg, *Lectures on the History of the Development of Chemistry since the time of Lavoisier*, trans. L. Dobbin, from the 2nd German edition, Edinburgh, 1900, p. 254n.
26 A. M. Butlerov, *Zeitsch. f. Chem.*, 1861, *4*, 559.
27 G. J. Hoijtink, *Chem. Weekblad.*, 1957, *53*, 164.
28 *Idem, ibid.*, 1958, *24*, 218.
29 A. M. Butlerov, *Annalen*, 1859, *110*, 58.

150
30 *Ibid.*, 1867, *144*, 9.
31 *Ibid.*, 1868, *145*, 124.
32 H. M. Leicester, *J. Chem. Educ.*, 1959, *36*, 329.
33 A. M. Butlerov, *Annalen*, 1868, *146*, 260.

151
34 *Idem, Zeitsch. f. Chem.*, 1861, *4*, 560.
35 *Cf.* F. Richter, *Ber.*, 1938, *71*, *A*, 35.

152
36 E. Frankland, *J. Chem. Soc.*, 1851, *3*, 322. See also pp. 35 *et seq.*
153
37 A. M. Butlerov, *Zeitsch. f. Chem.*, 1861, *4*, 556.
38 *Idem, ibid.*, 1862, *5*, 298.
39 *Ibid.*, p. 299.
40 *Ibid.*, p. 300.
41 Kekulé, *op. cit.* (see note 2), pp. 382, 410.
42 *Idem, Bull. Acad. roy. Belg.*, 1861, [2], *11*, 676.

154
43 R. A. C. E. Erlenmeyer, *Zeitsch. f. Chem.*, 1862, *5*, 30.
44 A. Crum Brown, *Trans. Roy. Soc. Edin.*, 1864, *23*, 715.
45 E. Frankland, *J. Chem. Soc.*, 1866, *19*, 381.

Page
155   46 C. Schorlemmer, *J. Chem. Soc.*, 1864, *17*, 262.
       47 E. Frankland, *Experimental Researches in Pure, Applied and Physical Chemistry*, London, 1877, pp. 47–8.
       48 C. Schorlemmer, *J. Chem. Soc.*, 1872, *25*, 429.
       49 See note 39.
       50 J. Thomsen, *Thermochemische Untersuchungen*, Leipzig, 1882, vol. i.
       51 See note 38.
156   52 A. W. Hofmann, *Phil. Trans.*, 1851, [*141*], 396.
       53 A. Ihde, *J. Chem. Educ.*, 1959, *36*, 333.
       54 J. A. Geuther, *Jahresberichte*, 1863, *32*, 323.
157   55 E. Frankland and B. F. Duppa, *J. Chem. Soc.*, 1866, *19*, 395.
       56 The problem of acetoacetic ester has been discussed from a historical standpoint by A. Lachmann in *The Spirit of Organic Chemistry*, New York, 1899, p. 80.
       57 A. M. Butlerov, *Annalen*, 1877, *189*, 76.
       58 R. A. C. E. Erlenmeyer, *Ber.*, 1880, *13*, 303.
       59 A. von Baeyer, *ibid.*, 1882, *15*, 2093; 1883, *16*, 2193.
       60 See note 57.
       61 A. Crum Brown, *Proc. Roy. Soc. Edin.*, 1866/7, *6*, 518.
       62 P. C. Laar, *Ber.*, 1885, *18*, 648; 1886, *19*, 730.
       63 A. R. Hantzsch and F. Hermann, *ibid.*, 1887, *20*, 3811.
       64 L. Knorr, O. H. Rothe and H. Averbeck, *ibid.*, 1911, *44*, 1138.
158   65 A much later ramification of the present subject is the phenomenon of "valence isomerism". Some aspects of this are considered in subsequent chapters (pp. 245, 249, 252).

CHAPTER IX

159   1 *E.g.*, E. M. L. Brown, "The History of Stereochemistry up to 1890", M.Sc. dissertation, London, 1949.
       2 J. H. van't Hoff, *La Chimie dans l'Espace*, Paris, 1875, an enlarged translation of *Voorstel tot Uitbreiding der . . . Structuur Formules in de Ruimte*, Utrecht, 1874.
       3 J. A. le Bel, *Bull. Soc. chim. France*, 1874, [2], *22*, 337.
160   4 W. H. Wollaston, *Phil. Trans.*, 1808, [*98*], 101.
       5 L. Gmelin, *Handbook of Chemistry*, London, trans. by H. Watts, vol. vii, 1852, p. 30.
       6 J. Wislicenus, *Annalen*, 1873, *167*, 343–4. Cf. *ibid.*, *166*, 3; *167*, 302; *Ber.*, 1869, *2*, 620.
       7 L. Pasteur, *e.g. Ann. Chim. Phys.*, 1848, [3], *24*, 456; 1851, [3], *31*, 72; 1853, [3], *38*, 437.
       8 *Idem, Leçons de chimie professées en 1860*, Paris, 1861.
       9 *Ibid.*, p. 25.
       10 *Ibid.*
161   11 See note 3.
       12 Much later he wrote "I used the greatest efforts in all my explanations to abstain from basing my ideas on the preliminary hypothesis that carbon compounds of the formula CR_4 have the shape of a regular tetrahedron", (*Bull. Soc. chim. France*, 1890, [3], *3*, 780).
       13 See note 6.

Page 161

14 J. H. van't Hoff, *The Arrangement of Atoms in Space*, trans. by A. Eiloart, London, 2nd ed., 1898, pp. 2–3.
15 *Idem, op. cit.* (see note 2), p. 7.

Page 162

16 F. A. Kekulé, *Zeitsch f. Chem.*, 1867, [2], *3*, 217.
17 R. Anschütz, *August Kekulé*, Berlin, 1929, vol. i, p. 359.
18 A. Ihde, *J. Chem. Educ.*, 1959, *36*, 330.
19 See note 2.
20 J. Wislicenus, *Abhandl. Königl. Sächs. Ges. Leipzig*, (Math. Phys.), 1887, *14*, 1, 27, 32, *etc.*

Page 163

21 Walter (no initials given), *Ber.*, 1873, *6*, 1402.
22 H. Goldschmidt, *ibid.*, 1883, *16*, 2176.
23 V. Meyer, *ibid.*, 1888, *21*, 784, 3510.
24 A. R. Hantzsch and A. Werner, *ibid.*, 1890, *23*, 11; Hantzsch claimed Werner originated the idea. On this see also G. B. Kauffman, *J. Chem. Educ.*, 1966, *43*, 155.
25 H. Goldschmidt, *ibid.*, 1889, *22*, 3113.
26 E. O. Beckmann, *ibid.*, 1877, *20*, 2766.
27 A. R. Hantzsch, *ibid.*, 1894, *27*, 701.
28 W. J. Pope and S. J. Peachey, *J. Chem. Soc.*, 1899, *75*, 1127.

Page 164

29 Baeyer himself improved Kekulé's model for sale in Munich; see van't Hoff, *op. cit.*, (see note 14), p. 8.
30 E.g., V. Meyer and J. J. Sudborough, *Ber.*, 1894, *27*, 1580. Full references in *J. Chem. Soc.*, 1899, *75*, 580.

Page 165

31 A. Crum Brown, *Trans. Roy. Soc. Edin.*, 1864, *23*, 710, 715, *etc.*
32 E. Frankland, *Lecture Notes for Chemical Students, embracing Mineral and Organic Chemistry*, London, 1st ed., 1866, p. 25.
33 *Ibid.*, p. vi.
34 It has been suggested that the failure of Ladenburg's "prism formula" for benzene (p. 244) to gain favour, *before* it had been definitely disproved, was partly owing to the "extreme distrust" of contemporary chemists in 3-dimensional models – *i.e.* stereochemistry (A. Lachmann, *The Spirit of Organic Chemistry*, New York, 1899, p. 45).
35 H. Kolbe, *J. prakt. Chem.*, 1877, [2], *15*, 474.

Page 166

36 A. Sementsov, *Amer. Scientist*, 1955, *43*, 97.
37 A. Claus, *Ber.*, 1881, *14*, 432.
38 R. Demuth and V. Meyer, *ibid.*, 1888, *21*, 265.
39 W. Lossen, *ibid.*, 1887, *20*, 3306. See also W. V. Farrar, " 'Chemistry in Space' and the Complex Atom", *Brit. J. Hist. Sci.*, 1968, *4*, 65.
40 J. A. le Bel, *Annalen*, 1880, *204*, 337.

Page 167

41 *Idem, Bull. Soc. chim. France*, 1882, [2], *37*, 300.
42 *Idem, ibid.*, 1892, [3], *7*, 164.
43 *Idem, ibid.*, p. 163.
44 *Idem, Annalen*, 1880, *204*, 337; *Bull. Soc. chim. France*, 1893, [3], *11*, 292.
45 K. Weissenberg, *Ber.*, 1926, *59*, 1526.
46 A. Wunderlich, *ibid.*, 1886, *19*, 592.
47 H. E. A. Knoevenagel, *Annalen*, 1900, *311*, 203.
48 L. Knorr, *ibid.*, 1894, *279*, 202.
49 K. Auwers, *Der Entwicklung der Stereochemie*, Heidelberg, 1890.
50 J. A. le Bel, *Bull. Soc. chim. France*, 1890, [3], *3*, 788.
51 N. V. Sidgwick, *Electronic Theory of Valency*, Oxford, 1927, p. 220.

CHAPTER X

52 W. H. Mills and E. H. Warren, *J. Chem. Soc.*, 1925, *127*, 2507.

1. H. Buff, *Ber.*, 1869, *2*, 144.
2. C. Schorlemmer, *The Rise and Development of Organic Chemistry*, Manchester, 1879, p. 45.
3. R. A. C. E. Erlenmeyer, *Zeitsch. f. Chem.*, 1862, *5*, 27.
4. J. P. Griess, *Proc. Roy. Soc.*, 1859, *10*, 591.
5. G. C. Foster, *Rép. de Chim. pure*, 1861, *3*, 273.
6. A. Wurtz, *Bull. Soc. chim. France*, 1864, [2], *2*, 247.
7. Letter dated 12 Feb., 1862, reproduced in R. Anschütz, *August Kekulé*, Berlin, 1929, vol. i, p. 218.
8. *Ibid.*
9. A. Naquet, *Compt. rend.*, 1864, *58*, 381.
10. F. A. Kekulé, *ibid.*, pp. 510–11.
11. A. Naquet, *ibid.*, p. 677.
12. F. R. Japp, *J. Chem. Soc.*, 1898, *73*, 118.
13. Reproduced in Anschütz, *op. cit.*, vol. ii, pp. 554–68.
14. F. Sestini, *Nuovo Cim.*, 1871, *20*, 274.
15. Account of the Congress reproduced in Anschütz, *op. cit.*, vol. ii, p. 687.
16. H. E. St-C. Deville and L. J. Troost, *Compt. rend.*, 1857, *45*, 824.
17. Anschütz, *op. cit.*, vol. ii, p. 687.
18. A. W. Williamson, *J. Chem. Soc.*, 1864, *17*, 216–17.
19. H. E. Roscoe, *Phil. Trans.*, 1868, *158*, 1.
20. T. E. Thorpe, *Phil. Mag.*, 1871, [4], *42*, 306.
21. L. Meyer, *Die modernen Theorien der Chemie*, Breslau, 1864, p. 100.
22. H. E. St-C. Deville and L. J. Troost, *Compt. rend.*, 1857, *45*, 821.
23. A. Scheurer-Kestner, *ibid.*, 1861, *53*, 653.
24. A. Wurtz, *Rép. de Chim. pure*, 1860, *2*, 451.
25. C. Friedel, *Bull. Soc. chim. France*, 1863, *5*, 202.
26. E. Frankland, *Lecture Notes for Chemical Students*, London, 1870, vol. i, *Inorganic Chemistry*, p. 22.
27. *Idem*, *Lecture Notes for Chemical Students embracing Mineral and Organic Chemistry*, London, 1866, p. 22.
28. W. Biltz [and J. Meyer], *Zeitsch. phys. Chem.*, 1902, *40*, 200.
29. L. F. Nilson and O. Pettersson, *J. Chem. Soc.*, 1888, *53*, 827.
30. L. Playfair and J. A. Wanklyn, *ibid.*, 1862, *15*, 153.
31. *E.g.*, W. Odling, *Phil. Mag.*, 1864, [4], *27*, 115.
32. C. H. Wichelhaus, *Ber.*, 1868, *1*, 77.
33. *Idem*, *Annalen*, 1868, supp. vol. vi, p. 1257.
34. See note 20.
35. H. Kopp, *Annalen*, 1855, *96*, 180.
36. C. A. A. Michaelis, *ibid.*, 1872, *164*, 9.
37. C. A. A. Michaelis and W. la Coste, *Ber.*, 1885, *18*, 2109.
38. C. A. A. Michaelis and L. Gleichmann, *ibid.*, 1882, *15*, 803; C. A. A. Michaelis and W. H. Sodeau, *Annalen*, 1885, *229*, 305.
39. C. A. A. Michaelis and W. la Coste, *Ber.*, 1885, *18*, 2118.
40. F. A. Kekulé (a) *Zeitsch. f. Chem.*, 1866, [2], *2*, 689; (b) *ibid.*, p. 700;

(c) *Lehrbuch der organischen Chemie*, Erlangen, 1866, vol. ii, pp. 715–23.
41 C. W. Blomstrand, *Der Chemie der Jetztzeit*, Heidelberg, 1869, p. 272.
42 Kekulé, *op. cit.* (see note 40c), p. 743.
43 A. Angeli, *Gazzetta*, 1916, *46*, 67. The history of these compounds is reviewed by H. E. Bigelow, *Chem. Rev.*, 1931, *9*, 117.
44 F. A. Kekulé, *Annalen*, 1854, *90*, 309; *Proc. Roy. Soc.*, 1854, *7*, 37.
45 *Ibid*, *Annalen*, 1857, *104*, 129.
46 *Ibid.*, 1858, *106*, 129.
47 *Idem*, *ibid.*, 1857, *104*, 129 and 1858, *106*, 129; *cf.* C. W. Blomstrand, *Ber.*, 1870, *3*, 958.
48 *Ibid.*
49 H. Muller, *Ber.*, 1873, *6*, 227.
50 G. A. Barbaglia and F. A. Kekulé, *Ber.*, 1872, *5*, 878.
51 W. V. Spring, *ibid.*, 1873, *6*, 1108.
52 Blomstrand, *op. cit.* (see note 41), pp. 160–5.
53 C. A. A. Michaelis, *Annalen*, 1873, *170*, 1.
54 G. A. Barbaglia, *Ber.*, 1872, *5*, 270.
55 *Ibid.*, p. 687.
56 C. Vogt, *Annalen*, 1861, *119*, 142.
57 A. Michael and A. Adair, *Ber.*, 1877, *10*, 583.
58 *Ibid.*, 1878, *11*, 116.
59 A. Wurtz, *The Atomic Theory*, London, 1880, trans. by E. Cleminshaw, p. 218.
60 A. W. Hofmann, *Proc. Roy. Inst.*, 1865, *4*, 414.
61 A. Wurtz, *Leçons de philosophie chimique*, Paris, 1864, p. 157.
62 Wurtz, *op. cit.* (see note 59), p. 229n.
63 *E.g.*, C. Graebe and C. T. Liebermann, *Ber.*, 1870, *3*, 634.
64 A. Wurtz, *Phil. Mag.*, 1869, [4], *38*, 455.
65 *Ibid.*
66 Wurtz, *op. cit.* (see note 59), p. 334.
67 Blomstrand, *op. cit.*, p. 283.
68 *Ibid.*, p. 337.
69 W. Lossen, *Ber.*, 1875, *8*, 47.
70 W. Odling, *Phil. Mag.*, 1864, [4], *27*, 115.
71 S. Cannizzaro, *Nuovo Cim.*, 1858, *7*, 352.
72 A. M. Butlerov, *Zeitsch. f. Chem.*, 1861, *4*, 556–7.
73 E. Frankland, *Experimental Researches in Pure, Applied and Physical Chemistry*, London, 1877, p. 145.
74 *Idem*, *J. Chem. Soc.*, 1866, *19*, 378; this is repeated in *Experimental Researches* (see note 73) with slight variations. Frankland's "law of variation" received quantum-mechanical justification from F. W. London (*Zeitsch. Phys.*, 1927, *46*, 455). *Cf.* p. 304.
75 Frankland, *op. cit.* (see note 27), pp. 19–22; *idem*, *op. cit.* (see note 26), pp. 18–21.
76 *Idem*, *op. cit.* (see note 27), p. 21.
77 C. R. A. Wright, *Phil. Mag.*, 1872, [4], *43*, 260.
78 H. E. Roscoe, *Phil. Trans.*, 1868, *158*, 1.
79 *Idem*, *Annalen*, 1872, *162*, 349.
80 J. C. G. Marignac, *ibid.*, 1866, supp. vol. iv, p. 350.
81 *Idem*, *ibid.*, p. 273.

[189–95] NOTES AND REFERENCES 349

Page
189 82 D. I. Mendeléef, *ibid.*, 1871, supp. vol. viii, p. 217.
190 83 See note 10.
     84 I. Freund, *The Study of Chemical Composition*, Cambridge, 1904, p. 532.
     85 F. A. Kekulé, *Compt. rend.*, 1864, *58*, 513.
     86 See note 9.
191 87 J. B. A. Dumas, *Ann. Chim. Phys.*, 1832, [2], *50*, 175.
     88 A. Bineau, *ibid.*, 1838, [2], *68*, 416.
     89 A. A. T. Cahours, *Compt. rend.*, 1845, *21*, 625.
     90 H. Kopp, *Annalen*, 1858, *105*, 290.
     91 A. F. Horstmann, *Ber.*, 1869, *2*, 140.
     92 *Idem, ibid.*, 1868, *1*, 213.
     93 L. von Pebal, *Annalen*, 1862, *123*, 199.
     94 J. A. Wanklyn and J. Robinson, *Proc. Roy. Soc.*, 1863, *12*, 507.
     95 H. E. St-C. Deville, *Compt. rend.*, 1857, *48*, 857; 1863, *56*, 730. In the second reference "dissociation" is limited to reversible changes.
     96 *Ibid.*, 1867, *64*, 242.
     97 *Ibid.*, 1865, *60*, 823.
     98 A. Wurtz, *ibid.*, 1866, *62*, 1182.
     99 *Idem, Ann. Chim. Phys.*, 1864, [4], *3*, 130.
    100 *Idem, Compt. rend.*, 1865, *60*, 728.
    101 S. Cannizzaro, *Nuovo Cim.*, 1857, *6*, 428.
    102 L. Playfair and J. A. Wanklyn, *J. Chem. Soc.*, 1862, *15*, 156.
    103 C. H. Wichelhaus, *Annalen*, 1868, supp. vol. vi, 257.
192 104 A. Wurtz, *Compt. rend.*, 1865, *60*, 731.
    105 A. F. Horstmann, *Ber.*, 1869, *2*, 299.
    106 *Ibid.*, p. 140.
    107 *Ibid.*, pp. 301–2.
    108 C. H. Wichelhaus, *ibid.*, p. 302.
    109 A. Wurtz, *Compt. rend.*, 1873, *76*, 609.
    110 T. E. Thorpe, *Proc. Roy. Soc.*, 1876, *25*, 122.
    111 *Ibid.*, 1875, *23*, 364.
    112 *Ibid.*, 1876, *25*, 123.
    113 H. E. Armstrong, *Phil. Mag.*, 1888, [5], *25*, 28.
193 114 J. W. Mallet, *Amer. Chem. J.*, 1881, *3*, 89.
    115 T. E. Thorpe and F. J. Hambly, *J. Chem. Soc.*, 1889, *55*, 163.
    116 H. E. Armstrong, *Phil. Mag.*, 1888, [5], *25*, 28.
    117 A. Naquet, *Compt. rend.*, 1864, *58*, 676.
    118 H. Buff, *Ber.*, 1869, *2*, 142.
    119 C. H. Wichelhaus, *ibid.*, p. 304.
    120 See note 116.
    121 C. H. Wichelhaus, *Annalen*, 1868, supp. vol. vi, 257.
194 122 V. Meyer and M. T. Lecco, *ibid.*, 1876, *180*, 173.
195 123 F. Krüger, *J. prakt. Chem.*, 1876, [2], *14*, 193.
    124 R. Anschütz, *Annalen*, 1889, *253*, 343.
    125 M. Traube, *Ber.*, 1886, *19*, 1115.
    126 *Ibid.*, p. 1111.
    127 J. A. Geuther and C. A. A. Michaelis, *Zeitsch. f. Chem.*, 1864, [2], *7*, 158.
    128 W. Lossen, *Annalen*, 1876, *181*, 364.
    129 H. E. Armstrong, *Phil. Mag.*, 1888, [5], *25*, 27.

Page
196 130 A. W. Williamson, *J. Chem. Soc.*, 1864, *17*, 216.
   131 Idem, *Rep. Brit. Assoc. Adv. Sci.*, 1881, *51*, 576.
   132 Letter to Williamson 23 Nov. 1856, reproduced in Anschütz, *op. cit.* (see note 7), vol. i, pp. 69–70.
   133 C. H. Wichelhaus, *Compt. rend.*, 1865, *60*, 304.
   134 Blomstrand, *op. cit.* (see note 41), p. 127.
197 135 F. R. Japp, *J. Chem. Soc.*, 1898, *73*, 118.
   136 H. Kolbe, *Ausführliches Lehrbuch der organischen Chemie*, Braunschweig, 1854, vol. i, p. 22.
   137 A. S. Couper, *Ann. Chim. Phys.*, 1858, [3], *53*, 469.
   138 See note 9.
   139 A. M. Butlerov, *Zeitsch. f. Chem.*, 1861, *4*, 556.
   140 R. A. C. E. Erlenmeyer, *ibid.*, 1864, *7*, 631–2.
198 141 E. Frankland, *J. Chem. Soc.*, 1861, *13*, 229 *et seq.*
   142 See note 41.
   143 J. F. Heyes, *Phil. Mag.*, 1888, [5], *25*, 227n.
   144 Wurtz, *op. cit.* (see note 59), p. 229.
   145 C. Friedel, *Compt. rend.*, 1875, *81*, 152.
199 146 *Ibid.*, p. 236.
   147 See note 66.
   148 Wurtz, *op. cit.* (see note 61), p. 80.
   149 Idem, *Elements of Modern Chemistry*, trans. by W. H. Greene, London and Philadelphia, 1881, p. 222.
   150 J. F. Heyes, *Phil. Mag.*, 1888, [5], *25*, 226.
   151 J. N. Collie and T. Tickle, *J. Chem. Soc.*, 1899, *75*, 710.
   152 J. U. Nef, *J. Amer. Chem. Soc.*, 1904, *26*, 1549.
200 153 J. A. Wanklyn, *Ber.*, 1869, *2*, 64.
   154 C. H. Wichelhaus, *ibid.*, p. 65.
   155 E. Frankland and B. F. Duppa, *Phil. Trans.*, 1866, *156*, 37.
   156 J. A. Wanklyn, *J. Chem. Soc.*, 1869, *22*, 199.
   157 *Ibid.*, pp. 200–2.
   158 A. W. Williamson, *Rep. Brit. Assoc. Adv. Sci.*, 1881, *51*, 577–8.
201 159 J. R. Partington (*A History of Chemistry*, London, 1964, vol. iv, p. 538) refers to some early insights into the related concept of "residual affinity" (Dalton, Higgins, Mercer, Playfair). It is not clear how important these tentative suggestions may have been for later workers, however.
   160 H. Kommrath, *Ber.*, 1876, *9*, 1394–5.
202 161 W. Lossen, *Annalen*, 1880, *204*, 284.
   162 *Ibid.*, pp. 288–9.
   163 A. Claus, *Ber.*, 1881, *14*, 432.
203 164 *Ibid.*, p. 433.
   165 *Ibid.*
   166 *Ibid.*, p. 435.
   167 W. Lossen, *ibid.*, pp. 764–5.
204 168 H. Klinger, *ibid.*, p. 783.
   169 S. U. Pickering, *Proc. Chem. Soc.*, 1885, *1*, 122.
205 170 Idem, *J. Chem. Soc.*, 1886, *49*, 411.
206 171 H. B. Baker, *ibid.*, 1885, *47*, 349.
   172 H. E. Armstrong, *Proc. Chem. Soc.*, 1885, *1*, 40.
   173 H. von Helmholtz, *J. Chem. Soc.*, 1881, *39*, 303.

Page		
206	174	H. E. Armstrong, *Rep. Brit. Assoc. Adv. Sci.*, 1885, *55*, 956.
	175	See note 158.
	176	H. E. Armstrong, *Rep. Brit. Assoc. Adv. Sci.*, 1885, *55*, 959.
	177	See note 167.
207	178	H. E. Armstrong, *Rep. Brit. Assoc. Adv. Sci.*, 1885, *55*, 960.
	179	Idem, *Proc. Roy. Soc.*, 1886, *40*, 274.
208	180	See note 114.
	181	H. E. Armstrong, *Proc. Roy. Soc.*, 1886, *40*, 285.
	182	Idem, *Phil. Mag.*, 1888, [5], *25*, 21.
209	183	*Ibid.*, p. 24.
210	184	*Ibid.*, p. 25.
	185	*Ibid.*, p. 26.
212	186	*Ibid.*, pp. 29–30.
	187	H. V. Eyre, *Henry Edward Armstrong, 1848–1937*, London, 1958.
	188	J. F. Heyes, *Phil. Mag.*, 1885, [5], *25*, 226.
	189	*Ibid.*, p. 297.
213	190	A. Werner, *Beitrage zur Theorie der Affinitat und Valenz*, Zurich, 1891. An English translation by G. B. Kauffman is in *Chymia*, 1967, *12*, 189.
	191	Idem, *Zeitsch. anorg. Chem.*, 1893, *3*, 267.
	192	Idem, (a) *op. cit.*, Braunschweig, 1905; (b) *New Ideas on Inorganic Chemistry*, trans. by E. P. Hedley, London, 1911.
	193	On Werner see H. A. Bent, *J. Chem. Educ.*, 1967, *44*, 512; G. B. Kauffman, *Educ. Chem.*, 1967, *4*, 11; idem, *Alfred Werner, Founder of Co-ordination Chemistry*, Berlin, Heidelberg and New York, 1966; the A.C.S. *Werner Centennial Symposium*, Washington, 1967. The sudden inspiration that came to Werner one night and resulted in the main outline of his co-ordination theory has been aptly compared with the "flash of genius" exemplified by Kekulé's "vision" (Kauffman, *op. cit.* [2], p. 30).
	194	J. Lewkowitsch, *Ber.*, 1883, *16*, 1575, 2722.
214	195	See note 162.
	196	See note 163.
	197	Werner, *op. cit.* (see note 192b), p. 58.
	198	P. T. Cleve, *Akad. Handl.*, 1871, *10*, no. 9.
215	199	Werner, *op. cit.* (see note 192b), pp. 29–30.
216	200	See note 161.
	201	See note 163.
	202	Werner, *op. cit.* (see note 192b), p. 35.
218	203	*Ibid.*, p. 43.
	204	A. Werner, *Zeitsch. anorg. Chem.*, 1893, *3*, 323; in the third example given modern chemistry would prefer a value of -1 (*i.e.*, $3-4$).
219	205	Werner, *op. cit.* (see note 192b), p. 56.
	206	*Ibid.*, p. 66.
	207	*Ibid.*, p. 69.
	208	L. Lindet, *Compt. rend.*, 1885, *101*, 492, 164.
220	209	A. Werner, *Zeitsch. anorg. Chem.*, 1893, [2], *3*, 298.
	210	S. M. Jörgensen, *J. prakt. Chem.*, 1889, [2], *39*, 1. On Jörgensen, see G. B. Kauffman, *J. Chem. Educ.*, 1959, *36*, 521, and *Chymia*, 1960, *6*, 180.
	211	A. Werner, *Zeitsch. anorg. Chem.*, 1893, *3*, 299.
221	212	Idem, *Ber.*, 1911, *44*, 2445.
	213	*Ibid.*, 1914, *47*, 3087.

M**

Page
222   214 *Idem, Zeitsch. anorg. Chem.*, 1893, *3*, 321.
      215 B. Flürscheim, *J. prakt. Chem.*, 1902, [2], *66*, 321.
      216 *Idem, J. prakt. Chem.*, 1907, [2], *76*, 185; *J. Chem. Soc.*, 1909, *95*, 718, *etc.*
223   217 A. Lapworth, *J. Chem. Soc.*, 1901, *79*, 1273.
      218 G. N. Lewis, *Valence and the Structure of Atoms and Molecules*, New York, 1923, p. 68.
      219 A. W. Stewart, *Stereochemistry*, London, 2nd ed., 1919, p. 158.

CHAPTER XI

224   1 A. Wurtz, *Compt. rend.*, 1856, *43*, 199.
      2 *Ibid.*, 1859, *49*, 813.
      3 *Ibid.*, p. 898.
      4 H. Buff, *Proc. Roy. Soc.*, 1857, *8*, 188.
      5 E. Frankland, *J. Chem. Soc.*, 1861, *13*, 231.
225   6 E. J. Mills, *Phil. Mag.*, 1867, [4], *33*, 8.
      7 F. A. Kekulé, (a) *Bull. Acad. roy. Belg.*, 1862, [2], *13*, 367; (b) *Annalen*, 1862, supp. vol. ii, 114–15.
226   8 A. W. Hofmann, *Proc. Roy. Inst.*, 1865, *4*, 414.
      9 T. Swarts, *Bull. Acad. roy. Belg.*, 1866, [2], *21*, 538.
      10 F. A. Kekulé, *ibid.*, 1867, [2], *24*, 8. Modern formulae are as follows:

$$\text{itaconic acid} \longrightarrow \underset{H}{\overset{H}{}}C = C\underset{CH_2COOH}{\overset{COOH}{}}$$

$$\text{citraconic acid} \longrightarrow \underset{HOOC}{\overset{CH_3}{}}C = C\underset{COOH}{\overset{H}{}}$$

$$\text{mesaconic acid} \longrightarrow \underset{HOOC}{\overset{CH_3}{}}C = C\underset{H}{\overset{COOH}{}}$$

227   11 *Idem, Bull. Acad. roy. Belg.*, 1862, [2], *13*, 364.
      12 *Ibid.*, 1861, [2], *11*, 84.
      13 V. von Richter, *Zeitsch. f. Chem.*, 1868, [2], *4*, 453.
      14 R. Anschütz, *August Kekulé*, Berlin, 1929, vol. i, pp. 490, 492.
228   15 A. M. Butlerov, *Zeitsch. f. Chem.*, 1861, *4*, 465.
      16 B. C. G. Tollens, *Annalen*, 1866, *140*, 244.
      17 F. A. Kekulé, *ibid*, 1865, *137*, 134n.
      18 A. Crum Brown, *Trans. Roy. Soc. Edin.*, 1864, *23*, 709n.
229   19 F. A. Kekulé and E. C. T. Zincke, *Ber.*, 1870, *3*, 129.
      20 *Idem, Annalen*, 1872, *162*, 125; *cf.* T. Harnitz-Harnitzky, *ibid.*, 1859, *111*, 192.
230   21 R. Fittig, *Ber.*, 1877, *10*, 513.
      22 See note 16.
      23 See note 15.
      24 V. Y. Krivaksin, *Zeitsch. f. Chem.*, 1871, [2], *7*, 265.
      25 C. Friedel and A. Ladenburg, *Annalen*, 1868, *145*, 190.
231   26 A. M. Butlerov, *Chem. Centr.*, 1871, 89.

Page
231   27 F. A. Kekulé, *Annalen*, 1858, *106*, 156–7.
      28 *Idem, Lehrbuch der organischen Chemie*, Erlangen, 1861, vol. i, p. 166.
      29 *Ibid.*, p. 159.
232   30 J. Loschmidt, *Chemische Studien*, Vienna, 1861.
      31 R. A. C. E. Erlenmeyer, *Zeitsch. f. Chem.*, 1862, *5*, 28.
      32 F. A. Kekulé, *Annalen*, 1862, supp. vol. ii, 31n.
      33 R. A. C. E. Erlenmeyer, *Zeitsch. f. Chem.*, 1862, *5*, 302.
      34 See note 15.
      35 A. M. Butlerov, *Zeitsch. f. Chem.*, 1861, *4*, 465.
233   36 See note 18.
      37 *Ibid.*
      38 H. Kolbe, *Annalen*, 1857, *101*, 257.
      39 Thus in a review of the constitutions of hydrocarbons, in 1872, C. Schorlemmer deals with most possibilities *except* this: *J. Chem. Soc.*, 1872, *25*, 430.
      40 A. A. Baker, jr., *Kekulé Centennial*, ed. O. T. Benfey, Washington, 1966, p. 90. Butlerov's first use of the double bond (*Zhur. obshchei Khim.*, 1870, *2*, 187; *J. Chem. Soc.*, 1871, *24*, 214) followed receipt of a letter from Beilstein giving news of Crum Brown's work (see G. V. Bykov and L. M. Bekassova, *Physis*, 1966, *8*, 277).
234   41 J. Wilbrand, *Zeitsch. f. Chem.*, 1865, *8*, 685.
      42 Kekulé, *op. cit.* (see note 28), vol. i, p. 163.
      43 See note 7.
235   44 F. A. Kekulé, *Zeitsch. f. Chem.*, 1867, [2], *3*, 218.
      45 *Idem, op. cit.* (see note 28), 1866, vol. ii, p. 496.
      46 *Idem, Bull. Acad. roy. Belg.*, 1867, [2], *24*, 10.
      47 *Idem, Ber.*, 1870, *3*, 125.
236   48 W. Lossen, *Annalen*, 1880, *204*, 288–9.
      49 *Ibid.*, p. 285n.
      50 *Ibid.*, p. 295.
      51 A. Claus, *Ber.*, 1881, *14*, 432.
237   52 F. W. Hinrichsen, *Ueber den gegenwärtigen Stand der Valenzlehre*, Stuttgart, 1902.
      53 A. Werner, *New Ideas on Inorganic Chemistry*, London, trans. by E. P. Hedley, 1911, pp. 72–3.
      54 Kekulé, *op. cit.* (see note 28), vol. i, p. 169.
      55 See note 46.
238   56 Anschütz, *op. cit.* (see note 14), vol. i, p. 169.
      57 A. von Baeyer, *Ber.*, 1885, *18*, 2269.
      58 V. V. Markovnikov and A. Krestovnikov, *Annalen*, 1881, *208*, 333.
      59 A. Freund, *J. prakt. Chem.*, 1882, [2], *26*, 367.
239   60 A. von Baeyer, *Ber.*, 1885, *18*, 2278.
      61 *Ibid.*
      62 W. H. Perkin, *ibid.*, 1885, *18*, 3246.
      63 H. Sachse, *ibid.*, 1890, *23*, 1363.
      64 E. W. M. Mohr, *J. prakt. Chem.*, 1918, [2], *98*, 349, *etc.*
      65 See also K. T. Finley, "The Synthesis of Carbocyclic compounds. A historical survey", *J. Chem. Educ.*, 1965, *42*, 536.
      66 E. W. M. Mohr, *J. prakt. Chem.*, 1918, [2], *98*, 315.
      67 V. Meyer, *Ber.*, 1890, *23*, 582.

Page 239

68 G. N. Lewis, *Valence and the Structure of Atoms and Molecules*, New York, 1923, pp. 93, 126.
69 *Cf.* A. Lapworth, *J. Chem. Soc.*, 1901, *79*, 1265, etc.

Page 240

70 Work by Baeyer on the incomplete reduction of terephthalic acid (with Herb, *Annalen*, 1890, *258*, 1) and dichloromuconic acid (with Rupe, *ibid.*, 1890, *256*, 1) had anticipated Thiele's recognition of 1:4 addition.
71 F. K. J. Thiele, *Annalen*, 1899, *306*, 89.
72 *Ibid.*, pp. 89–90.

Page 241

73 See note 51.
74 See note 48.
75 S. U. Pickering, *Proc. Chem. Soc.*, 1885, *1*, 122.
76 H. E. Armstrong, *Rep. Brit. Assoc. Adv. Sci.*, 1885, *55*, 945; *Proc. Roy. Soc.*, 1886, *40*, 268; *Phil. Mag.*, 1888, [5], *25*, 21.
77 See note 53.
78 See note 71.
79 H. Klinger, *Ber.*, 1881, *14*, 783.
80 M. J. S. Dewar, *The Electronic Theory of Organic Chemistry*, Oxford, 1949, p. 9.
81 R. Huisgen, *Proc. Chem. Soc.*, 1958, 210.

CHAPTER XII

Page 242

1 Recent articles on this matter include E. E. van Tamelen, *Chemistry*, 1965, *38*, 6; C. A. Russell, *Chem. Brit.*, 1965, *1*, 141. See n. 5.
2 F. A. Kekulé, (a) *Bull. Soc. chim. France*, 1865, [2], *3*, 98; (b) *Annalen*, 1866, *137*, 129 (an expanded version of this paper and the one cited in note 3).

Page 243

3 F. A. Kekulé, *Bull. Acad. roy. Belg.*, 1865, [2], *19*, 551.
4 *Idem*, *Lehrbuch der organischen Chemie*, Erlangen, 1866, vol. ii, p. 496.
5 *Ber.*, 1890, *25*, 1265. (English translation in *J. Chem. Educ.*, 1958. *35*, 21.) See also J. Gillis (trans. R. E. Oesper), "Kekulé's life at Ghent (1858–1867)", *J. Chem. Educ.*, 1961, *38*, 118; *idem*, "Naissance et étapes de la Théorie Benzénique de Kekulé", *Actes of XIth International Congress of the History of Science*, Warsaw and Cracow, 1965, *4*, 104; the A.C.S. *Kekulé Centennial Symposium*, Washington, 1966.
6 "R.S." in *M. & B. Lab. Bull.*, 1961, *4*, 78.
7 H. E. Armstrong, *J. Soc. Chem. Ind.*, 1929, *48*, 914; for reactions to Kekulé's formula see also J. Read, *Humour and Humanism in Chemistry*, London, 1947, p. 217.

Page 244

8 A. Ladenburg, *Ber.*, 1869, *2*, 140; *Annalen*, 1874, *172*, 344.
9 H. Hübner, *Annalen*, 1869, *149*, 130; Z. F. von Wroblewsky, *Ber.*, 1872, *5*, 30, etc.
10 W. Körner, *Gazzetta*, 1874, *4*, 305.
11 A. Ladenburg, *Ber.*, 1869, *2*, 140, 272.
12 F. R. Japp, *J. Chem. Soc.*, 1898, *73*, 135.

Page 245

13 *E.g.*, A. von Baeyer, *Annalen*, 1892, *269*, 145.
14 *E.g.*, *idem*, *Ber.*, 1886, *19*, 1797.
15 *E.g.*, K. Lonsdale, *Trans. Farad. Soc.*, 1935, *31*, 1036.

Page	
245	16 K. E. Wilzbach and L. Kaplan, *J. Amer. Chem. Soc.*, 1965, *87*, 4004.
	17 R. Criegee and R. Askani, *Angew. Chem., Internat. Edn.*, 1966, *5*, 519.
246	18 R. A. C. E. Erlenmeyer, *Annalen*, 1866, *137*, 327.
	19 C. Graebe and C. T. Liebermann, *Ber.*, 1868, *1*, 49; *Annalen*, 1870, supp. vol. vii, p. 257.
	20 C. Graebe and C. Glaser, *Annalen*, 1873, *167*, 131.
	21 R. Fittig and E. Ostermeyer, *ibid.*, 1873, *166*, 361.
	22 W. Körner, *J. nat. econ. Sci. Palermo*, 1869.
	23 J. Dewar, *Trans. Roy. Soc. Edin.*, 1872, *26*, 195; *Proc. Roy. Soc. Edin.*, 1872, *7*, 192 (paper read 6 June 1870). The Dewar–Körner controversy has been fully discussed by L. Dobbin in *J. Chem. Educ.*, 1934, *11*, 596.
	24 A. von Baeyer and A. Emmerling, *Ber.*, 1869, *2*, 679.
	25 *Ibid.*, 1870, *3*, 517.
247	26 A. Ladenburg, *Ber.*, 1869, *2*, 140.
	27 V. Meyer, *Annalen*, 1870, *156*, 293.
	28 H. E. Armstrong, *J. Soc. Chem. Ind.*, 1929, *48*, 916.
248	29 J. Dewar, *Proc. Roy. Soc. Edin.*, 1866/7, *6*, 82. On this see also A. Sementsov, *J. Chem. Educ.*, 1966, *43*, 151.
	30 J. Dewar, *Proc. Chem. Soc.*, 1897, *13*, 239.
	31 C. H. Wichelhaus, *Ber.*, 1869, *2*, 199.
	32 W. Körner, *Gazzetta*, 1874, *4*, 305; abstract in *J. Chem. Soc.*, 1876, *29*, 204.
	33 G. Städeler, *J. prakt. Chem.*, 1868, *103*, 105.
	34 C. Graebe and C. T. Liebermann, *Annalen*, 1870, supp. vol. vii, p. 315.
	35 C. Riedel, *Ber.*, 1883, *16*, 1612.
249	36 See note 23.
	37 K. E. Wilzbach and D. J. Rausch, *J. Amer. Chem. Soc.*, 1970, *92*, 2178; M. G. Darlow, J. G. Dingwall and R. N. Haszeldine, *Chem. Comm.*, 1970, 1580.
	38 A. Ladenburg, *Ber.*, 1869, *2*, 272.
	39 F. A. Kekulé, *e.g., ibid.*, 362; memoir, 1890, published by R. Anschütz in his *August Kekulé*, Berlin, 1929, vol. ii, p. 769.
	40 As shown by Ladenburg (note 8), Hübner (note 9) and others.
	41 A. von Baeyer, *Ber.*, 1890, *25*, 12.
	42 H. E. Armstrong, *J. Chem. Soc.*, 1887, *51*, 263.
	43 C. K. Ingold, *ibid.*, 1922, *121*, 1133, 1143; *cf.* R. Robinson, *Ann. Rep. Chem. Soc.*, 1922, *29*, 86.
	44 E. E. van Tamelen and S. S. Pappas, *J. Amer. Chem. Soc.*, 1962, *84*, 3789; 1963, *85*, 3297.
	45 For reviews see W. Baker, *Chem. Brit.*, 1965, *1*, 191; and E. E. van Tamelen, "Benzene: the story of its formulas, 1865–1965", *Chemistry*, 1965, *38*, 6. Recent references are in W. Schäfer and H. Hellmann, *Angew. Chem., Internat. Edn.*, 1967, *6*, 518; I. O. Sutherland, *Ann. Rep. Chem. Soc.*, 1967, *64*, 281; H. Heaney, *ibid.*, 1968, *65*, 351; 1969, *66*, 336, *etc.*
	46 W. Schäfer, *Angew. Chem., Internat. Edn.*, 1966, *5*, 668.
250	47 See note 17.
	48 H. R. Ward and J. S. Wishnok, *J. Amer. Chem. Soc.*, 1968, *70*, 1085.
	49 A. Claus, *Theoretische Betrachtungen und deren Anwendung zur Systematik der organischen Chemie*, Freiburg, 1867, p. 207.

50 F. R. Japp, *J. Chem. Soc.*, 1898, *73*, 136–7.
51 *Ibid.*, p. 136.
52 Wichelhaus, *op. cit.* (see note 31), p. 198.
53 A. Claus, *Ber.*, 1882, *15*, 1405.
54 A. von Baeyer, *Annalen*, 1888, *245*, 118.
55 A. Claus, *J. prakt. Chem.*, 1890, [2], *42*, 24, 260.
56 A. von Baeyer and H. Rupe, *Annalen*, 1890, *256*, 1.
57 A. von Baeyer and J. Herb, *ibid.*, 1890, *258*, 1.
58 A. von Baeyer, *Ber.*, 1890, *23*, 1279.
59 A. Claus, *J. prakt. Chem.*, 1890, [2], *42*, 24, 260, 458; 1891, [2], *43*, 321.
60 See note 50.
61 L. Pauling, *J. Amer. Chem. Soc.*, 1926, *48*, 1132.
62 *Idem*, *J. Chem. Phys.*, 1933, *1*, 362.
63 L. Meyer, *Die modernen Theorien der Chemie*, Breslau, 4th ed., 1883, p. 262.
64 J. Thomsen, *Thermochemische Untersuchungen*, Leipzig, 1886, vol. iv.
65 H. E. Armstrong, *Phil. Mag.*, 1887, [5], *23*, 108.
66 Thomsen, *op. cit.* (see note 64), p. 353.
67 See note 11.
68 See note 65.
69 See note 65.
70 A. von Baeyer, *Annalen*, 1888, *245*, 12 et seq.
71 *Ibid.*, p. 126.
72 See note 56.
73 E. Bamberger, *Ber.*, 1891, *24*, 1758.
74 F. A. Kekulé, *Annalen*, 1872, *162*, 77.
75 *Idem*, *Compt. rend.*, 1865, *60*, 174.
76 *Idem*, *Annalen*, 1872, *162*, 86.
77 *Ibid.*, p. 87.
78 *Ibid.*
79 *Ibid.*, p. 88.
80 *Ibid.*, p. 89.
81 A. Ladenburg, *Ber.*, 1872, *5*, 322.
82 H. E. Armstrong, *J. Soc. Chem. Ind.*, 1929, *48*, 916.
83 However, in an article on "Concepts of time in Chemistry" (*J. Chem. Educ.*, 1963, *40*, 574), O. T. Benfey has observed that "the idea of a transition state or intermediate state predates by many decades the mathematical procedure for obtaining its energy" (*ibid.*, p. 575).
84 L. Fieser in *Organic Chemistry, an advanced Treatise*, ed. H. Gilman, New York, 2nd ed., 1943, vol. i, p. 122.
85 J. N. Collie, *J. Chem. Soc.*, 1897, *71*, 1013.
86 Γ. G. Arndt, letter quoted by E. Campaigne in *J. Chem. Educ.*, 1959, *36*, 336.
87 H. C. Longuet-Higgins, *Proc. Chem. Soc.*, 1957, 157.
88 F. K. J. Thiele, *Annalen*, 1899, *306*, 125.
89 Formation of free cyclobutadiene was first demonstrated by L. Watts, J. D. Fitzpatrick and R. Pettit, *J. Amer. Chem. Soc.*, 1965, *87*, 3253. For historical aspects see M. P. Cava and M. J. Mitchell, *Cyclobutadiene and Related Compounds*, New York and London, 1967, pp. 3–10.

Page		
257	90	R. M. Willstäter, *Ber.*, 1911, *44*, 3423.
	91	H. Kaufmann, *Phys. Zeitsch.*, 1908, *9*, 311.
	92	J. J. Thomson, *Phil. Mag.*, 1914, [6], *27*, 757.
	93	L. S. Efros, *Russ. Chem. Rev.*, 1960, *29*, 66.

CHAPTER XIII

261	1	For a review see, *e.g.*, C. A. Russell, *Ann. Sci.*, 1963, *19*, 117, 127.
	2	*Idem, ibid.*, 1959, *15*, 1, 15; 1963, *19*, 255.
	3	J. J. Berzelius, *J. de Phys.*, 1811, *73*, 253.
	4	*Ibid.*, p. 257, etc.
	5	*Ibid.*, p. 283, etc.
	6	*Idem, Essai sur la Théorie des Proportions Chimiques*, Paris, 1819, pp. 95–6.
262	7	*E.g., ibid.*, pp. 54–5.
	8	*Idem, Nic. J.*, 1813, *34*, 155.
	9	*Idem, Traité de Chimie*, Paris, 2nd ed., 1845, vol. i, p. 106.
	10	W. G. Palmer, *Valency, Classical and Modern*, Cambridge, 1929, p. 6.
263	11	J. A. A. Ketelaar, *Chemical Constitution*, Amsterdam, 2nd ed., 1958, p. 19n.
	12	J. Clerk Maxwell, *Electricity and Magnetism*, Oxford, 1873, vol. i, pp. 313–15.
	13	H. Kolbe, *Short Textbook of Inorganic Chemistry*, trans. by T. S. Humpidge, London, 1884, p. 45.
264	14	E. Frankland, *Rep. Brit. Assoc. Adv. Sci., (Trans. of Sections)*, 1868, *38*, 34.
	15	*Idem, Lecture Notes for Chemical Students embracing Mineral and Organic Chemistry*, London, 1866, p. 25; *J. Chem. Soc.*, 1866, *19*, 377–8. For a commentary on both the background to this statement and the development of gravitational "explanations", see D. M. Knight, *Atoms and Elements, A Study of Theories of Matter in England in the Nineteenth Century*, London, 1967.
	16	E. Frankland, *Lecture Notes for Chemical Students*, London, 1870, 2nd ed., vol. i, (*Inorganic Chemistry*), p. 25; *Experimental Researches in Pure, Applied and Physical Chemistry*, London, 1877, p. 9.
	17	J. Loschmidt, *Sitzungsber. Akad. Wiss. Wien*, 1865, *52*, 395.
	18	H. Buff, *Ber.*, 1869, *2*, 145.
	19	C. A. A. Michaelis, *ibid.*, 1872, *5*, 145.
265	20	J. P. Cooke, *The New Chemistry*, London, 1874, p. 265.
	21	C. W. Blomstrand, *Die Chemie der Jetztzeit vom Standpunkte der electrochemischen Auffassung, aus Berzelius Lehre entwickelt*, Heidelberg, 1869.
	22	M. Faraday, *Experimental Researches in Electricity*, London, 1839, vol. i, p. 314.
	23	W. Whewell, *History of the Inductive Sciences*, London, 1847, vol. iii, pp. 194–5.
266	24	H. Helmholtz, *J. Chem. Soc.*, 1881, *39*, 284.
	25	*Ibid.*, pp. 289–90.
	26	*Ibid.*, pp. 302–3.
	27	*Ibid.*, p. 303.

Page
266   28 L. Koenigsberger, *Hermann von Helmholtz*, trans. by F. A. Welby, Oxford, 1906, p. 332.
267   29 O. Lodge, *Rep. Brit. Assoc. Adv. Sci.*, 1885, *55*, 763.
      30 F. M. Raoult, *Ann. Chim. Phys.*, 1883, [5], *28*, 133, *etc.*
      31 J. H. van't Hoff, *Zeitsch. phys. Chem.*, 1887, *1*, 481.
      32 F. Kohlrausch, *Ann. Physik.*, 1879, *6*, 167.
      33 S. Arrhenius, *Zeitsch. phys. Chem.*, 1887, *1*, 631.
      34 F. G. Donnan, *J. Chem. Soc.*, 1933, 324 (Ostwald Memorial Lecture).
      35 H. W. Nernst, *Theoretical Chemistry from the Standpoint of Avogadro's Rule and Thermodynamics*, trans. C. S. Palmer, London, 1895, from the German, Stuttgart, 1893.
      36 H. E. Armstrong, *Proc. Chem. Soc.*, 1885, *1*, 40.
      37 *Idem, Rep. Brit. Assoc. Adv. Sci.*, 1885, *55*, 945; *Proc. Roy. Soc.*, 1886, *40*, 268; *Phil. Mag.*, 1888, [5], *25*, 21.
268   38 W. Crookes, *Proc. Roy. Soc.*, 1880, *30*, 469.
      39 J. J. Thomson, *Phil. Mag.*, 1897, [5], *44*, 293.
      40 The name "electron" was introduced by G. Johnstone Stoney (*Trans. Roy. Soc. Dublin*, 1891, *4*, 582.) See also C. T. Walter and G. A. Slack, *Amer. J. Phys.*, 1970, *38*, 1380.
      41 E. Rutherford, *Phil. Mag.*, 1902, [6], *4*, 315 (gives further references).
269   42 Sir H. Hartley, *J. Chem. Soc.*, 1947, 1281.
      43 Sir A. Todd, "The Development of Organic Chemistry since Perkin's Discovery", in *The Perkin Centenary, London*, London, 1958, p. 91.

CHAPTER XIV

271   1 G. N. Lewis, *Valence and the Structure of Atoms and Molecules*, New York, 1923, pp. 29–30.
      2 An account of Lewis's contribution by R. Kohler is due to appear in *Hist. Stud. Phys. Sci.*, 1971, *3*.
      3 R. Abegg and G. Bodländer, *Zeitsch. anorg. Chem.*, 1899, *20*, 453.
      4 *Ibid.*, 1904, *39*, 330.
273   5 *Ibid.*, p. 380.
      6 J. J. Thomson, *Phil. Mag.*, 1904, [6], *7*, 237.
      7 *Ibid.*, p. 262.
      8 W. Ramsay, *J. Chem. Soc.*, 1908, *93*, 774.
      9 W. L. J. P. H. Kossel, *Ann. Physik.*, 1916, [4], *49*, 229.
      10 G. N. Lewis, *J. Amer. Chem. Soc.*, 1916, *38*, 762.
274   11 E. C. C. Baly and C. H. Desch, *J. Chem. Soc.*, 1905, *87*, 784.
      12 E. Rutherford, *Phil. Mag.*, 1911, [6], *21*, 669; this was a revival of Nagaoka's hypothesis (*ibid.*, 1904, [6], *7*, 445). On Nagaoka see E. Yagi, *Jap. Stud. Hist. Sci.*, 1964, 29.
      13 H. G. J. Moseley, *Phil. Mag.*, 1913, [6], *26*, 102; 1914, [6], *27*, 703.
275   14 A. van den Broek, *Phys. Zeitsch.*, 1914, *14*, 32.
      15 J. Rydberg, e.g., *Phil. Mag.*, 1914, [6], *28*, 144; on Rydberg see St J. Nepomucene, *Chymia*, 1960, *6*, 127.
      16 *Phil. Mag.*, 1913, [6], *26*, 1, 27.
      17 *Ann. Physik*, 1901, [4], *4*, 553.

Page	
275	18 See M. Planck, "Origin and Development of the Quantum Theory", *J. Chem. Educ.*, 1963, *40*, 262 (address given 1920).
	19 See T. Hirosige and Sigeku Nisio, "Formation of Bohr's Theory of Atomic Constitution", *Jap. Stud. Hist. Sci.*, 1964, 6; also Sir G. Thomson, "Niels Bohr", *Proc. Chem. Soc.*, 1964, 351.
	20 It is curious that some of the most stubborn exceptions to the "rule of eight" had been known for a very long time: PCl_5 (H. Davy, *Phil. Trans.*, 1809, [99], 39), ICl_3 (J. L. Gay-Lussac, *Ann. Chim.*, 1814, *91*,5) and $SeCl_4$ (J. J. Berzelius, *Akad. Handl.*, 1818, *39*, 13). These and other "hypervalent" molecules perplexed chemists for a century and more. See J. L. Musher, *Angew. Chem., Internat. Edn.*, 1969, *8*, 59.

CHAPTER XV

276	1 G. N. Lewis, *Valence and the Structure of Atoms and Molecules*, New York, 1923, p. 30.
	2 J. J. Thomson, *Electricity and Matter*, London, 1904, p. 90. The book contains all six lectures, given by Thomson to inaugurate those given under the Silliman foundation.
277	3 *Ibid.*, pp. 132–4.
	4 P. Walden, *Zeitsch. phys. Chem.*, 1903, *43*, 385.
	5 J. J. Thomson, *The Corpuscular Theory of Matter*, London, 1907, pp. 130–1.
278	6 E. C. C. Baly and C. H. Desch, *J. Chem. Soc.*, 1905, *87*, 784.
	7 H. S. Fry, *Zeitsch. phys. Chem.*, 1911, *76*, 385, 398. Translations of papers read to the Cincinatti Section of the American Chemical Society, 15 Jan. 1908 and 19 Oct. 1909.
	8 A. Crum Brown and J. Gibson, *J. Chem Soc.*, 1892, *61*, 367.
279	9 K. G. Falk and J. M. Nelson, *J. Amer. Chem. Soc.*, 1910, *33*, 1641n.
	10 K. G. Falk, *ibid.*, 1911, *33*, 1140, *etc.*
	11 London, 1921. This gives a comprehensive bibliography up to 1920 (p. 272).
280	12 W. C. Bray and G. E. K. Branch, *J. Amer. Chem. Soc.*, 1913, *35*, 1440.
	13 G. N. Lewis, *ibid.*, p. 1451.
	14 J. J. Thomson, *Phil. Mag.*, 1914, [6], *27*, 747.
	15 S. J. Bates, *J. Amer. Chem. Soc.*, 1914, *36*, 789.
	16 Fry, *op. cit.* (see note 11), p. 283.
	17 See note 12.
281	18 A. A. Noyes, *J. Amer. Chem. Soc.*, 1913, *35*, 767.
	19 Thomson, *op. cit.* (see note 5), p. 121; *cf.* p. 133.
	20 W. Ramsay, *J. Chem. Soc.*, 1908, *93*, 774.
282	21 J. Stark, *Phys. Zeitsch.*, 1908, *9*, 85.
	22 H. Kaufmann, *ibid.*, p. 311.
	23 See note 14.
	24 A. L. Parson, *Smithsonian Inst. Publ.*, 1915, *65*, no. 11.
	25 Lewis, *op. cit.* (see note 1), pp. 29–30. See Chapter XIV, p. 271.
	26 W. L. J. P. H. Kossel, *Ann. Physik*, 1916, [4], *49*, 229.
	27 G. N. Lewis, *J. Amer. Chem. Soc.*, 1916, *38*, 762.
283	28 *Idem, op. cit.* (see note 1), p. 68.

Page
284 29 Idem, *J. Amer. Chem. Soc.*, 1916, *38*, 778.
     30 See note 1.
285 31 Lewis, *op. cit.* (see note 1), p. 90.
     32 *Ibid.*, p. 91.
     33 *Ibid.*
286 34 *Ibid.*, p. 87.
     35 I. Langmuir, *J. Amer. Chem. Soc.*, 1919, *41*, 868.
     36 *Elektrovalenz* had been used by R. Abegg, *Zeitsch. anorg. Chem.*, 1904, *39*, 335.
     37 I. Langmuir, *Science*, 1921, *54*, 59.
     38 N. Bohr, *The Theory of Spectra and Atomic Constitution*, Cambridge, 1922.
     39 C. R. Bury, *J. Amer. Chem. Soc.*, 1921, *43*, 1602.
     40 G. A. Perkins, *Philippine J. Sci.*, 1921, *19*, 7. Perkins symbolized this as ∞.
     41 S. Sugden, J. B. Reed and H. Wilkins, *J. Chem. Soc.*, 1925, *127*, 1528.
     42 N. V. Sidgwick, *The Electronic Theory of Valency*, Oxford, 1927, p. 60.
287 43 J. J. Berzelius, *Nic. J.*, 1813, *34*, 155.
288 44 G. N. Lewis, *J. Amer. Chem. Soc.*, 1916, *38*, 782.
     45 *Idem, op. cit.* (see note 1), p. 139.
     46 C. K. Ingold, *Ann. Rep. Chem. Soc.*, 1926, *23*, 140.
     47 C. K. Ingold and C. C. N. Vass, *J. Chem. Soc.*, 1928, 417.
     48 This phrase uses *conjugation* in its modern sense of a system of alternating double and single bonds.
     49 A. Lapworth, *Mem. Proc. Manchester Lit. Phil. Soc.*, 1920, *64*, no. 3, 16.
     50 *Ibid.*, p. 1.
289 51 *Ibid.*, p. 2.
     52 *Idem, J. Chem. Soc.*, 1903, *83*, 995.
     53 *Idem, Mem. Proc. Manchester Lit. Phil. Soc.*, 1920, *64*, no. 3, 3.
     54 *Ibid.*, p. 5.
290 55 R. Robinson *et al.*, *J. Chem. Soc.*, 1926, 404n.
     56 W. O. Kermack and R. Robinson, *ibid.*, 1922, *121*, 427.
     57 T. M. Lowry, *ibid.*, 1923, *123*, 822.
     58 C. K. and E. H. Ingold, *ibid.*, 1926, 1310.
     59 R. Robinson *et al.*, *J. Chem. Soc.*, 1926, 401.
     60 C. K. Ingold, *ibid.*, 1933, 1120.
     61 R. Robinson, *An Outline of an Electrochemical (Electronic) Theory of the Course of Organic Reactions*, London, 1932.
     62 J. W. Armit and R. Robinson, *J. Chem. Soc.*, 1925, *127*, 1604.
     63 Robinson, *op. cit.* (see note 61), p. 10.
     64 *E.g.*, Ingold, *op. cit.* (see note 60), p. 1127.
     65 A modern view on "The Stability of the Aromatic Sextet" is given by R. Waack, *J. Chem. Educ.*, 1962, *39*, 469.

CHAPTER XVI

292 1 On this subject generally, see B. Hoffmann, *The Strange Story of the Quantum*, London, 1963; W. Wilson, "The Origin and Nature of Wave Mechanics", *Sci. Prog.*, 1937, *32*, 209; F. Hund, "Paths to Quantum Theory Historically Viewed", *Phys. Today*, 1966, *19*, no. 8,

23; J. Gerber, "Geschichte der Wellenmechanik" *Arch. Hist. Ex. Sci.*, 1969, *5*, 350; A. Hermann, "From Planck to Bohr; the first 15 years in the development of the quantum theory", *Angew. Chem., Internat. Edn.*, 1970, *9*, 34.
2 A. Einstein, *Ann. Physik*, 1905, [4], *17*, 132.
3 A. H. Compton, *Phys. Rev.*, 1923, *21*, 483, 715; *22*, 409.
4 L. de Broglie, *Phil. Mag.*, 1924, [6], *47*, 446. See also M.-A. Tonnelet, *L. de Broglie et la Méchanique Ondulatoire*, Paris, 1966.
5 L. de Broglie, *Wave Mechanics*, trans. H. T. Flint, London, 1930, p. 3.
6 C. J. Davisson and L. H. Germer, *Nature*, 1927, *119*, 558.
7 G. P. Thomson, *ibid.*, 1927, *122*, 279.
8 P. H. E. Rupp, *Zeitsch. Phys.*, 1928, *52*, 8.
9 A. Eddington, *The Nature of the Physical World*, Everyman ed., London, 1947, p. 204; the text of the Gifford Lectures was prepared in 1926 and delivered at Edinburgh in 1927.
10 See also H. Cluny, *Werner Heisenberg et la Méchanique Quantique*, Paris, 1966.
11 W. Wilson, *Sci. Prog.*, 1937, *32*, 227.
12 W. Heisenberg, *Zeitsch. Phys.*, 1925, *33*, 879.
13 M. Born and E. P. Jordan, *ibid.*, 1925, *34*, 858.
14 P. A. M. Dirac, *Proc. Roy. Soc.*, 1925, *109*, 642; 1926, *110*, 561.
15 See note 9.
16 E. Schrödinger, *Ann. Physik*, 1926, [4], *79*, 360.
17 *Collected Papers on Wave Mechanics*, trans. W. F. Shearer and W. M. Deans, London, 1928.
18 "My theory was inspired by L. de Broglie . . . and by brief, yet infinitely far-seeing remarks of A. Einstein, *Berl. Ber.*, 1925, p. 9 *et seq*. I did not at all suspect any relation to Heisenberg's theory at the beginning. I naturally knew about his theory, but was discouraged, if not repelled, by what appeared to me as very difficult methods of transcendental algebra, and by the want of perspicuity", *ibid.*, p. 46n.
19 See note 16.
20 W. Heisenberg, *Zeitsch. Phys.*, 1927, *43*, 172.
21 M. Born, *ibid.*, 1926, *38*, 803.
22 A. S. Coolidge and H. M. James, *J. Chem. Phys.*, 1933, *1*, 825.
23 E. Schrödinger, *Ann. Physik.*, 1926, [4], *79*, 734.
24 D. P. Craig, *J. Roy. Inst. Chem.*, 1959, *83*, 141.
25 W. Pauli, *Zeitsch. Phys.*, 1925, *31*, 765.
26 F. G. Arndt, E. Scholz and P. Nachtwey, *Ber.*, 1924, *57*, 1003.
27 J. N. Collie, *J. Chem. Soc.*, 1904, *85*, 971.
28 F. P. Pfeffer, *Ber.*, 1922, *55B*, 1762.
29 F. G. Arndt, *ibid.*, 1924, *57*, 1906.
30 *Idem*, private communication to E. Campaigne, *J. Chem. Educ.*, 1959, *36*, 337.
31 *Idem, Ber.*, 1930, *63*, 2963.
32 *Idem, J. Chem. Educ.*, 1959, *36*, 338.
33 H. Wren, *Brit. Chem. Abs.*, 1931, *A*, 234.
34 E. Campaigne, *J. Chem. Educ.*, 1959, *36*, 336.
35 See note 32.
36 F. G. Arndt and B. Eistert, *Zeitsch. phys. Chem.*, 1935, *31*, 125; these names are now associated in the well-known Arndt–Eistert reaction of diazo-ketones.

Page	
299	37 *Cf.* the observation by Todd quoted on p. 269.
	38 L. E. Sutton, *Proc. Chem. Soc.*, 1958, 312.
	39 *Ibid.*, p. 314.
	40 S. Sugden, *J. Chem. Soc.*, 1924, *125*, 32.
	41 See note 26.
	42 Quoted by F. Challenger, *J. Roy. Inst. Chem.*, 1953, *77*, 166.
300	43 See note 42.
	44 C. K. and E. H. Ingold, *J. Chem. Soc.*, 1926, 1312. They also wrote of molecules possessing "a betaine-like phase" in a reaction, as

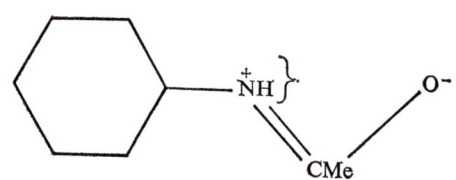

adding in a footnote that they might be "in a condition corresponding with partial conversion into such a phase" (p. 1311).

	45 C. K. Ingold, *Ann. Rep. Chem. Soc.*, 1926, *23*, 149.
301	46 (a) K. Höjendahl, *Studies of Dipole Moments*, Copenhagen, 1928; (b) L. E. Sutton, *Proc. Roy. Soc.*, 1931, *A*, *133*, 668, and on.
	47 See note 39.
	48 C. K. Ingold and F. Shaw, *J. Chem. Soc.*, 1927, 2918.
	49 C. K. Ingold, *ibid.*, 1933, 1120.
	50 *Idem*, *Nature*, 1934, *133*, 946 (letter dated 24 May).
	51 *Idem*, *Chem. Rev.*, 1934, *15*, 225.
302	52 L. E. Sutton, *Proc. Chem. Soc.*, 1958, 315.
	53 (a) N. V. Sidgwick, L. E. Sutton and W. Thomas, *J. Chem. Soc.*, 1933, 406; (b) N. V. Sidgwick, *Trans. Farad. Soc.*, 1934, *30*, 801; (c) *idem*, *J. Chem. Soc.*, 1936, 533; 1937, 694.
	54 N. V. Sidgwick, *Ann. Rep. Chem. Soc.*, 1934, *31*, 37.
	55 *Ibid.*, p. 40.
	56 *Ibid.*
	57 See note 52.
	58 See note 53c.
304	59 W. Heitler and F. W. London, *Zeitsch. Phys.*, 1927, *44*, 455. A useful analysis of the contributions of Heitler and London is given by W. G. Palmer, *A History of the Concept of Valency to 1930*, Cambridge, 1965, pp. 155–66.
	60 A. D. McLean, A. Weiss and M. Yoshimine, *Rev. Mod. Phys.*, 1960, *32*, 211.
305	61 L. Pauling, *J. Amer. Chem. Soc.*, 1931, *53*, 1367.
	62 *Idem, ibid.*, p. 3225.
	63 *Idem, ibid.*, 1932, *54*, 988.
	64 *Idem, J. Chem. Phys.*, 1933, *1*, 362 (with G. W. Wheland).
	65 *Idem, Proc. Acad. Nat. Sci.*, 1933, *19*, 860 (with L. O. Brockway).
	66 *Ibid.*, 1932, *18*, 293, 498.
	67 *Ibid.*, p. 414 (with D. M. Yost).
306	68 H. C. Longuet-Higgins, *Proc. Chem. Soc.*, 1957, 158.
	69 C. N. Hinshelwood, *Ann. Rep. Chem. Soc.*, 1932, *29*, 20.

NOTES AND REFERENCES

70 L. Pauling and G. W. Wheland, *J. Chem. Phys.*, 1933, *1*, 606.
71 G. W. Wheland, *op. cit.*, New York, 1944.
72 *Idem, op. cit.*, New York, 1955.
73 L. Pauling, *The Nature of the Chemical Bond and the Structure of Molecules and Crystals*, New York, 1st ed., 1939; 2nd ed., 1940.
74 C. N. Hinshelwood, *Ann. Rep. Chem. Soc.*, 1932, *29*, 17–18.
75 N. V. Sidgwick, *ibid.*, 1934, *31*, 37.
76 W. G. Penney and G. J. Kynch, *ibid.*, 1936, *33*, 44.
77 M. J. S. Dewar, *Electronic Theory of Organic Chemistry*, Oxford, 1949.
78 *Ibid.*, pp. ix–x.
79 G. V. Chelintsev, *Ocherki po Teorii Organicheskoi Khimii*, Moscow, 1949.
80 V. M. Tatevskii and M. I. Shakhparanov, *Voprosy Filosofii*, 1949, no. 3, 176.
81 See L. R. Graham, "A Soviet Marxist View of Structural Chemistry: The Theory of Resonance Controversy", *Isis*, 1964, *55*, 20.
82 *Ibid.*, p. 27.
83 Graham, *op. cit.*
84 J. M. Hunt, M. P. Wisherd and L. C. Bonham, *Anal. Chem.*, 1950, *22*, 1478.
85 F. A. Miller and C. H. Wilkins, *ibid.*, 1952, *24*, 1253.
86 A. D. Liehr, *J. Chem. Educ.*, 1962, *39*, 135.
87 L. Pauling, *ibid.*, p. 461.
88 On Pauli and Hund see W. G. Palmer, *op. cit.* (see note 59), pp. 143–5 and 166–71 respectively. An account of Mulliken's work in quantum chemistry by C. A. Coulson, and of his life, by J. C. Slater, appears in *Molecular Orbitals in Chemistry, Physics and Biology, A Tribute to R. S. Mulliken*, ed. P.-O. Lowdin and B. Pullman, New York, 1964.
89 Letter to C. A. Coulson, *ibid.*, p. 1.
90 L. Pauling, *Phys. Rev.*, 1928, *32*, 186, 761.
91 See note 88.
92 F. Bloch, *Zeitsch. Phys.*, 1928, *52*, 555.
93 L. Pauling, *Chem. Rev.*, 1928, *5*, 173. For a review of recent progress in L.C.A.O. calculations for small molecules see R. G. Clark and E. T. Stewart, *Qu. Rev.*, 1970, *24*, 95.
94 J. E. Lennard-Jones, *Trans. Farad. Soc.*, 1929, *25*, 668.
95 E. Hückel, *Zeitsch. Phys.*, 1931, *70*, 204; 1932, *76*, 628.
96 H. A. Bethe, *Ann. Physik*, 1929, [5], *3*, 133.
97 Lowdin and Pullman (eds.), *op. cit.* (see note 88), p. 9.
98 E. Hückel, *Zeitsch. Phys.*, 1930, *60*, 423.
99 L. Pauling and G. W. Wheland, *J. Amer. Chem. Soc.*, 1935, *57*, 2086.
100 R. Mulliken, *J. Chem. Phys.*, 1939, *7*, 339.
101 J. H. van Vleck, *ibid.*, 1935, *3*, 807; see also P. W. Anderson, "Van Vleck and magnetism", *Phys. Today*, 1968, *21*, no. 10, 23.
102 See, *e.g.*, J. S. Griffith and L. E. Orgel, *Qu. Rev.*, 1957, *11*, 381, and F. A. Cotton, *J. Chem. Educ.*, 1964, *41*, 466.
103 M. J. S. Dewar and H. C. Longuet-Higgins, *Proc. Roy. Soc.*, 1952, *A, 214*, 482; N. S. Ham, *J. Chem. Phys.*, 1958, *29*, 1229.
104 E. Heilbronner, in Lowdin and Pullman (eds.), *op. cit.*, (see note 88), p. 356. *Cf.* W. T. Simpson, *ibid.*, p. 385.
105 C. A. Coulson, *Valence*, Oxford, 2nd ed., 1961, p. 71.

CHAPTER XVII

1 E. F. Caldin, *Proc. Chem. Soc.*, 1959, 271.
2 E. Frankland, *Phil. Trans.*, 1852, [*142*], 439.
3 M. B. Hesse, *Models and Analogies in Science*, London, 1963, pp. 97–8.
4 He says, *e.g.*, "nor is it necessary to insist upon the accuracy of all these compounds, both in number and weight", *New System of Chemical Philosophy*, Manchester, 1808, p. 220.
5 J. J. Berzelius, *Lehrbuch der Chemie*, Dresden, 5th ed., 1843, vol. i, p. 10.
6 See note 3.
7 G. Buchdahl, *Brit. J. Phil. Sci.*, 1959, *10*, 130.
8 A. G. N. Flew, *New Biol.*, 1959, *28*, 37–8.
9 J. B. Conant, *Science and Common Sense*, London, 1951, p. 292.
10 See, *e.g.*, G. M. Fleck, "Atomism in late Nineteenth Century Physical Chemistry", *J. Hist. Ideas*, 1963, *24*, 106; W. H. Brock and D. M. Knight, "The Atomic Debates", *Isis*, 1965, *56*, 5, and ch. I of the book of the same title (ed. W. H. Brock), Leicester, 1967.
11 W. Ostwald, *Grundlinien der anorganischen Chemie*, Leipzig, 1900.
12 C. A. Coulson, *J. Chem. Soc.*, 1955, 2084.
13 W. F. Cannon, *Hist. Sci.*, 1964, *3*, 27.
14 *Report of the Consultative Committee on Secondary Education with Special Reference to Grammar Schools and Technical High Schools* (Spens Report), London, H.M.S.O., 1938, pp. 51–2.
15 *The New Cambridge Modern History*, ed. by F. H. Huisby, Cambridge, 1962, vol. xi, pp. 59–60.
16 W. F. Cannon, *Hist. Sci.*, 1964, *3*, 34.
17 E. Mendelsohn, *ibid.*, p. 45.
18 H. E. Armstrong, *Nature*, 1930, *127*, 807.
19 *Cf.* L. Pearce Williams, *Hist. Sci.*, 1962, *1*, 12.
20 P. Cook, *Science News*, 1959, *53*, 35.
21 C. Gerhardt, *Ann. Chim. Phys.*, 1845, [3], *14*, 107.
22 F. J. Kilgour, *Actes of the Xth International Congress of the History of Science*, Ithaca, 1962, p. 329.
23 J. Jacques, *Rev. d'Hist. Sci.*, 1950, *3*, 32.
24 On vitalism see particularly J. H. Brooke, "Wöhler's Urea, and its Vital Force?—A Verdict from the Chemists", *Ambix*, 1968, *15*, 84.

Indexes

INDEX OF SECONDARY SOURCES

Anderson, P. W., 363
Bailey, D. and K. C., 87, 336
Baker, A. A., 353; Baker, W., 355
Bekassova, L. M., 353
Benfey, O. T., 338, 353, 356
Bent, H. A., 351
Bentley, J., 328
Bigelow, H. E., 348
Braun, J., 329
Brock, W., 328, 330, 343, 364
Brooke, J. H., 364
Brown, E. M. L., 345
Brown, H. C., 338, 342
Buchdahl, G., 315, 364
Butterfield, H., 327
Bykov, G. V., 148, 327, 344, 353
Caldin, E. F., 313, 364
Campaigne, E., 356, 361
Cannon, W. F., 319, 329, 364
Cardwell, D. S. L., 328
Carrière, J., 328
Cava, M. P., 356
Challenger, F., 362
Cluny, H., 361
Cohen, I. B., 328
Conant, J. B., 364
Cook, P., 323, 333, 364
Coulson, C. A., 311, 363
Craig, D. P., 296, 361
Crosland, M. P., 330
Dobbin, L., 355
Efros, L. S., 357
Eyre, H. V., 351
Farrar, K. R., 70, 334, 344
Farrar, W. V., 70, 328, 334, 344, 346
Fieser, L., 356
Findlay, A., 87, 336
Finley, K. T., 353
Fleck, G. M., 364
Flew, A. G. N., 316, 364
Freund, I., 190, 349
Fullmer, J. Z., 327
Gerber, J., 361
Gillis, J., 354
Gilman, H., 356
Graham, L. R., 308, 363
Grant, A. J., 328
Graves, N. J., 330
Green, J. H. S., 338
Gregory, J. C., 7, 327, 336
Haber, L., 327

Hammond, J. L. and B., 339
Hartley, H., 328, 358
Heaney, H., 355
Heilbronner, E., 363
Hennemann, G., 328, 329
Hermann, A., 361
Hesse, M. B., 314, 364
Hiebert, E. W., 57, 63–4, 334
Hirosige, T., 359
Hoffmann, B., 360
Hoijtink, G. J., 344
Huisby, F. H., 364
Huisgen, R., 354
Hund, F., 360
Ihde, A., 332, 341, 345, 346
Jacques, J., 364
Jones, H. M., 328
Kauffman, G. B., 346, 351
Kazansky, B. A., 327, 344
Kendall, J., 327, 339
Ketelaar, J. A. A., 263, 357
Kilgour, F. J., 364
Kingzett, C. T., 83, 335
Knight, D. M., 327, 328, 329, 357, 364
Koenigsberger, L., 358
Kohler, R., 358
Kolodkin, P., 343
Lachmann, A., 345, 346
Larder, D. F., 335, 338
Leicester, H. M., 150, 344
Levere, T. H., 229, 333
Linstead, P., 11, 328
Lowdin, P.-O., 363
Mellanby, J., 337
Mendelsohn, E., 321, 328, 364
Meyer, E. von, 64, 102, 128, 130, 339, 342
Mitchell, M. J., 356
Musher, J. L., 359
Nepomucene, St J., 358
Nisio, S., 359
Palmer, W. G., 23, 262, 330, 357, 362, 363
Partington, J. R., 329, 331, 338, 350
Porter, G., 337
Prandtl, W., 329
Pullman, B., 363
"R. S.", 354
Read, J., 354
Richter, F., 344
Russell, C. A., 328, 330, 331, 337, 339, 342, 354, 357

INDEX OF NAMES

Russell, S. P., 331, 337
Scott, A., 331, 339
Sementsov, A., 166, 346
Sharrock, R., 329
Sidgwick, N. V., 23, 330
Simpson, W. T., 363
Slack, G. A., 358
Slater, J. C., 363
Smeaton, W. A., 330, 343
Söderbaum, H. G., 327
Spronsen, J. W. van, 343
Stubbs, W., 336
Sutherland, I. O., 355
Tamelen, E. E. van, 354, 355
Temperley, H., 328
Thayer, J. S., 331
Thompson, D., 331
Tilden, W. A., 101, 338

Todd, A., 358, 362
Tonnelet, M.-A., 361
Treneer, A., 327, 329
Verkade, P. E., 333, 338
Vernon, K. D. C., 327
Walker, J., 341
Wallach, O., 329
Walter, C. T., 358
Wan, J. S., 330
Webb, K. R., 333
Wheeler, T. S., 100, 338
Wightman, W. P. D., 343
Williams, L. P., 329, 364
Wilson, C., 320, 364
Wilson, W., 294, 361
Woodward, L., 8, 327
Yagi, E., 358

INDEX OF NAMES

Abegg, R., 271, 358, 360
Adair, A., 185, 348
Ampère, A. M., 24, 330
Anderson, P. W., 363
Angeli, A., 348
Anschütz, R., 72, 78–80, 98, 102, 108, 133, 162, 174, 195, 227, 238, 333, 335–6, 338, 339, 340, 341, 342, 346, 347, 349, 350, 352, 353, 355
Armit, A. W., 360
Armstrong, H. E., 106, 108, 112, 128, 129, 193, 195, 205–14, 216, 241, 243, 247, 249, 252–3, 254, 256, 267, 269, 321, 339, 340, 342, 349, 350, 351, 354, 355, 356, 358, 364
Arndt, F. G., 297–9, 356, 361
Arrhenius, S., 205, 212, 267, 274, 358
Askani, R., 355
Atherton, F. R., 72–3, 335
Auwers, K., 167, 346
Averbeck, H., 345
Avogadro, A., 9, 23, 37, 38, 263, 268, 315, 328, 358

Baeyer, A. von, 54, 71, 110, 111, 116, 131, 157, 161, 164, 235, 236, 237–9, 241, 245–6, 251, 253–4, 257, 317, 333, 340, 341, 342, 345, 346, 354, 356
Baker, H. B., 206, 208, 350
Baly, E. C. C., 274, 277–8, 358, 359
Bamberger, E., 254, 356
Barbaglia, G. A., 183, 184, 348
Bates, S. J., 280, 359
Baumé, A., 17
Beckmann, E. O., 163, 346
Beethoven, L. von, 328, 329
Beilstein, F. K., 353
Bergman, T., 18
Berthelot, M., 145, 146, 330, 344
Berthollet, C. L., 14, 17–18, 46, 118, 330, 332

Berzelius, J. J., 3, 4, 12, 14, 15, 16, 18, 22–31, 44–5, 47, 58, 83, 128, 137, 198, 261–3, 265, 267, 268, 269, 270, 272, 276, 277, 281, 287, 301, 311, 315, 323, 327, 328, 329, 330, 332, 342, 357, 359, 360, 364
Bethe, H. A., 311, 363
Biltz, W., 347
Bineau, A., 191, 349
Black, J., 9
Bloch, F., 310–11, 363
Blomstrand, C. W., 100, 182, 184, 186–7, 196, 198, 214, 265, 268, 338, 348, 350, 357
Bodländer, G., 271, 358
Bohr, N., 275, 286, 292–5, 360, 361
Bonham, L. C., 363
Born, M., 293–5, 361
Boscovitch, R. J., 7
Bottone, S. R., 336
Boullay, P., 23, 24, 330
Boyle, R., 116
Branch, G. E. K., 280, 359
Bray, W. C., 280, 359
Brockway, L. O., 362
Brodie, B. C., 39, 84, 145, 332, 336, 344
Broek, A. van den, 275, 358
Brough, J. C., 106, 339
Brown, A. Crum, 9, 80, 99–107, 127, 130, 133, 143, 151–2, 154, 157, 165, 228–9, 233–4, 248, 279, 328, 335, 344, 345, 346, 352, 353, 359, Plate 4
Buff, H. L., 172, 224, 264, 341, 347, 352, 357
Buffon, G. L. L., 17
Bunsen, R. W., 25, 28, 31, 34–6, 38, 57, 111, 130, 238, 331
Bury, C. R., 286, 360
Butlerov, A. M., 3, 34, 80, 133, 144, 146–55, 157, 187, 197, 228, 230, 232, 234, 327, 335, 342, 343, 344, 345, 348, 350, 352, 353

INDEX OF NAMES

Cadet, L. C., 25, 331
Cahours, A. A. T., 24, 191, 331, 349
Cannizzaro, S., 127, 129, 138, 187, 191, 268, 348, 349
Carlisle, A., 327
Chelintsev, G. V., 308, 363
Chevreul, M. E., 14, 19
Clark, R. G., 363
Claus, A., 166, 202–3, 207, 214, 216, 236, 241, 249–52, 254, 346, 350, 353, 356
Clausius, R. J. E., 317
Cleve, P. T., 214, 351
Cockeram, H., 336
Cohen, J. B., 299
Coleridge, S. T., 13, 329
Collie, J. N., 199, 256, 297, 350, 356, 361
Collier, W. J., 73, 335
Compton, A. H., 292, 361
Comte, A., 19, 48, 49, 51, 76, 330
Cooke, J. P., 85, 105, 265, 336, 339, 357
Coolidge, A. S., 296, 361
Cotton, F. A., 363
Coulson, C. A., 312, 318, 337, 363, 364
Couper, A. S., 62, 71–80, 83, 97, 100–2, 109, 111, 113, 120, 124–5, 133, 142, 146–52, 197–9, 263, 335, 341, 350, Plate 3
Criegee, R., 355
Crookes, W., 268, 269, 358

Dalton, J., 6, 9, 12, 13, 117, 174–5, 262, 264, 313–16, 323, 344, 350
Darlow, M. G., 355
Darwin, C., 316, 317
Davies, G, E., 88
Davisson, C. J., 361
Davy, E., 12
Davy, H., 5–17, 22, 46, 145, 261–2, 265, 268–70, 277, 315, 327, 330, 332, 359
Davy, J., 7
Dawson, H. M., 299–300
De Broglie, L., 293–6, 361
Debye, P., 301
De Chancourtois, A. E. B., 137, 343
Democritos, 323
De Morveau, Guyton, 17, 22, 330
Demuth, R., 346
Desch, C. H., 274, 277–8, 358, 359
Deville, H. E. St C., 178, 191, 347, 349
Dewar, J., 107, 162, 246, 247–50, 252, 339, 354–5, Plate 8
Dewar, M. J. S., 241, 307, 309, 354, 363, 363
Dingwall, J. G., 355
Dirac, P. A. M., 293–4, 361
Divers, E., 54, 118
Döbereiner, J. W., 137, 343
Donnan, F. G., 358
Dulong, P. L., 22, 46, 330, 333
Dumas, J. B. A., 19, 23, 24, 25, 46–8, 52, 63, 73, 137, 191, 315, 330, 331, 333, 343, 349
Duppa, B. F., 120, 157, 200, 345, 350, 353

Eddington, A., 293–4, 361
Einstein, A., 292, 361

Eistert, B., 298, 361
Emmerling, A., 246, 355
Erdmann, O. L. E., 173
Erlenmeyer, R. A. C. E., 71, 84, 99, 147, 154, 157, 173–4, 197, 232, 246, 335, 336, 344, 345, 347, 350, 353, 355

Falk, K. G., 279, 359
Faraday, M., 9, 12, 205, 273–8, 329
Fehling, H., von, 87, 335, 336
Fisher, E. G., 14
Fittig, R., 230, 243, 246, 352, 355
Fitzpatrick, J. D., 356
Flürscheim, B., 222, 288, 352
Foster, G. C., 172–4, 176, 342
Fourcroy, A. F., 17
Frankland, E., 9, 10, 15, 27–43, 48, 50, 52–3, 62, 65, 79, 83–5, 89, 90–1, 93, 94, 95, 104–6, 109–17, 119–25, 127–33, 138–9, 152–3, 155, 157, 165, 175–6, 188–9, 197–8, 200, 202, 224, 229, 255, 261–2, 264–6, 268, 313–14, 320–1, 332, 333, 334, 335, 336, 337, 338, 339, 340, 341, 342, 343, 344, 345, 346, 348, 350, 352, 353, 357, 364, Plates 1 and 12
Freund, A., 353
Friedel, C., 198, 230, 347, 350, 352
Friend, J. N., 337
Fry, H. S., 276, 278–82, 287, 289–90, 359

Gay-Lussac, J. L., 14, 19, 22, 27, 330, 359
Geoffroy, E. F., 17
Gerhardt, C., 15, 19–20, 28, 33, 38–9, 46–5, 53, 55–7, 63–4, 66, 70–1, 79, 95–6, 119, 121–2, 126–7, 129–31, 144–5, 315, 323, 330, 331, 333, 338, 341, 364
Germer, L. H., 361
Geuther, J. A., 87, 156, 195, 337, 345, 349
Gibbs, J. W., 317
Gibson, J., 279, 359
Glaser, C., 246, 355
Gleichmann, L., 347
Gmelin, L., 47, 160, 345
Goethe, J. W., 13, 328
Goldschmidt, H., 163, 346
Gomberg, M., 332
Graebe, C., 186, 246, 248, 348, 355
Graham, T., 46, 333
Griess, J. P., 173, 347
Griffith, J. S., 363
Grignard, F. A. V., 38
Guldberg, C. M., 18, 330

Ham, N. S., 363
Hambly, F. J., 193, 349
Hamilton, W., 71, 76
Hantzsch, A. R., 157, 163, 288, 345, 346
Harcourt, A. G. V., 10, 88–9, 200, 328, 337
Harnitz-Harnitzky, T., 229, 352
Haszeldine, R. N., 355
Heisenberg, W., 293–5, 361
Heitler, W., 304, 362
Hellmann, H., 355
Helmholtz H. von, 205–6, 208, 265–9, 277, 350, 358

INDEX OF NAMES

Herb, J., 354, 356
Hermann, F., 157, 345
Herschel, W., 322
Heslop, R. B., 337
Heyes, J. F., 84, 199, 212, 336, 350
Higgins, W., 100–1, 338, 350
Hinrichs, J. C., 121
Hinrichsen, F. W., 236–7, 353
Hinshelwood, C. N., 306–7, 363
Hoff, J. H. van 't, 159, 161–2, 164–6, 227, 250, 267, 317, 345, 346, 358, Plate 10
Hofmann, A. W., 4, 10, 15, 38, 47, 50, 82–3, 85–6, 94, 104, 155–6, 186, 226, 229, 328, 332, 333, 335, 336, 338, 339, 345, 348, 352, Plate 7
Höjendal, K., 362
Holzmann, M., 98
Horstmann, A. F., 192, 349
Hübner, H., 244, 354
Hückel, E., 298, 311–12, 363
Humboldt, A. von, 16, 329
Hund, F., 309, 363
Hunt, J. M., 363

Ingold, C. K., 249, 288, 298–302, 306, 308, 355, 360, 362
Ingold, E. H., 290, 300, 360, 362

James, H. M., 296, 361
Japp, F. R., 45, 57, 64, 89, 109, 112, 117, 131, 134, 197, 250, 252, 332, 334, 337, 340, 341, 342, 347, 350, 356
Jordan, E. P., 293–4, 361
Jörgensen, S. M., 220, 351

Kant, I., 13, 71, 76, 329
Kaplan, L., 355
Kaufmann, H., 257, 282, 357, 359
Kay, G., 53
Kay-Shuttleworth, J., 106
Kekulé, F. A., 4, 15, 24, 27, 30–1, 49–50, 54–72, 78–82, 84, 86–7, 90, 93–105, 107–34, 138, 142–50, 161–2, 164, 166, 172–8, 180, 182–4, 190–3, 196–7, 200, 219, 225–32, 242–6, 248–51, 254–7, 263–4, 314, 321, 330, 333, 334, 335, 336, 337, 338, 339, 340, 341, 342, 343, 344, 345, 346, 347, 348, 349, 350, 351, 353, 354, 355, 356, Plates 2 and 9
Kermack, W. O., 290, 360
Kharasch, M. S., 332
Klinger, H., 203–4, 241, 350, 354
Knoevenagel, H. E. A., 167, 346
Knorr, L., 167, 345, 346
Köhlrausch, F., 267, 358
Kolbe, H., 27–34, 39, 41, 44, 47, 56, 64, 72, 91, 94–6, 100, 109, 111, 113–14, 117, 119–23, 125, 127–8, 130–2, 145–6, 165–6, 175–6, 197, 233, 250, 261, 263, 321–2, 331, 332, 333, 334, 337, 338, 341, 344, 346, 350, 353, 357
Kommrath, H., 201–2, 350
Kopp, H., 71, 111, 144, 181, 191, 340, 347, 349
Körner, W., 244, 246, 248, 355

Kossel, W. L. J. P. H., 273–5, 282, 358, 359
Krestovnikov, A., 353
Krivaskin, V. Y., 352
Krüger, F., 194, 349
Kynch, G. J., 307, 363

Laar, P. C., 157, 345
La Coste, W., 181, 211, 347
Ladenburg, A., 102, 111, 123, 128, 132–3, 149, 230, 244–5, 253, 255–6, 339, 340, 341, 342, 344, 346, 352, 356
Langmuir, I., 285–6, 296, 360
Lapworth, A., 222, 288–90, 299, 352, 354, 360
Laurent, A., 19–20, 38, 46–9, 51, 66, 82, 330, 332, 333, 335
Lavoisier, A. L., 16. 22, 330
Le Bel, J. A., 159, 161, 166–7, 345, 346
Lecco, M. T., 194–5, 211, 349
Lennard-Jones, J. E., 306, 310–11, 363
Lewis, G. N., 141, 223, 270–7, 280, 282–8, 296, 304, 352, 354, 359, 360
Lewkowitsch, J., 351
Liebermann, C. T., 186, 246, 248, 348, 355
Liebig, J. von, 8, 14–15, 23–5, 27–8, 31–2, 38, 44, 46–7, 49–50, 71, 79, 128, 238, 320, 323, 330, 331, 332
Liehr, A. D., 363
Limpricht, H., 58, 61, 66, 71, 330, 334
Lindet, L., 351
Lodge, O., 266, 358
London, F. W., 304, 348, 362
Longuet-Higgins, H. C., 89, 305–6, 337, 356, 362, 363
Lonsdale, K., 354
Loschmidt, J., 97–100, 231–2, 242, 264, 338, 353, 357
Lossen, W., 166, 195, 202–3, 206, 214, 216, 236–7, 241, 346, 348, 349, 350, 353
Lowry, T. M., 290, 360

Mach, E., 294
McLean, A. D., 362
Macquer, P. J., 17
Mallet, J. W., 192–3, 349
Marignac, J. C. G., 348
Markovnikov, V. V., 353
Marsh, J. E., 341
Maxwell, J. Clerk, 263, 317, 357
Melsens, L. H. F., 47, 332, 333
Mendel, J. G., 323
Mendeléef, D. I., 139–40, 189, 198, 271, 343, 349
Mendius, O., 58, 334
Mercer, J., 350
Meyer, J., 347
Meyer, J. Lothar, 139–41, 150, 252–3, 272, 343, 344, 356
Meyer, Victor, 164, 194–5, 211, 247, 346, 349, 353, 355
Michael, A., 185, 348
Michaelis, C. A. A., 172, 181–2, 184, 195, 211, 264, 347, 349, 357
Miller, F. A., 363

INDEX OF NAMES

Miller, G. W., 331
Miller, W. A., 106, 129, 138–9, 339, 342, 343
Mills, E. J., 225, 352
Mills, W. H., 167, 347
Mohr, C. F., 331
Mohr, E. W. M., 239, 353
Moissan, H., 192
Moseley, H. G. J., 358
Müller, H., 70, 183, 348
Mulliken, R. S., 309–11, 363

Nachtwey, P., 361
Nagaoka, H., 358
Napoleon I, 6, 12, 17
Naquet, A., 100, 146, 172, 174, 190, 193, 197, 338, 347, 349
Nef, J. U., 199, 350
Nelson, J. M., 279, 359
Nernst, H. W., 267, 358
Newlands, J. A. R., 137, 343
Nicholson, W., 327
Nilson, L, F., 347
Noyes, A. A., 281, 359

Odling, W., 49–50, 54–5, 61, 63, 66, 82, 84, 92–3, 96, 111, 118–19, 121, 124–7, 129, 131, 137–9, 145, 178, 187, 314, 333, 335, 336, 337, 338, 341, 343, 347, 348, Plate 6
Oersted, H. C., 13, 328
Orgel, L. E., 363
Ostermeyer, E., 246, 355
Ostwald, W., 267, 317–18, 329, 357, 364

Paneth, F. A., 332
Pappas, S. S., 249, 355
Parson, A. L., 282, 359
Pasteur, L., 160–2, 323, 345
Pauli, W., 296, 309, 361, 363
Pauling, L., 252, 298, 302, 304–11, 356, 363
Peachey, S. J., 163, 346
Pebal, L. von, 191, 349
Péligot, E. M., 24, 331
Penny, W. G., 307, 363
Perkin, W. H., 239, 299, 352
Perkins, G. A., 286, 360
Pettersson, O., 347
Pettit, R., 257, 356
Pfeffer, F. P., 361
Pickering, S. U., 204–6, 241, 350, 354
Pike, W. M., 106, 339
Pinkus, A. G., 72, 335
Planck, M., 275, 292–4, 323, 359, 361
Playfair, L., 11, 15, 27, 34, 80, 144, 191, 248, 343, 347, 349, 350
Poggendorf, J. C., 25
Pope, W. J., 163, 346
Prince Consort, 11
Proust, L. J., 18, 178, 330, 347

Ramsay, W., 273, 281, 358, 359
Raoult, F. M., 267, 358
Rausch, D. J., 355
Reed, J. B., 360

Regnault, H. V., 46, 333
Richter, J. B., 13–14
Richter, V. von, 227, 352
Riedel, C., 248–9, 355
Ritter, J. W., 16, 329
Robinson, J., 191, 349
Robinson, R., 288, 290, 299–300, 355, 360
Rochleder, F., 70, 334
Roscoe, H. E., 27, 85–6, 139, 155, 178, 331, 343, 347, 348
Rose, F. L., 331
Rothe, O. H., 345
Rumford, Count, 5
Rupe, H., 354, 356
Rupp, P. H. E., 360
Rutherford, E., 269, 274–5, 292, 358
Rydberg, J., 275, 358

Sachse, H., 239, 353
Schäfer, W., 355
Schelling, F. W. J., 13, 329
Scheurer-Kestner, A., 178–9, 347
Schischkoff, L., 63, 334
Schorlemmer, C., 39, 82, 155, 172, 335, 345, 347, 353
Scholz, E., 361
Schreib, H., 335
Schrödinger, E., 293–6, 302, 304, 361
Schwann, T., 323
Seebeck, T. J., 13, 328
Sestini, F., 177, 347
Shakhparanov, M. I., 308, 363
Shaw, F., 301, 362
Sidgwick, N. V., 23, 167, 286, 299, 301–2, 306–7, 330, 346, 360, 363
Sodeau, W. H., 347
Sprague, J. T., 88, 337
Spring, W. V., 183, 348
Städeler, G., 248, 355
Stallo, J. B., 85, 336
Stark, J., 282, 359
Stenhouse, J., 49–50, 54–5
Stewart, A. W., 223, 352
Stewart, E. T., 363
Stokes, G. G., 40, 109
Stoney, G. J., 358
Sudborough, J. J., 346
Sugden, S., 286, 299, 360, 362
Sutton, L. E., 299, 301–2, 362
Swartz, T., 226, 230, 352
Swedenborg, E., 315

Tamelen, E. van, 249, 355
Tatevskii, V. M., 308, 363
Thiele, F. K. J., 213, 239–41, 256–7, 354, 356
Thomas, W., 362
Thomsen. J., 253, 345, 356
Thomson, G. P., 359, 361
Thomson, J. J., 257, 269, 273–4, 276–82, 357, 358, 359
Thomson, T., 6, 327
Thorpe, J. F., 299
Thorpe, T. E., 178, 180, 182, 192–3, 347, 349

Tickle, T., 199, 350
Tilden, W. A., 126–7, 341
Tillman, S. D., 344
Tollens, B. C. G., 352
Traube, M., 195, 349
Troost, L. J., 178, 347
Tyndall, J., 36

"Urban", 91, 337
Uslar, L. W. J. von, 58, 334

Valentin, H. G., 105, 339
Vass, C. C. N., 288, 360
Virchov, R. L. C., 321
Vleck, J. H. van, 363
Vogt, K., 323
Volhard, J., 109, 111, 113
Volta, A., 327

Waack, R., 360
Waage, P., 18, 330
Waldrop, P. G., 73, 335
Walter, [no initials given], 163, 346
Wanklyn, J. A., 172, 191, 200, 347, 349, 350
Ward, F. O., 336
Warren, E. H., 167, 347
Watts, H., 343
Watts, L., 356
Weiss, A., 362
Weissenberg, K., 167, 346
Werner, A., 159, 163, 167, 199, 213–23, 237, 241, 265, 283–4, 288, 346, 351, 353
Wheland, G. W., 306–8, 311, 363
Whewell, W., 16, 329, 357

Wichelhaus, C. H., 85–6, 172, 180, 191–6, 200, 248, 336, 347, 349, 350, 356
Wilbrand, J., 106, 234, 339, 353
Wilkins, C. H., 363
Wilkins, H., 360
Williamson, A. W., 10, 15, 20, 47, 49–58, 61, 66, 79, 81, 83–4, 91, 94, 96, 111–12, 118–19, 125–7, 129, 131, 138–9, 145, 178, 188, 196, 200, 206, 214, 314, 330, 333, 335, 336, 337, 338, 341, 343, 344, 347, 348, 350, Plate 5
Willstätter, R. M., 257, 357
Wilzbach, K. E., 355
Wisherd, M. P., 363
Wislicenus, J., 160–2, 345
Wöhler, F., 15–16, 21, 24, 27, 329, 331, 334, 364
Wollaston, W. H., 159–60, 345
Wordsworth, D., 329
Wordsworth, W., 13–14
Wren, H., 361
Wright, C. R. A., 188–9, 348
Wroblewsky, Z. F. von, 354
Wunderlich, A., 167, 346
Wurtz, A., 15–16, 57, 66, 71, 79–80, 84–5, 93, 100, 109, 111, 118–19, 125, 129, 131, 133–4, 138, 145, 161, 173, 179, 185–7, 191–2, 198–9, 224, 329, 334, 335, 336, 338, 341, 342, 344, 347, 348, 349, 350

Yoshimine, Y. M., 362
Yost, D. M., 362

Zincke, E. C. T., 229, 235, 352

SUBJECT INDEX

acetoacetic ester, 156–7, 200, 345
acetylenes, 232, 236, 238, 248, 285
acid strength, 288
acids, 22, 28–9, 46, 52, 54, 58, 72–3, 75, 129; unsaturated, 225–7, 235, 352
acridine, 248–9
addition, 1–4, 240, 251, 354
affinity, 17–18, 73, 83, 125, 197, 202, 329; residual, 204–5, 207–13, 219, 222, 241, 267, 350; zones of, 217–20
"agnosticism", 20, 33, 127, 143, 146, 250, 330
alcohol, 23–4, 200
alkalis, 11
allyl compounds, 224–5, 232
amateurism, 8
American Chemical Society, 183
ammines, 186–7, 214–15, 217–22
ammonium, 24, 284, 285, 330; compounds, 191, 193; quaternary, 155–6, 164, 167, 194–5, 208–10
anthracene, 246, 248
Arceuil, 17, 330
aromatic compounds, 242–57, 278–9, 282,
306–8; sextet, 290, 360
atom-fixing power, 83, 85
atomic models, 317; Dewar's, 107, 248; Hofmann's, 104, 186, 226, Plate 7; Kekulé's, 107, 162, 234–5, 238–9, 346, Plate 9
atomic number, 274; weight, 114–17, 122, 130, 133, 137, 139–41, 176
atomicity, 81, 84–5, 88, 111; latent, 93, 104, 188–9, 229
atomism, 6–9, 21, 145, 262, 276, 313–18, 323–4, 328, 294
azides, 302, 305
azo compounds, 182
azoxy compounds, 183

Baconianism, 322–3
basicity, 81, 84, 111
Belgian Academy, 226
benzene, 111–12, 134, 213, 234, 241–57, 278–9, 282, 305–6, 311; hexagon, 242–7; substitution, 47, 222–3
benzoyl, 24–5, 27
Berlin, 238, 294

SUBJECT INDEX

Berthollides, 18
bond, diagonal, 247–52; double, 162–4, 166–7, 228–9, 234–6, 239, 249, 253, 279, 290, 353; length, 278, 303–5; triple, 162, 166–7, 234–6, 239, 284–5
"bonds", 89–91, 100–7, 114, 313, 318
Bonn, 72, 183
British Association, 10, 88–9, 129, 133, 139, 264, 266
butadiene, 241, 285

cacodyl, 25–7, 31, 41, 111, 131, 238
California Institute of Technology, 304
Cambridge, 306
carbon atom, tetrahedral, 159–62, 166–7, 239, 291, 345; tetravalent, 61–80, 119–24, 150
carbon oxides, 73, 197, 199, 203, 236
carbonyl, C_2, 122–3; addition, 288–90
catenation, 69–70, 125, 150, 179–87
"cathode rays", 268
centric formulae, 252–4
chain theory, 179–87
Chemical Society, 34, 38, 54, 94, 126, 133, 200, 204–6, 265
chlorine, oxyacids, 186, 198
"combining power", 42–3, 83
conductivity, 267–277
conformational analysis, 239
conjugation, 222–3, 240, 288–91, 360
co-ordination, 214–17, 285, 351; co-ordination number, 218
copulae, 28, 30–2, 45, 58–9, 122, 172
Couper's compound, 72–3
coupling of polyvalent atoms, 117–19
cryoscopy, 179, 267
crystal field theory, 311–12
crystallography, 166–7
cyanogen, 25–7
cyanuric acid, 78
cycloalkanes, 231, 238–9
cyclobutadiene, 257, 356
cyclo-octatetrene, 257

"Dewar structures", 247–50
diamond, 239
diazo compounds, 163–4, 182, 302
diazonium compounds, 182
dipoles, 300–1
direct effect, 288
dissociation, 191–2
dualism, 16, 129–30, 262–4, 276–82, 287

Edinburgh, 71, 80, 101–2, 107, 247–8
education, 9–11, 14, 19, 103–6, 319–21, 339
electrochemical theory, 5, 11–12, 16, 22–3 44–6, 50, 132, 137, 172, 198, 205–7, 261–9, 287, 311
electrolysis, 6, 11, 29–30, 265–8, 270
electromeric effect, 290
electromerism, 278–81
electron, 268, 292–6, 358; electron-pair, 283–5, 287, 304; electron-sharing, 281–7; electron spin resonance, 312
electrovalency, 270–5, 287

elements, classifications of, 137–41
Encyclopaedia Britannica, 89, 337
England, 4, 265–9, 320
English Mechanic, The, 84, 88, 91, 103, 105, 336–7, 339
etherin, 23–4
ethers, 23–4, 50–3, 126, 198
"ethyl", 23–4, 35–7
ethylene, 228–30, 232–7, 236, 239, 284; ethylene radical, 224
ethylidene, 228, 230
equivalence, 82, 85, 87–8, 335
exclusion principle, 296

Faraday's laws, 205
formulae, rational, 143–4, 149
France, 16–20, 48
fulminates, 62–3, 65, 199

"Gap-theory", 225–31
Germany, 3–4, 12–16, 122, 265, 268–9, 297–9, 320–1, 334
Ghent, 86, 107, 238, 243, 248
Giessen, 14–15, 49, 144, 320
Glasgow, 71
glycols, 57, 71, 224
Göttingen, 309, 334
gravitation, 264, 357

H.M.O. theory, 312
halogen, 11
Heidelberg, 50, 57–8, 62, 130–1, 147, 238
homologous series, 323
hybridization, 305
hydrates, 186, 188, 199, 201, 204, 208, 216
hydrogen bond, 302; hydrogen cyanide, 157; hydrogen fluoride, 192–3, 207–8; hydrogen molecule, 304; hydrogen molecule-ion, 305, 310; hydrogen peroxide, 179, 185, 195, 199, 204
hyperconjugation, 311
hypervalency, 359

indole, 246
inductive effect, 288
iodine chlorides, 175, 190, 193, 196, 225, 358; iodine positive, 277, 283
ions, 262, 266, 270–5; ion-pairs, 274
iron chlorides, 105, 178–9, 188
isatin, 157
isomerism, 21, 149, 154–5; geometrical, 162–4, 221; optical, 160–1, 163–4, 167, 221

Karlsruhe, 115, 341
"key atom", 289
kinetics, 299–300
King's College, London, 10, 138

L.C.A.O., 310, 363
Latin etymology, 82, 86–8
Leeds, 299–300
ligand field theory, 221, 311 12
London (see also institutions), 5–6, 10, 57, 70, 129–30, 138–9
Lysenkoism, 307

SUBJECT INDEX

magnetism, 203–4, 241, 261, 282, 302, 311
Manchester, Owen's College, 39, 50, 130
Marburg, 25–8, 34–8, 130
Marxism, 308
mass action, 17–18
mathematization, 13–14, 318
matrix mechanics, 294, 296
Mechanics' Institutes, 10, 103, 105
mercurous chloride, 177–8, 187–8
mesomerism, 299, 301
metal atom, octahedral, 220–1; planar, 221–2
metals, transition, 173, 186–7, 189, 214–22
methyl, 24, 30, 35, 37, 130, 152–3
methylene, 228, 230
molecular compounds, 175, 177, 189–97, 266, 271–2, 285
molecular orbital approach, 306, 307, 309–12
monads, 82
Munich, 238
naphthalene, 246, 252
nationalism, 3 *et seq.*
Naturphilosophie, 13–16, 165, 317, 329
nitric oxide, 177–8, 188, 283
nitriles, 28–9, 34–6, 155
nitrogen atom, stereochemistry, 163–4, 167
nitrogen, molecular, 202–3
"nitrogen peroxide", 191, 283
nitrogen trichloride, 281
non-electrolytes, 270–91
notation, 9, 50, 55, 74–5, 77–8, 80, 92–107, 127, 130, 133, 151, 165, 233–4
nucleii, 46

octet theory, 271–5, 277, 282–3, 286–7, 290
one-electron bonds, 305
orbitals, 302–12
organic compounds, classification, 70
organo-metallic compounds, 38–43, 114, 120, 130–1, 200, 224
organic syntheses, 12, 14–15, 269
oscillation hypotheses, 157, 244, 254–6, 264, 307
Oxford, 301, 306
oxidation state, 280
oximes, 163–4
oxy-acids, 283
oxygen, 237

parachor, 299
Paris, 17, 46, 49, 71, 129–30, 133, 147, 160–2, 197
Periodic Law, 137, 140–2, 270–4
Periodic Table, 140–1, 198
phenanthrene, 246, 248
phlogiston, 12, 328
phosphorus atom, stereochemistry, 163
phosphorus compounds, 54–7, 180–2, 191–6, 209, 211, 359
physical chemistry, 18, 269, 299, 301–2
picrates, 195
polarizability, 290
polarization, 285, 287–91, 300–1

polymeric theory, 177–9
positivism, 20, 48, 76
priority, 108–34
prismanes, 244–5, 250, 346
propanols, 99, 102–3
Putney, 34, 39
pyridine, 246, 249
pyrimidines, 35–6
pyrones, 199, 297
pyrrole, 246

quantivalence, 85–6
quantum theory, 275, 292–6, 298
Queenwood College, 34–6, 331
quinoline, 252
quinones, 186

racemization, 213
radicals, 20–43, 109, 117, 129–30, 287; basicity, 60; free, 332; polyatomic, 49–61
radioactivity, 268, 274, 292
reflux condenser, 28, 331
refraction, 299
relativity, theory of, 292, 318
research training, 11, 14, 134
resonance, 298, 301–2, 305–8
romanticism, 14
Royal College of Chemistry, 10–11, 50, 96
Royal Institute of Chemistry, 126, 290
Royal Institution, 5, 9, 49, 83, 120, 124, 126, 138
Royal School of Mines, 11
Royal Society, 5, 10, 40
Royal Society of Arts, 10
Russia, 3, 307–8

St Bartholomew's Hospital, 49, 138
salts, 51, 54; double, 206–7, 215
saturation capacity, 110, 113–17, 129, 131
selenium compounds, 211, 358
solid state, 302, 310–11
spectroscopy, 72–3, 274–5, 286, 302, 308–11
stanethylium, 40–2
stereochemistry, 13, 21, 101–2, 107, 144 159–67, 239, 339; inorganic, 220–2
strain theory, 235, 237–9, 254
substitution, 45–6, 58, 263; aromatic, 47, 60, 279, 289, 299
sulphones, 185
sulphonic acids, 28, 58–9, 184–5, 280
sulphonium salts, 188, 198, 218
sulphur halides, 179, 184, 190, 197
sulphuric acid, 53, 59, 183, 216
sulphuryl chloride, 68, 183–4, 300
superscript dashes, 92–4

tautomerism, 156–8, 256, 278, 299, 301
tellurium halides, 190, 211
thermochemistry, 155, 253, 305
thermodynamics, 317, 324
types: ammonia, 46, 50; carbonic acid, 46; H₂S, 56; marsh gas, 63–5, 120–2, 124; mechanical, 46, 64–5, 122; mixed, 55, 58–9; multiple, 59; water, 46, 50–1

SUBJECT INDEX

Types, Theory of, 19, 21, 44–80, 84, 91, 94–6, 100, 109, 117, 127, 129–31, 142–6, 148, 172, 223, 287, 323

uncertainty principle, 295
University College, London, 10, 49, 54, 126, 138, 306
unsaturation, 224–41, 266
U.S.A., 105, 265, 268–9, 276–81
Utrecht, 161

"valence", 85–9
valence bond approach, 303–12
valence isomerism, 245, 249, 252, 345
valencies, secondary, 151–5; equivalence of, 152–6
valency, auxiliary, 201–3; fixed, 117, 176–89; "non-directional", 223–4; paired, 187–9; partial, 241, 257; variable, 130, 171–223, 236–7

valyl, 29–30
vanadyl compounds, 178, 180
vapour densities, 177–9, 190–3
vinyl compounds, 157, 229
vitalism, 323, 364

war, 321–2; Napoleonic War, 5, 17–18
water, 196, 204, 206, 208
wave equation, 295, 302
wave function, 304; mechanics, 292–312, 318
"wavicle", 293
Wertigkeit, 87–8, 336

X-rays, 274

Zurich, 57, 223, 294
Zwitterion, 297